Progress in Colloid & Polymer Science, Vol. 101 (1996)

PROGRESS IN COLLOID & POLYMER SCIENCE
Editors: F. Kremer (Leipzig) and G. Lagaly (Kiel)

Volume 101 (1996)

Interfaces, Surfactants and Colloids in Engineering

Guest Editor:
H.-J. Jacobasch (Dresden)

ISBN 978-3-662-15698-8
ISSN 0340-255 X

Die Deutsche Bibliothek –
CIP-Einheitsaufnahme

Interfaces, surfactants and colloids in engineering
/ guest ed.: H.-J. Jacobasch.
(Progress in colloid & polymer science ; Vol. 101)
ISBN 978-3-662-15698-8
ISBN 978-3-7985-1664-9 (eBook)
DOI 10.1007/978-3-7985-1664-9

NE: Jacobasch, Hans-Jörg [Hrsg.] ; GT

Originally published by Dr. Dietrich Steinkopff Verlag GmbH & Co. KG, Darmstadt in 1996
Softcover reprint of the hardcover 1st edition 1996

Chemistry editor: Dr. Maria Magdalene Nabbe; English editor: James C. Willis; Production: Holger Frey, Bärbel Flauaus.

Type-Setting: Macmillan Ltd., Bangalore, India

Progr Colloid Polym Sci (1996) V

PREFACE

The 37th General Meeting of the German Colloid Society was held in Dresden from September 26 to 29, 1995. The main subject was "Interfaces, Surfactants and Colloids in Engineering." The organizers – the Institute of Polymer Research Dresden and the Dresden University of Technology – had chosen this particular title to point to the important role that interface phenomena play in technological processes.

The meeting was attended by about 230 participants from 11 countries. About 20 % of them came from industry, mainly from the chemical industry.

The meeting was inaugurated by M. J. Schwuger, chairman of the German Colloid Society. Greetings were expressed by E. Noack, undersecretary of state in the Saxon State Ministry of Science and Arts, and by A. Mehlhorn, rector of the Dresden University of Technology.

Some general thoughts on the position of science and research in society, on the tremendous challenges for research, and on the organization of scientific research in Germany were reflected in lectures by J. Treusch, president of the Hermann-von-Helmholtz Society of German Research Centres, and by V. Hertel, president of the Science Association of the Blue List Institutions (WBL).

The highlight of the opening session was the presentation of three prizes of the German Colloid Society. The Wolfgang Ostwald Prize was conferred to H. Sonntag, Berlin, for extraordinary lifetime achievement in colloid science.

The Thomas Graham Medal was bestowed upon J. Lyklema, Wageningen, for his outstading contributions to colloid science and international cooperation in the field. The Richard Zsigmondy Scholarship for successful young colloid scientists was awarded to F. Simon, Dresden.

In the meeting 14 invited lectures, 33 contributed lectures and 80 posters were presented. They dealt mainly with theory and application aspects in the fields of preparation and stability of dispersions and emulsions, adsorption of surfactants, polyelectrolytes and neutral polymers on solid surfaces, wetting processes and electrokinetic phenomena.

Investigations on new methods to characterize interfaces, surfactants and colloids were also presented.

Quite a number of the contributions dealt with applications of basic colloid-chemical findings and methods on practical problems, such as utilization of surfactants and polymeric auxiliaries in washing processes, flocculation and duplication processes, emulsion polymerization, film formation, stability and demulsification of petroleum emulsions, polymer blends and fiber-reinforced plastics, the development of materials for medical applications (microcapsules, artificial organs), solvent extraction of heavy metals, e.g., from effluents, application of sensors, and membrane separation methods.

The papers and discussions have shown that colloid and interface science may make valuable contributions to control and optimization of processes and material properties in the above fields, and that cooperation between natural scientists and engineers as well as between academic research institutions and industrial enterprises is imperative.

However, the meeting has also shown that there remains a lot to do in the field of application-oriented colloid and interface science, e.g., development of methods to describe interface properties in multiphase materials, such as blends and reinforced plastics, application of established methods of interface characterization to real systems, i.e., physically and morphologically heterogeneous surfaces, multi-component systems, description of the effect of surfactants, polyelectrolytes and other auxiliaries on adhesion and wetting in technological processes, and investigation of process stages in the chemical and processing industries by means of methods and approaches of interface science.

German industry has a great backlog demand, especially in the last-mentioned field, and this applies both to conventional technologies such as paper and textile technology, plastics and chemical processing, and to high-tech branches like computer technology. Major advance is also necessary in the field of development of materials and processes to be used in medicine.

The 37th General Meeting of the German Colloid Society has helped to recognize problems to be solved and to start discussing approaches offered by colloid science.

I would like to thank all those who contributed to making the meeting successful. For the preparation of the scientific program, I am grateful to the members of the scientific committee: M. J. Schwuger, Jülich; Th. Wolff, Dresden; H.-J. Schulze, Freiberg; and H. Stechemesser, Freiburg. For all their efforts in the organization of the meeting, I wish to thank K. Wustrack, M. Creutz, and all the involved coworkers of the Institute of Polymer Research and of the Conference Service of the Dresden University of Technology.

Last but not least, I would like to express my thanks to all the scientists who actively contributed to the scientific program and, in particular, to those who prepared their manuscripts for publication in this volume.

Prof. Dr. H.-J. Jacobasch

CONTENTS

Surfactants

Interfaces

Progr Colloid Polym Sci (1996) 101:1–3

M. J. Schwuger

Opening address at the 37th General Meeting in Dresden

Prof. Dr. M. J. Schwuger (✉)
Forschungszentrum Jülich GmH
Institut für Angewandte Physikalische Chemie,
52425 Jülich, FRG

Ladies and Gentlemen,

On behalf of the Executive Committee of the Colloid Society, I would like to welcome you most warmly to the opening of this 37th General Meeting. We are gathering here in Dresden for several reasons. After the German reunification, this is the first meeting to be held in eastern Germany. The last General Meeting of the Colloid Society before the worst catastrophe in German history took place in Dresden, in 1941, which is why we consciously decided to come here again. Furthermore, this year we celebrate the 125th anniversary of Professor Lottermooser's birth, who for many years shaped interfacial and colloid chemistry in Germany from his base in Dresden. At the same time, we commemorate the 50th anniversary of his death.

Colloid chemistry has a great tradition in Saxony. It was molded for many years by a number of respected researchers – first and foremost, undoubtedly, Wolfgang Ostwald – and later also by a number of world famous institutions. In prewar Germany, this great tradition led to colloid science becoming a major pillar to German science. The last big conference of the Colloid Society held in Germany before the war, in Mainz, attracted more than 1000 participants. Today, like other large societies, we can only look back enviously upon this large number of delegates.

After the partitioning of Germany, developments took a different course in the eastern and the western parts. A number of colloid chemistry institutes continued their work in the East. Classical colloid chemistry was promoted both at the universities and at the Academy of Sciences. In the Western part of Germany, the introduction of new methods and new research areas in physical chemistry led to a slow decline in the traditional colloid chemistry departments. This development went so far that from the mid-1960s and early 1970s, we no longer had a single institute in this field in West Germany. All activities were subsumed within the broader framework of physical chemistry. Unification has brought a great surge of innovations in colloid chemistry. We have taken an important step forward with the establishment of new institutes. Special credit among the large societies must be given to the Max Planck Society for the establishment of the Institute of Colloid research in Potsdam, near Berlin. Even if the entire scientific structure is considered, there has been no comparable colloid chemistry book in German. [since the excellent book "Colloid Chemistry" by Joachim Staufff in 1960]. I am particularly pleased to point out that last year an equally good modern book was published by Professor Dörfler from Dresden.

Colloid chemistry continued to decline in the West until about the mid-1960s, early 1970s. Its significance for technology and also for the environment has now been recognized and a new phase to establish fields of work concerned with interfacial and colloid chemistry has begun. Figure 1 shows the locations where work is now being carried out in the field of interfacial and colloid chemistry and a clear upward trend can be seen. Considering that there are several institutions involved in this field in large cities such as Berlin or Munich, even though we cannot be entirely satisfied we can welcome the positive development in interfacial and colloid chemistry. However, a further point must be emphasized, namely that the relationship between university research and research in industry and large research establishments is not yet well balanced. The large centers, and the large societies such as the Max Planck Society, recognized the significance at a relatively early stage so that research on interfacial and colloid chemistry is being carried out there at several institutes. This is also true for the national research centers, in particular the Research Centre Jülich, where two large institutes have been established – one with a more basic orientation and the other rather in the field of environmental chemistry.

The situation looks worse at the universities since all activities are concentrated on physical chemistry. Due to new appointments to chairs and a number of retirements, there is a great danger of the discipline dying out, as for example in Regensburg at the moment.

In the future, we must ensure that once again chairs of interfacial and colloid chemistry are established separately in Germany in addition to the existing major disciplines. Irrespective of this, I am able to report that the number of members of the Colloid Society has risen perceptibly in the course of the last few years. Interest has also clearly increased so that we are able to welcome about 250 to 300 delegates. This is a significant success.

Unfortunately, not all colleagues are able to be with us. I regret the death of Prof. Kratky, Graz; Dr. Herzog, Munich; Dr. Strnad, Regensburg; Dr. Zembrod, Leverkusen; Prof. Scheludko, Sofia; and Prof. Peterlin and Dr. Junghans.

I have already mentioned the significance of interfacial chemistry. I would now like to expand on this and quantify it more precisely. There are a number of governmental and private institutions concerned with future developments in science. As examples, I would like to mention three reports. In 1987, the American government commissioned the so-called Amundson Report, dealing with "Frontiers in Chemical Engineering". In 1988, a DECHEMA study was presented with the title 'Studies on the Evaluation of Modern Trends in Basic Chemical Research with a High Technical Innovation Potential' and finally in 1992 the report 'Interindustrial Chemical and Technical Research Priorities with a High Innovation Potential' was isued, supported jointly by DECHEMA, FCI and VCI. The quintessence of all these studies is that interfacial chemistry is emphasized everywhere and recognized as particularly innovative. Thus, for example, the DECHEMA study, designed to cover the requirements of innovative basic research for relevant

Fig. 1 Centres of Excellence in Colloid and Interface Science

branches of industry, emphasizes "interfaces and membranes" among five particularly innovative fields. Furthermore, aspects of interfacial sciences are to be found in all five fields, and in some cases are important as in the material sciences or in connection with the new principles of substrate and energy conversion. The second study by DECHEMA, VCI and FCI comes to a similar conclusion, assessing interfacial chemistry as one of ten priority fields with a great innovation potential. There is no existing or planned technology in which surface processes, and particularly chemical processes, do not play a decisive role. Chemical research in this central field is therefore of great importance in order to avoid or overcome current obstacles to present and future developments in many technologies. It is therefore an extremely innovative field with a great potential. Nevertheless, I would like to add that the environmental discussion, which is particularly intensive here in Germany, has mainly concentrated on biological degradation in the past. However, all processes in soils and sediments, on aerosols, in water and suspended matter are processes of colloid and interfacial chemistry, and the understanding and further development of new technologies based on the present state of the art can only proceed by means of basic research in interfacial and colloid chemistry.

A certainly important sign for our promising field of work is also the acceptance accorded to graduates from our discipline. You are all familiar with the problems experienced in the past few years by chemistry graduates trying to find their first job. In spite of these difficulties, practically all good graduates from departments of interfacial and colloid chemistry found an interesting position in industry or research institutions relatively easily; this is a very important point which we must also keep in mind. For this reason, we have intensified the work of the Colloid Society and this year, in addition to the general meetings, we have initiated a series of Wolfgang Ostwald colloquia. These colloquia are designed in such a way that for 1 or 2 days, we deal with a closely defined topic from interfacial and colloid chemistry involving particularly distinguished specialists in the respective fields. The first two colloquia have been extremely well received and I hope that in the future we shall be able to establish ourselves in German science. On this note, I would like to wish the organizers every success for the conference and a fruitful scientific atmosphere for the delegates.

Progr Colloid Polym Sci (1996) 101:4–6

G. Kretzschmar

Wolfgang-Ostwald-Prize recipient 1995: Prof. Dr. Hans Sonntag

Prof. Dr. Hans Sonntag

Prof. Dr. G. Kretzschmar (✉)
MPI für Kolloid
und Grenzflächenforschung
Rudower Chaussee 5
12489 Berlin, FRG

In Hans Sonntag we pay homage to a scientist whose dedicated work has deeply and lastingly enriched the field of colloid chemistry. His scientific achievements do honor to the outstanding academic teachers and leading colloid scientists of other countries who helped to inspire them.

Professor Sonntag was born January 25, 1932, in Heidenau, a small town to the south of Dresden. It was in Heidenau that he received his first schooling, and in nearby Pirna he completed his secondary school certificate. After a subsequent apprenticeship in a pulp factory, Sonntag went on in 1951 to enroll as a student of chemistry at the Technical University of Dresden. Here he carried out his first scientific work under the direction of Prof. K. Schwabe. Undoubtedly, Prof. Schwabe's great diligence and fastidious methods of research did not fail to make a deep impression upon the young Sonntag.

Sonntag's final thesis and later doctoral dissertation both dealt with issues of electrochemistry.

From 1956 till 1960, Sonntag was a research assistant at the Institute of Physical chemistry at the Technical University in Dresden. There he was entrusted with a teaching position in the field of colloid chemistry. The tradition of colloid chemistry had been eagerly cultivated in Dresden since the time of A. Lottermoser and E. Wolfram. In retrospect, we must say that it was a fortunate decision and perhaps also an instance of farsightedness on the part of Prof. Schwabe to charge Hans Sonntag with precisely this task.

It was in Moscow, during a 6 month academic residency under Prof. Rehbinder, that Sonntag chose colloid chemistry as his field of known Prof. Rehbinder personally will recall the wealth of ideas and intellectual enthusiasm which this impressive scientist was able to impart to his students. After Schwabe it was Rehbinder who exercised the most profound influence on Sonntag as an ambitious young scientist. It was during this stay in Moscow as well that Mr. Sonntag established his first contacts to B.V. Derjaguin and his school.

In Moscow Sonntag became intimately familiar with the basic problems of collochemical processes in the coalescence of dispersed particles. While Rehbinder and his colleagues were focusing on the steric-mechanical barrier as a stabilizing factor in the coalescence of dispersed particles, Derjaguin, Landau, Verwey and Overbeek emphasized the overlapping of Van der Waal's attraction with electrostatic repulsion as the truly decisive factor. At the time of Sonntag's arrival in Moscow these two competing conceptions were considered to be irreconciliable. After he had prepared him sufficiently for the task, Rehbinder asked Sonntag to take advantage of the more advanced experimental facilities in East Germany, in order to contribute to a future solution of these basic problems. As we know today, the application of the "pure DLVO-theory to a series of real systems underwent corrections based in part on Rehbinder's concepts.

When he arrived back in East Germany, Sonntag set to work on this task immediately. Professor Schwabe was the first to offer his support; soon after he was joined by P.A. Thiessen, whose colleague and co-researcher Sonntag became in 1960 when he joined the newly founded Institute of Physical Chemistry of the Academy of Sciences in East Berlin. Sonntag was entrusted here with putting together a department of colloid chemistry. As a former colleague of R. Szigmondy and former director of the Kaiser Wilhelm Institute of Physical Chemistry in Dahlem, Thiessen had a special relationship with colloid chemistry. In his concern to bring an active, yound colloid scientist to the Institute to work at his side, his chose Sonntag.

In one of his first Adlershof publications, Sonntag was concerned with observing the coalescence rate of two mercury hemispheres. This classical problem of diffusion controlled (fast) coagulation as developed by v. Smoluchovski entails a series of fundamental questions as to the stability of colloid dispersions. Upto today Mr. Sonntag has remained true to this complex of problems – both in his constant refinement of experimental techniques and in his related theoretical papers. Next to his new ideas for experimental work, intensive collaboration on the explanation of crucial theoretical questions has played an equally important role in his work on coagulation kinetics.

Under Sonntag's leadership, the Department of Colloid Chemistry attracted many talented scientists; as representative of his many equally gifted fellow researchers let us mention K. Strenge in particular. In 1963, Sonntag habilitated with a thesis on "Forces of Interaction between Dispersed Particles and the Influence of Surface Active Substances on Stability in Non-Polar Media".

Many of his own scientific results are part of the textbook **Coagulation and Stability of Dispersed Systems**, written by H. Sonntag and K. Strenge (Berlin 1970). This fundamental book has editions in English (1973) and in Russian (1973). Again with Strenge he wrote the textbook **Coagulation Kinetics and Structure Formation** (New York 1987).

Alongside his work on the coagulation kinetics of homogeneous particles and on heterocoagulation stand wide-ranging contributions on interparticular forces of interaction and experiments carried out on thin fluid-films using pressure/distance functions between solid surfaces.

As is well known, interparticular interaction in concentrated dispersions leads to a number of remarkable phenomena, for instance rheological behavior. The investigations on magnetic fluids, which Sonntag carried out with N. Buske, were an especially interesting case. Sonntag expanded the foundations of our understanding of dispersed system stability by incorporating a variety of theoretical and experimental methods, such as the pressure-distance function of thin fluid-films. Especially impressive was his employment of emulsion films, which for optical reasons are exceedingly difficult to measure. Several of the techniques which he developed in order to overcome these difficulties deserve special emphasis: The measurement of equilibrium thickness in emulsion films as a function of the ion-strength; the measurement of film tension by means of contact-angle measurements; and the investigation of solid dispersions using electro-optical methods.

These investigations were carried out in cooperation with research groups in Sofia supported by the scientists A. Sheludko, D. Platikanov, D. Exerova and S. Stoylov. Particularly A. Sheludko's stimulating relationship with Sonntag should be mentioned here. After Thiessen became professor emeritus, W. Schirmer succeeded him as the Institute's new director; he provided Sonntag with constant support in the following years.

Over the course of several decades, the fundamental insights of SONNTAG's research group bore fruit for the industrial sector, especially in cooperation with the Chemical Plant Buna. In 1975, Sonntag was appointed professor. As mentioned above, the outstanding merit of SONNTAG and his co-researchers consisted in large part in their elucidation of peculiarities of the different dispersed systems.

After the study of foam properties SONNTAG's attention was focused on emuslion films. This topic was a weak point in the scientific literature. The reasons for that may be the fact that emulsion films possess a lower degree of light reflection at the oil/water interface.

Consequently, up until the mid-Seventies, Sonntag's publications with H. Klare, J. Netzel, and K. Strenge focused on problems of emulsion stability and emulsification; for the most part, a thin-film model was employed.

In his research and in many personal discussions, SONNTAG was always deeply interested in problems of heterocoalescence. Early on he had already made a quantitative study of the wash-

Progr Colloid & Polymer Sci (1996) 101: 4–6

process together with Strenge, working on the basis of theoretical conceptions of heterocoalescence.

Further steps in his research related to the stability of polymer dispersions with a transition to oxidic solid dispersions. For these investigations, Sonntag developed interesting procedures for the production of monodisperse particles. I shall only alude to the interesting methods for measuring interfacial tension, which he developed in cooperation with Fieber.

In spite of East Germany's restrictive policies, Sonntag was able to maintain fruitful contact with the leading centers of colloid chemistry. In accord with the esteem in which his work is held, Sonntag has been a sought-after guest at many significant international conferences.

His textbook, **The Science of Colloids**, appeared in 1977.

The unquestionable relevance and high research standards of Sonntag's department have led to a close working relationship with leading scientists lasting over many years. This intensive cooperation has in part taken the form of shared publications with the research teams of B.V. Derjaguin and E. Shchukin in Moscow and V. Shilov and S.S. Duchin in Kievw; a friendly relationship of equal long standing has likewise been maintained with Hans Lyklema in Wageningen. Members of all these research teams have often enjoyed lengthy academic stays with SONNTAG in Berlin.

Considering the overwhelming intensity and productivity of his scientific research, it was inevitable that Sonntag would find recognition in a whole series of academic distinctions. In 1971, he was awarded the Friedrich-Wöhler Prize by the Chemistry Society of the GDR, in 1982 the Jacobus Henricus van't Hoff Medal by the Academy of Sciences of the GDR and a further distinction by the Rudjer Boscovich Institute in Zagreb.

In order to underline Sonntag's indefatigable energy, one may recall a situation in the early 1980s, 1990s, when Sonntag had been stricken by a grave illness. For rest he had no patience; as soon as he had recovered from the worst of his health problems he was back at work, giving a lecture at a conference in Australia!

After the German Reunification, the former East German Academcy of Sciences had to be restructured and its sundry departments reintegrated. This process was placed in the hands of a Council on Science appointed by the federal government. In the course of evaluation it was recommended that the department of colloid chemistry and the area of surface active substances research, both in Adlershof and housed respectively in the central institutes of physical and organic chemistry, be partially adopted by the Max Planck Society. The founding committee of the new Max Planck Institute for Colloid and Interface Research had an active and competent member in SONNTAG. The Insitute offered him a remarkably well-suited environment in which to further develop his thoughts on interparticular interaction. The laser equipment available here made it possible to measure single-particle light-scattering with improved accuracy by means of hydrodynamic focusing. Substantial new results, achieved in cooperation with H. Lichtenfeld and L. Knapschinsky, are at present being prepared for publication. This paper will show which kinetic laws are at play in the formation of single-particle aggregates. In this way it will be possible to distinguish diffusion-controlled (fast) coagulation and slow coagulation in terms of velocity constants in the formation and decay of dispersed single-particle aggregates. As a consequence of these measurements and calculations, a new understanding of force effects between dispersed particles has become possible. The results of this research are therefore of great practical significance for the employment of dispersed systems.

So many are Sonntag's contributions to colloid chemistry that it is not possible to adequately each in so short a space. His life's work has been marked throughout by assiduousness and determination in the overcoming of obstacles – especially of those placed until recently in the way of scientific research in the Eastern Bloc countries. Obstacles such as the limits placed on world-wide communication, an often insufficient financial and technical basis for the conducting of experiments, and the crass personal restrictions so familiar in East Germany in the latter half of our century. It is therefore not only his scientific achievements that we honor, but Sonntag's unswerving path. He displayed not only an unrelinquishing dedication in the pursuit of scientific goals, but possessed as well the uncompromising outlook necessary for an achiever living under the constraints of a dictatorial regime. Those who have known can well appreciate the sensitivity and understanding he has displayed toward his friends. Toward government paternalism and excessive bureaucracy, on the other hand, his rejection was absolute and unyielding. That Sonntag remained true to this fundamental principle over the course of decades, even when dealing with his would-be superiors, is a sign of determination and a mark of distinction.

His colleagues here at the Institute and many other of his colleagues elsewhere fondly remember the long years of friendship and cooperation in good times and in bad, and are glad of this special opportunity to honor Hans Sonntag.

Progr Colloid Polym Sci (1996) 101:7–8

B. Bijsterbosch
G.J. Fleer

Thomas-Graham-Medal recipient 1995: Prof. Dr. Johannes Lyklema

Prof. Dr. Johannes Lyklema

Prof. Dr. B.H. Bijsterbosch
Prof. Dr. G.J. Fleer (✉)
Department of Physical
and Colloid Chemistry
Wageningen University
Postbus 8038
6700 EK Wageningen, The Netherlands

Hans Lyklema was born in the Dutch town of Apeldoorn on November 23, 1930, and his physical-chemical life started when he enrolled in the chemistry program of Utrecht University in September 1948. He majored in physical and colloid chemistry in 1955. Just two and a half years later he obtained his PhD, which, by all then prevailing standards, was remarkably fast. His thesis dealt with the adsorption of counterions in the electrical double layer on silver iodide and was supervised by Prof. Theo Overbeek. He then became associate professor in physical chemistry in Utrecht, and worked as a visiting associate professor with professor Karol Mysels at the University of Southern California in 1961–1962. In 1962, Lyklema was appointed full professor in physical and colloid chemistry at Wageningen Agricultural University.

Taking up this chair in Wageningen was actually quite a challenge, since at the time colloid chemistry was rather undeveloped there. However, Lyklema discerned great potential for incorporating disciplinary approaches into agricultural science. After more than 30 years, we may safely say that his accomplishments have shown this vision to be correct. The interplay between fundamental research on the one hand and more applied areas (as in food, soil and environmental sciences) on the other has proved to be both practically useful and scientifically rewarding.

With his thesis topic and his collaboration with Mysels as a basis, Lyklema initially focused on electrical double layers and thin liquid films, but soon diversified to a much wider range of topics. Characteristically, every new topic always quite logically originated from a preceding one. For example, agricultural implications prompted the double-layer work to be extended to oxide surfaces, and the organic molecules (simple at first, but gradually more complex) that were introduced in the double-layer projects were also used in studies on the stability of hydrophobic colloids and emulsions. In the latter case thin liquid films came into the picture again. An important aspect in this work (but also in the later activities) was a close relation between theoretical developments and experimental verification. Moreover, new domains were only entered if preceding ones had become properly understood.

Lyklema has by now published over 250 papers and supervised more than 40 dissertations which were pioneering in such diverse areas as adsorption of phosphates, of humic acids, of polymers, and of polyelectrolytes, ion binding by natural and synthetic polyelectrolytes, adhesion of proteins and micro-organisms at various interfaces, self-association of amphiphilic molecules (expecially in relation to membrane formation), dynamics of double layers, stability of colloids, statistical-thermodynamical description of the water structure, and the importance of undulations and curvature in liquid films and micro-emulsions. Even this long list is far from complete.

Many of those research fields have now become well-established areas, with implications in agricultural and environmental applications (e.g., bacterial adhesion and metal speciation), fundamental physics and material science (e.g., the theory for chain molecules at interfaces developed in Wageningen) and biology (e.g., membrane permeation). It is obvious that Lyklema has not accomplished all these results on his own. Many co-workers have been involved in these projects, but his inspiration, supervision and continuous personal interest have always been invaluable. Characteristically, Lyklema keeps emphasizing the interrelations between the various areas and disciplines, and the need to learn from each other and to exchange ideas and knowledge.

Several of his co-workers have now become internationally renowned experts in their own right, but they all wholeheartedly acknowledge the great influence Lyklema has had (and still has) on their scientific development. Three of his former PhD students have become full professors, and some others hold industrial research positions of comparable rank. Under Lyklema's guidance an originally very small group of three staff members and three PhD students in the early 1960s has evolved into a department of 14 permanent scientific staff and some 20 PhD students, about half of which are sponsored by industry. On average, annually some 10 foreign guests are active in his department for periods ranging between 1 month and 1 year.

On the national level, Lyklema's merits have been well-recognized, the best proof of which is that, for his scientific and organizational accomplishments, in 1988 he was awarded the title of "Knight of the Order of the Dutch Lion". So, in a British context, we could call him "Sir Hans"!

Internationally his fame has become of equal measure, to say the least. Every year he receives numerous invitations for plenary lectures at international meetings, of which he can accept only a small part. He is (or was) a member of the Editorial Board of virtually all leading journals in colloid and interface science and electrochemistry, is for the second time a Council member of the "Faraday Division of the Chemical Society", has long functioned as deputy chairman of the "Ausschuss Grenzflächen" of the Verein Deutscher Ingenieure, is a member of the Advisory Board of the Fraunhofer Institute since 1981, and has been chairman of the "European Chemistry at Interfaces" committee for over 20 years. Between 1976 and 1983 he was first secretary and then chairman of the "Commission for Colloid and Surface Chemistry, including Catalysis" of IUPAC. In 1978 he founded, together with Professors Weiss and Parfitt, the "International Association of Colloid and Interfacial Scientists", for which he has since edited the "Newsletter", appearing twice a year.

After having received the "Nightingale Award for Medical Electronics" as early as 1963, and the gold medal of the Center for Marine Research "Ruder Boskovic" in Zagreb in 1986, he was awarded a honorary degree from the University of Louvain-la Neuve in Belgium in 1988, which explicitly honoured the physical chemical impact he created in addressing agricultural and related problems.

It is a rare phenomenon that such an impressive scientific status goes along with an outstanding proficiency to lecture. Lyklema is not only renowned for his presentations during conferences and symposia, but over the years has shown to be a gifted lecturer in a variety of courses. Many of these courses he initiated and structured himself, also at the introductory level. Wageningen Agricultural University has greatly benefited from his teaching capacities and also from those on the organization level.

About half of the dissertations in Lyklema's department have been financed by industry, illustrating once more his philosophy that fundamental research may be made beneficial to applied and industrial purposes. As a logical consequence of this attitude he has been a consultant for many Dutch and foreign companies.

In conclusion, Hans Lyklema has built up a most impressive reputation in the area of colloid and interface science. This holds both for his teaching and for the impressive amount of fundamental research. This research has proved to be of considerable importance for agricultural and industrial applications.

Last but not least, we would be amiss if not mentioning that Hans Lyklema is a most amiable, agreeable and humorous personality.

Progr Colloid Polym Sci (1996) 101:9–17

J. Lyklema
W. Norde

Interfacial behaviour of biomacromolecules

Prof. Dr. J. Lyklema (✉) · W. Norde
Wageningen Agricultural University,
Department of Physical and Colloid
Chemistry, Deijenplein 6 6703 HB
Wageningen, The Netherlands

Abstract Biological macromolecules, proteins in particular, belong to the most interesting and challenging surfactants. Many of these molecules are endowed with very special properties, required to carry out specific biological processes. Structure-functioning relationships have been studies for decades by biologists and biochemists, and many applications are now known.

A colloid scientist would interpret biomacromolecules as supramolecular structures that only to a limited extent obey classical colloidal laws. The delicate balance between energetically and entropically driven interactions that determine the special structures and the stability of these molecules is a particularly interesting problem. Offering a surface onto which these molecules can adsorb and monitoring the incurred structural alterations is one approach to obtain more information. Results will be discussed for a number of proteins, including γ-globulins and constituent parts of them. In addition, some new information will be presented about the biopolymers in the surfaces of bacterial cells.

Key words Protein adsorption – proteins at interfaces, structure – immunoglobulins at interfaces – bacterial cell walls – dielectric dispersions

Introduction

Biological macromolecules belong to the most challenging model colloids that exist. Over epochs that have to be counted in millions of years they have evolved to their present-day compositions and functions, many of these being very specific. Resolution of structure-function relationship is one of the topical objects of study for biochemists and biologists. What can a colloid scientist contribute? What can we learn from studies involving interfacial aspects? And are there applications?

Let us first define the systems and the challenges. Biomacromolecules are large molecules of biological origin, including proteins, ribonucleic and desoxynucleic acids, and polysaccharides. In aqueous solution they may have molecular masses ranging from a few thousands to over a million daltons. The molecular sizes are such that they may be categorized as "colloidal". We have called them "model colloids" because nature is capable of producing exact replicas that remain invariant over the ages. Certainly, they are no model colloids in the more classical sense of "simple systems with simple properties". Classical colloid science has contributed by applying its techniques to determine sizes, shapes, electrokinetic charges etc., and the influences of electrolytes, pH, temperature on these parameters. More recently, these measurements have been supplemented by a variety of more advanced optical and calorimetric techniques.

More challenging are studies aiming at alterations that these molecules might undergo when subjected to external forces, or when brought in a different environment.

Denaturation by changes in temperature or by adding certain reagents are the most familiar examples. In the present work we are instead interested in the consequences of offering an interface. Will the biomacromolecule adsorb, and if yes, by what forces? Are there changes in the structure, and if yes, are these reversible or irreversible? Offering an interface makes good sense for at least three reasons:

i) For academic reasons it is noted that the external force is relatively mild, and can be modified by changing the surface charge and its relative bydrophilicity or hydrophobicity. In this way, subtle changes in the structure may be provoked. Adequate means are now available to also determine such structural alterations in the adsorbed state, including circular dichroism (CD)-, fluorescence-, infrared spectroscopy, differential scanning calorimetry and isothermal (micro-) calorimetry.

ii) In nature itself examples are known of biomacromolecules that display specific interfacial features. For instance, lipases tend to be active only if adsorbed at an oil-water interface, and γ-globulins tend to adsorb with their so-called Fc-fragments to surfaces (of cells, etc.) so that the parts, carrying the antigen-specific receptor sites (the $F(ab')_2$-fragments), can turn toward the (aqueous) solution side. Besides these examples, there are also biological systems where biomacromolecules intrinsically enrich surface layers, for instance in bacterial cell walls.

iii) Studying interfacial properties of biomacromolecules is also very relevant for practice, for instance in dealing with biofouling, dental plaque removal and the preparation of specific biosensors.

The above considerations motivate the present paper. We shall divide it into three parts: i) a brief review of our knowledge on structural changes upon adsorption. ii) presentation of some recent developments and iii) introduction of a new approach to obtain information on the composition and electrochemistry of bacterial cell walls. In parts i) and ii) only proteins will be considered. Given the subtlety and intricacy of some of the trends to be investigated, it is desirable to apply several techniques to study a specific system.

Inference on structural stability of proteins from adsorption studies

Over the last decade a multitude of protein adsorption studies have been published. Given the wide choice in adsorbate, adsorptive, pH, salt concentration, conditions of measurement and parameters studied, a vast literature, which has been reviewed several times, has become available [1–6]. Considering the basic mechanisms, it appeared expedient to make a first distinction between structurally "hard" and structurally "soft" proteins as the extremes of a scale of varying "hardness". Morphologically speaking, a protein is called (ideally) hard when upon adsorption it maintains the structure it has in the dissolved state. Less than ideally hard molecules may alter their structures somewhat, the more so the lower their internal coherence. On the other end of the range the (ideally) soft molecules are found; upon adsorption they spread completely. This distinction between "hard" and "soft" proteins is not absolute because the structural resilience against adsorption is not necessarily identical to that against changes in temperature or against addition of denaturant.

Figures 1 and 2 exemplify how from adsorption studies evidence for structural changes can be obtained. In both graphs the plateau adsorption Γ(max) is plotted as a function of pH, comparing different adsorbates. Figure 1 is representative of a hard protein (ribonuclease, RNase),

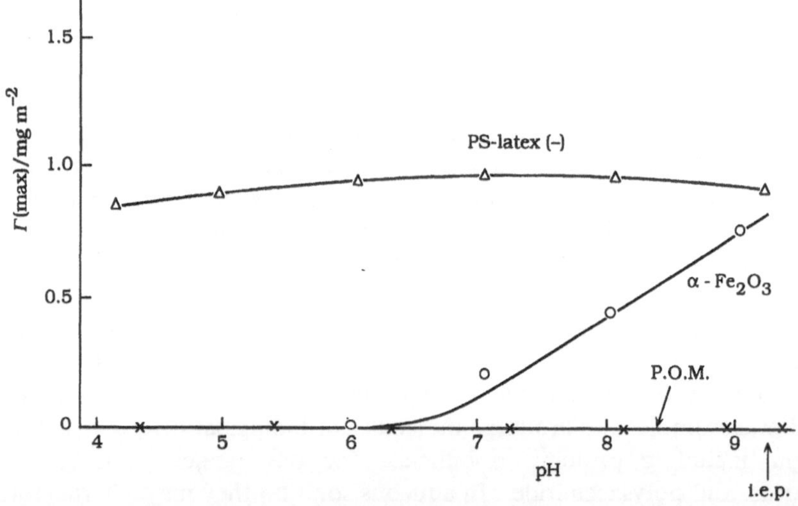

Fig. 1 Plateau adsorption of RNase as a function of pH on a negatively charged poly(styrene) latex, haematite (α-Fe_2O_3) and poly(oxymethylene) (P.O.M.) crystals. Electrolyte, 0.01 or 0.05 M KNO_3. Temperature 25 °C. The arrow indicates the isoelectric point of RNase

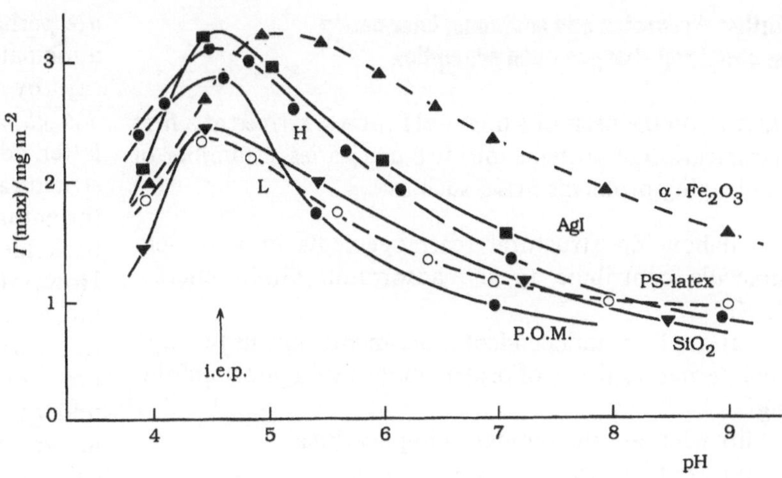

Fig. 2 Plateau adsorption of human serum albumin (HSA) as a function of pH on haematite (α-Fe_2O_3), silver iodide, (AgI) high (H)- and low (L)-charge poly(styrene) latex, silica (SiO_2) and poly(oxymethylene) (P.O.M.) crystals. Electrolyte, 0.01 or 0.05 M KNO_3. Temperature 25 °C. The arrow indicates the isoelectric point of HSA

Fig. 2 is typical for a soft protein (human serum albumin, HSA).

The curves of Fig. 1 are readily interpreted in terms of a simple adsorption picture, determined by electrostatic interaction and hydrophobic bonding. The isoelectric point (i.e.p.) of RNase is at pH $\approx$ 9.5, slightly depending on the concentration and nature of the indifferent electrolyte. Hence, on negatively charged poly(styrene) latex, RNase binds electrostatically at low pH. Moreover, RNase molecules have hydrophobic patches at their surfaces [7] and as the latex is also largely hydrophobic, adsorption also takes place by hydrophobic bonding and this attraction is strong enough to ensure adsorption till the i.e.p. and even beyond it. The result is a monolayer coverage, corresponding with about 0.9 mg m^{-2} which is more or less independent of pH. Poly(oxymethylene) is the other extreme. Such crystals do not permit hydrophobic bonding and are uncharged so that there is no attraction for RNase. Hence, on this adsorbate no adsorption takes place. Haematite (α-Fe_2O_3) takes an intermediate position: its surface is hydrophilic, but at sufficiently high pH this mineral is negatively charged so that adsorption of RNase is possible for electrostatic reasons. The precise transition between complete adsorption and no attachment at all depends on the surface treatment of the mineral, especially on the amount of low molecular mass ions that is adsorbed.

Figure 2 exhibits quite different trends. Neither is a clear relationship observed with the hydrophobicity of the adsorbate (AgI and the two latices are hydrophobic, haematite and silica are hydrophilic and poly(oxyethylene) takes an intermediate position), nor with the surface charge (the two latices carry a constant negative charge caused by strong sulfate groups, H and L standing for a high- and low-charge sample, respectively, silica bears a pH-dependent negative charge, AgI a pH-independent negative charge and haematite is positively charged over the pH range studied). Rather the curves appear more or less symmetrical with respect to the isoelectric point of the albumin, or at least close to it. Considering that there is some uncertainty in the determination of the specific surface area of the adsorbates, the overall impression is that the curves are primarily determined by the properties of the protein and only to a minor extent by the protein-adsorbent interaction. The amount adsorbed at the maximum (2–3 mg m^{-2}) corresponds more or less with a saturated monolayer of close-packed, side-on albumin molecules. Increasing the electrokinetic charge on the protein molecule, irrespective of the direction, leads to the reduction of the amount adsorbed. Of the two options to interpret this decrease i) the molecules retain their side-on mode of attachment but the lateral distances increase because of protein-protein repulsion and ii) the molecules unfold and spread to an extent that increases with their electrokinetic charges) only the latter was found tenable. One of the arguments stemmed from potentiometric titrations, from which it could be inferred that the tendency of the protein to direct carboxyl groups to the surface increases with increasing charge on the protein, be it negative or positive [9, 10]. In conclusion, the trends of Fig. 2 are typical for a "soft" protein.

Other examples of relatively structure-stable proteins are lysozyme, cytochrome-C and subtilisin. Relatively "soft" proteins include immunoglobulins, myoglobulin, fibrinogen, α-lactalbumin and casein. There is a trend that small protein molecules tend to be more hydrophobic than larger ones because they have a large area/volume ratio, so that it is more difficult for them to bury all the hydrophobic groups in the interior of the molecule.

Further discussion and additional information on structural changes upon adsorption

Having, on the basis of Γ(max)-pH curves, arrived at a first classification of proteins into two categories, a number of challenging problems arise, such as:

i) how do structural rearrangements in a protein molecule contribute to the adsorption Gibbs energy, $\Delta_{ads}G$?

ii) are there independent experimental means to study the internal resilience of protein molecules against unfolding?

iii) what are the biological implications?

iv) are there technical applications?

Issue i) first requires us to define the notion of structure. This is easy because the secondary and tertiary structure may serve for that. Practically, it means that the structure is determined by the spatial positions of all atoms, by the fractions of amino acids in α-helices or β-sheets, etc. More difficult is the definition of structural alteration, because minor or more drastic changes may be incurred, depending on the nature of the perturbation (mild or substantial temperature variations, offering a hydrophilic or hydrophobic adsorbent, etc.). Experimentally, the problem is that by different techniques different aspects of the structure are measured. Closely related is the issue of reversibility: can adsorbed proteins be desorbed, and if yes, does the desorbed protein regain its native conformation? The biological implications of this issue are obvious.

It is in line with the "holistic" approach of colloid science to consider the thermodynamics of the adsorption of the molecule as a whole. Spontaneous adsorption requires $\Delta_{ads}G$ to be negative, or

$$\Delta_{ads}H - T\Delta_{ads}S < 0. \tag{1}$$

For "hard" molecules the most prominent driving forces are:

- electrostatic attraction, for which $\Delta_{ads}H < 0$; $T\Delta_{ads}S$ may play an additional role, depending on the redistribution of charges that accompanies the attachment.
- hydrophobic bonding, for which $\Delta_{ads}H$ is small, it may be positive or negative; $T\Delta_{ads}S > 0$ and this is the driving force.
- Van der Waals attraction usually plays a minor role, so does the loss in kinetic energy upon adsorption (the ensuing entropy loss is negligible against the gain in entropy by hydrophobic bonding).
- Sometimes specific binding may occur.

For "soft" molecules $\Delta_{ads}H_{str}$ and $T\Delta_{ads}S_{str}$ have to be added, where the subscript str refers to "structural". These are perhaps the most interesting parameters. One would anticipate that breakdown of structure upon adsorption, say by the loss of part of the α-helices, requires $T\Delta_{ads}S_{str} > 0$ and $\Delta_{ads}H_{str} > 0$. If the former outweighs the latter, adsorption can be promoted by an entropy gain of structural origin, whereby the structural contribution to the enthalpy is endothermic. Indeed, such situations have been identified for serum albumins on latices [11, 12]. However, nature has its idiosyncrasies and cases are now known where the α-helix content increases upon adsorption (see below). Obviously, structural alterations never occur on their own because otherwise the protein would undergo spontaneous reconfirmations without the need for an adsorbent. However, such changes do contribute substantially to the $\Delta_{ads}G$ if the different groups that become exposed in this way can make favourable interactions with the absorbent.

Regarding issue ii), information about the structure and its resilience against changes mostly stems from thermodynamics and spectroscopy. Thermodynamically by differential scanning calorimetry (DSC) phase transitions as a function of temperature, and the enthalpy change of these transitions can be measured [13, 14]. Direct determination of $\Delta_{ads}H$ in a sensitive calorimeter may serve as another starting point; if the contributions to $\Delta_{ads}H$ can be estimated $\Delta_{ads}H_{str}$ is obtainable by subtraction. Using this procedure for the well-studied systems of serum albumin and RNase on poly(styrene) latices, it was indeed found that $\Delta_{ads}H_{str}$ was more positive for the latter. Further information is obtainable from isothermal proton titration microcalorimetry [14].

Of the spectroscopies, circular dichroism (CD) presents itself in the first place because by this technique the secondary structural elements can be determined. For free proteins in solution this is an established technique, but for adsorbates application requires solving some experimental problems. One of these is that situations should be avoided where the adsorbent absorbs (UV) light at the same wavelengths where the dichroism is to be measured. For this reason poly(styrene) latices are not convenient. Moreover, the refractive indices of sorbent and water should be close, to minimize scattering. Recently, measurements have been reported for lysozyme and albumin on silica [15] and for subtilisin on Teflon [16]. Structural information can also be obtained by fluorescence of intrinsic labels (such as the tryptophan) [16], or from infrared absorption, to which we shall return in the following section [17].

Some trends, obtained for albumin and lysozyme on silica [15] are illustrative. For both proteins the fraction of α-helix in the adsorbed state tends to be lower than that of the free protein, the more so, the lower the degree of occupancy. The implication is that upon adsorption these

Table 1 α-helix content of bovine serum albumin (top) and lysozyme (bottom), before, during and after adsorption on silica [12]. The degree of occupancy $\theta = \Gamma/\Gamma(\max)$. Temperature 20 °C, 10^{-2} M, phosphate buffer.

	pH	Before adsorption	In the adsorbed state		Desorbed
	4.0	69	–	–	50
BSA	4.7	70	–	–	51
	7.0	74	28 ($\theta = 0.24$)	38 ($\theta = 1.00$)	55
	4.0	32	13 ($\theta = 0.23$)	25 ($\theta = 1.00$)	31
LYS	4.7	33	22 ($\theta = 0.10$)	30 ($\theta = 1.00$)	31
	7.0	32	–	–	33

molecules can unfold more extensively when the surface is emptier. In addition, the earlier made contention that lysozyme is "hard" has to be qualified. For subtilisin, also relatively hard, something similar was found, but on Teflon the α-helix content increases upon adsorption, the more so, the lower the occupation [16]. Generally speaking, offering a hydrophobic surface to a protein implies a stronger demand on the structural integrity. Most proteins functioning in aqueous environments tend to screen their hydrophobic groups from the solution by burying them inside the molecule. Disrupture of the subtle entropy-enthalpy balance that keeps the molecule intact is more probable when hydrophobic bonding of the liberated hydrophobic groups is possible. So it is likely that there are proteins that adsorb in the "hard" fashion on hydrophilic surfaces but behave as if they were "soft" on hydrophobic adsorbents. Another observation along these lines [14] refers to the difference in behaviour between lysozyme (relatively hard) and α-lactalbumin (relatively soft) on the hydrophobic poly(styrene) and the hydrophilic haematite surface: both proteins lose substantial parts of their structure on the former but on haematite they are discriminated: lysozyme only loses a fraction of its ordered secondary structure, whereas α-lactoglobulin denatures almost completely.

Table 1 collects some interesting data on the reversibility of lysozyme and (bovine serum) albumin adsorption on silica. Besides the trends discussed above it is also seen that after desorption BSA molecules do not completely recover their native states, whereas lysozyme does. Otherwise stated, lysozyme can be fully recycled, in contrast to BSA. This observation may also be used to redefine the notion of structural resilience.

Reflectometry is another useful technique, which can give information on the dynamics of adsorption and desorption if it is combined with controlled flow supply. The principle of the method as applied by us [18] is that the protein solution approaches the flat adsorbent surface as a narrow, perpendicular jet. Arriving convectively at the surface, it spreads laterally, except in its centre, where the liquid is stagnant. The amount adsorbed at this stagnation point is monitored reflectometrically. The rate of supply can be computed [19]. The advantages of this method are the relatively simple construction and the possibility of following the adsorption as a function of time. By comparing the computed rate of supply with the reflectometrically measured rate of adsorption, information can be obtained on adsorption barriers, etc. Desorption and replacement by other adsorptives can also be investigated. One drawback is the translation of the signal in terms of adsorbed amounts, for which a standardization step is required: there is always some uncertainty in the interpretation of signal variations since they can be either caused by changes in adsorbed amounts or changes in conformation. Another restriction is that ideally flat, macroscopic surfaces, so-called Fresnel surfaces, are needed as the adsorbents; their surface properties may differ from those of dispersed particles.

Figure 3 gives an illustration of the potentialities of this method [20]. It mimics the various steps taken in practice in the preparation of immunochemical tests (on pregnancies in this example). Such tests are based on the action of immunoglobulins (IgG's) as antibodies against antigens. The IgG's are adsorbed on carrier surfaces, after flushing any remaining open patches on the surface are covered by bovine serum albumin (BSA), and finally, after another rinse, the antigen is bound, (in this example human chorionic gonadotropin (HCG)).

From the figure is can be inferred that after 30 min of adsorption most of the IgG on the surface cannot be desorbed by the solvent; some additional after-coating with BSA takes places and from the amount of HCG, bound in the last step, the ratio between the number of adsorbed HCG and IgG molecules can be determined, that is: the immunological activity of the IgG molecules in the adsorbate. We return to the structural properties of IgG molecules in the following section.

This discussions and the illustrations given take us automatically to the fascinating query iii): why do these molecules behave biologically as they do? Biologically

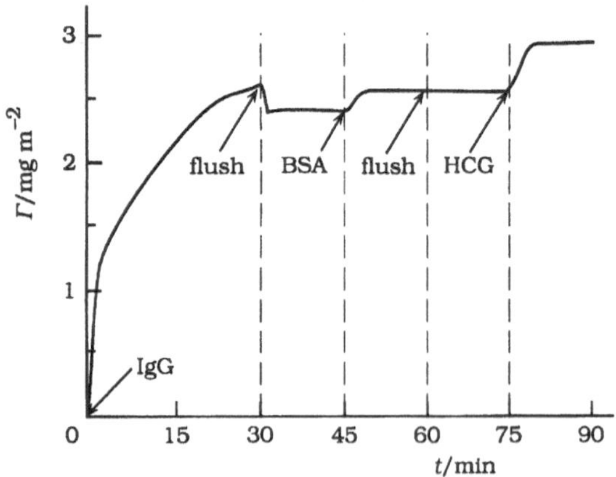

Fig. 3 Reflectometric study of the successive adsorption of immunoglobulin, bovine serum albumin and human chorionic gonadotropin on a methylated silica wafer. Concentrations of the three proteins $7.5\,\mathrm{g\,m^{-3}}$, $30\,\mathrm{g\,m^{-3}}$ and $1\,\mathrm{g\,m^{-3}}$, respectively. pH = 5, ionic strength 0.1 M

speaking, a certain hardness, or, for that matter, resilience against conformational alterations, may be conducive for the functioning of enzymes: this requires very specific structures. RNase, lysozyme and subtilisin belong to this category. Albumin is multifunctional; it may be advantageous that the molecules of this protein readily adjust their structures. Immunoglobulins have special properties that will be addressed in the following section. In connection with the observation that on hydrophobic Teflon surfaces the fraction of α-helices in subtilisin increases upon adsorption it may be noted that a number of membrane proteins also contain relatively large fraction of α-helix. For instance, bacteriorhodopsin (which uses light energy to pump protons across the membrane of *Halobacterium halobium*) contains seven trans-membrane helices [21]. Colicin A, the protein that kills *E. coli* cells, spontaneously forms ten α-helices when it enters the membrane of its host. Perhaps this transition is related to its antibiotic action. Upon adsorption, proteolytic enzymes may become more or less susceptible to "cannibalistic" attack by their colleagues in solution, depending on the nature of the surface [34]. This observation is relevant for the storage of enzymes in disperse systems, for example in liquid detergents.

Practical applications are abundant. Rather than summing them up, we shall discuss some background for preparing biosensors in the following section. This issue was already addressed in Fig. 3.

Adsorption of immunoglobulin fragments

Immunoglobulins constitute interesting biomacromolecular models because their molecules are bifunctional; they should attach to receptors on a variety of cell surfaces and at the same time expose their immuno-active parts to the solution in order to bind antigens. The experiments of Fig. 3 were motivated by this dual function. IgG molecules have, roughly speaking, the shape of the letter Y. The lower part (with which the molecule binds to receptors) is the Fc-fragment, the two upper parts, containing the antigen-binding sites at their tips, are the $F(ab')_2$-fragments. The hinges, connecting the three domains are flexible, allowing the distance between the two binding sites to vary. Given the different functionalities of the domains the question arises as to what extent the structural stabilities also differ.

The adsorption of (monoclonal) IgG has received due attention. In broad lines, these molecules behave as relatively soft. Part of the evidence stems from the Γ(max)-pH curves, which are similar to those of Fig. 2 [22]. The maxima correspond roughly with end-on adsorption. Such studies, even if extended by thermodynamic and some spectroscopic data, do not lend themselves very well to establishing the structural properties of the individual domains because only sum-effects are measured. Incorporation of fluorescence labels may proceed domain-specifically, but the drawback is that such guest molecules may themselves affect the structure.

Today it is possible to cleave IgG-molecules enzymatically into their Fc and $F(ab')_2$ parts and obtain sufficient amounts of monoclonal fragments for adsorption and spectroscopic studies. We shall now present some findings, obtained by Buijs [17]. As before [22] he found that on hydrophobic latices IgG and its $F(ab')_2$ fragments adsorb strongly, with maxima in the Γ(max)-pH curves. However, on hydrophilic latices the adsorption depends on electrostatics: neither IgG nor $F(ab')_2$ adsorbs under conditions of electrostatic repulsion and if adsorption does take place, it is in a flat orientation. Although on all surfaces the adsorption trends to IgG and $F(ab')_2$ are similar, there is some evidence that the $F(ab')_2$ fragments are more sensitive to changes in the electrostatic interaction. This trend was confirmed reflectometrically and from Fourier-transform infrared (FTIR) analysis. The inference is that as a whole complete IgG molecules are softer than their $F(ab')_2$ fragments. This, in turn, would mean that the Fc domain is the softer part and this would be in line with the trend for IgG molecules to adsorb with the Fc moieties towards surfaces.

From the second derivative of the infrared spectrum of the so-called amide-I region it is possible to obtain quantitative information on the fraction of β-sheets. In free IgG molecules these constitute the dominant structure elements. The domains of the different polypeptide chains form layers of antiparallel β-sheets, comprising 50–60% of the amino acid residues, enclosing a predominantly

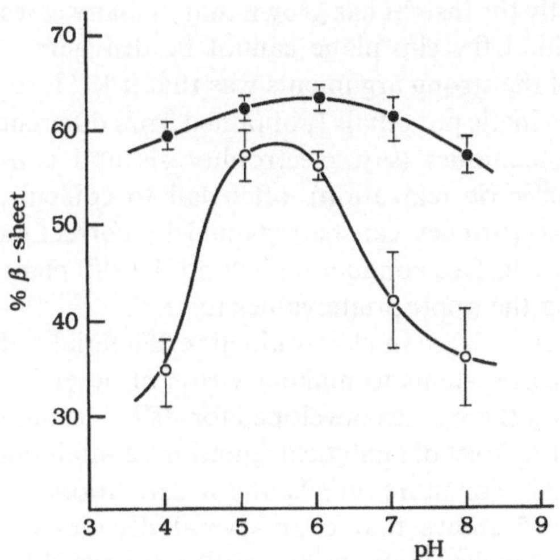

Fig. 4 Fraction of amino acids in β-sheets of (○), monoclonal IgG and (●), its corresponding $F(ab')_2$ fragments, adsorbed on silica. (Data by Buijs.)

hydrophobic interior [23–24]. FTIR-determined fractions of β-sheets in the absorbed state are shown in Fig. 4. The IgG studied was monoclonal and directed against the pregnancy hormone HCG, as before; its isoelectric point was 5.6–6.0, as determined by isoelectric focusing. The fragments were obtained from the intact IgG's by pepsin digestion. The trend is convincing: the β-sheet content in the adsorbed $F(ab')_2$ fragment is much less pH-sensitive, which is in line with the earlier conclusion that the IgG's as a whole are structurally weaker than the $F(ab')_2$ because of the relative softness of the Fc-domains. Similar trends were observed with other IgG's and their fragments and on other surfaces [17].

The consequence of this observation for the understanding of IgG attachment is that the tendency to do so with the Fc's directed to the surfaces may be determined by the relative softness of this fragment. In the literature this tendency is more commonly attributed to the putative hydrophobic nature of the Fc. However, in Buijs' experiments no support for this hydrophobicity could be found.

Composition and ionic mobilities in bacterial cell walls. A dielectric relaxation study

The last example concerns the composition of bacterial cell walls. In this case we are dealing with biomacromolecules that are intrinsically present in a biological system. Unexpectedly, these systems appear to be excellent models for electrochemical and electrokinetic studies: not only can useful information be obtained by these electrical techniques, but the models also serve as a check for new developments in electrokinetics. Anticipating more extensive publications elsewhere [25], the potentialities of systems and methods are illustrated by taking dielectric spectroscopy of Gram-positive bacteria as the example.

Bacterial cells are covered by a porous, three-dimensional biomacromolecular layer, the cell wall, or cell envelope. An important constituent of this wall is peptidoglycan, a protein-carbohydrate. In Gram-positive cell walls peptidoglycans are present in high concentrations. Other biopolyelectrolytes include teichuronic acid, (lipo-)teichoic acid, (lipo-)polysaccharides, (lipo-)proteins, enzymes and mycolic acids. Charge-carrying units are mostly carboxyl, phosphate and amino groups. Hence, the bacterial cell wall is amphoteric: its surface charge depends on pH and an isoelectric point can be assigned to it.

At their outer sides the cell walls are in open contact with the solution in which they are immersed. One of the consequences is that they can be titrated [26, 27]. Such titrations, if carried out at different electrolyte concentrations, yield useful information on the nature and numbers of the various groups; from the salt dependence ion binding can be studied. It is gratifying that these results are very well in line with chemical analyses [26]. It may be concluded that, at least for a number of species, the cell wall composition is well established.

At their inner sides, the walls of Gram-positive bacteria are separated from the cytoplasma by the bacterial membrane. The permeability of this membrane is an issue on its own but for the relaxation studies to be discussed, the membrane may be considered impenetrable on the time scale of one cycle in the a.c. electric field to be applied.

Besides the static information (how many charges, what kinds of groups, etc.) the dynamics of these charges (how mobile are they?) are also relevant for a number of purposes, for instance the interpretation of bacterial adhesion. It is here that electrokinetic methods become useful. When the cell wall is subjected to an electric field, the ions in it start to move. Ionic mobilities, or at least averaged values, can be assessed from the conductivity of this surface layer, $K^{\sigma i}$, Conductivities of colloids can be obtained in d.c. or a.c. fields. If the field is alternating, the total conductivity becomes complex, written as $\hat{K}(\omega)$ if ω is the frequency of the applied field. This complex conductivity consists of a real and an imaginary part

$$\hat{K}(\omega) = K(\omega) - \mathrm{i}\omega\varepsilon_0\varepsilon'(\omega), \qquad (2)$$

where ε_0 is the dielectric permittivity of vacuum and ε' the (real) relative dielectric constant. The two parts can be measured. In principle they give the same information because they are related through so-called Kramers-Kronig relations. In the present case we consider the dielectric dispersions $\Delta\varepsilon'(\omega)$, where $\Delta\varepsilon'$ is the increase in the relative dielectric constant of the medium caused by the polarization of the double layer around the colloidal

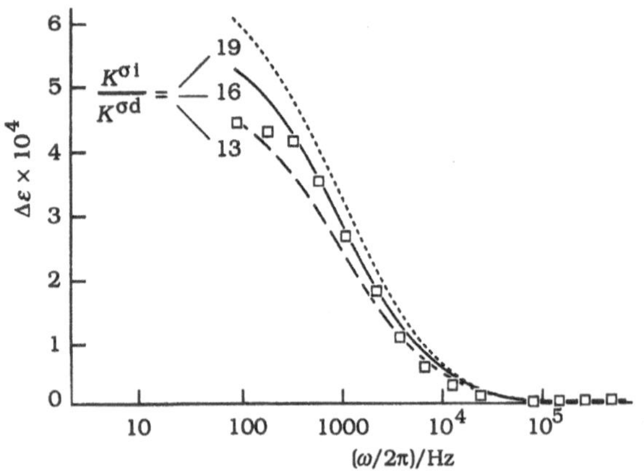

Fig. 5 Dielectric relaxation of a *Corynebacterium* species, strain 44016 suspension. $\zeta = 70$ mV. Electrolyte, 10^{-3} KNO_3, temperature 25 °C, pH = 6.5, volume fraction between 0.01 and 0.1. (Data van der Wal, theory by Kijlstra and Minor.) Discussion in the text

particles. The experimental points in Fig. 5 show such a decrease for a *Corynebacterium.*

To analyze these curves existing theory had to be extended to account for the conduction in the cell wall. Generally speaking, advanced electrokinetic theories take surface conduction in the diffuse part of the double layer (conductivity $K^{\sigma d}$) into account, but not conduction behind the slip plane. The reason is that in these theories the hydrodynamic and electric field equations are solved for the double-layer part outside the slip plane, the particle plus the hydrodynamically stagnant layer are considered impervious to ions and non-conducting. This is for instance the case with the often used O'Brien-White model for electrophoresis [28] and for the Dukhin-Shilov [29] and Fixman [30] theories for dielectric relaxation.

More recently the insight has grown that in many cases conduction behind the slip plane cannot be disregarded [31, 32]. One of the strong arguments was that if $K^{\sigma i}$ is set zero the electrokinetic potentials ζ, obtained from different electrokinetic techniques (say, electrophoresis and conductivity or dielectric relaxation), often fail to coincide. Electrokinetic consistency can be regained by correcting these theories for surface conduction behind the slip plane and substituting the appropriate values for $K^{\sigma i}$.

This is a current issue in electrokinetics. Bacterial cell walls are excellent systems to make a virtue of necessity. With M. Minor a theory was developed for $\Delta\varepsilon'(\omega)$ leading to a relatively simple set of analytical equations containing the ratio $K^{\sigma i}/K^{\sigma d}$. For more details and a derivation, see [25, 33]. Figure 5 shows that over several decades the frequency of the experiments is very well represented by the theory, setting $K^{\sigma i}/K^{\sigma d} = 16$. As $K^{\sigma d}$ is known from Bikerman's theory [33], $K^{\sigma i}$ can in this way be obtained. Since the numbers of charge carriers are known from titration, the ionic mobilities in the cell wall can be found. They appear to move about a factor of 3 slower than free ions would be in aqueous solution. Similar results have been obtained with other bacteria and extension appears appropriate.

Thus this example shows that the application of modern techniques to bacterial cells as model systems is not only conducive for obtaining valuable dynamic information, but also serves as a means to validate new theoretical developments.

Acknowledgments The authors thank M. Maste, J.A.G. Buijs, and A. van der Wal for allowing the use of some of their unpublished results.

References

1. Norde W (1986) Adv Colloid Interface Sci 25:267–340
2. Norde W, Lyklema J (1991) J Biomat Sci, Polymer Edn 2:183–202
3. Haynes CA, Norde W (1994) Colloids Surf B, Biointerfaces 2:517–566
4. Norde W (1995) Cells and Materials 5:97–112
5. Andrade JD, Hlady V (1986) Adv Polym Sci 79:1–63
6. Sadana A (1992) Chem Revs 92:1799–1818
7. Lee B, Richards FM (1971) J Mol Biol 55:379–400
8. Papenhuysen J, Fleer GJ (1984) J Colloid Interface Sci 100:561–570
9. Norde W, Lyklema J (1978) J Colloid Interface Sci 66:266–276
10. Lyklema J, Norde W (1995) In: Goodwin JW, Buscal R (eds) Colloidal polymer particles. Academic Press, London, pp 233–243
11. Norde W, Lyklema J (1978) J Colloid Interface Sci 66:295–302
12. Norde W, Lyklema J (1979) J Colloid Interface Sci 71:350–366
13. Privalov (1979) Adv Protein Chem 33:167–241
14. Haynes CA, Norde W (1995) J Colloid Interface Sci 169:313–328
15. Norde W, Favier JP (1992) Colloids Surf 64:87–93
16. Master MCL, Pap EHW, van Hoek A, Norde W and Visser AJWG (1995) J Colloid Interface Sci: submitted
17. Buijs JAG, (1995) PhD thesis Agricultural University Wageningen NL
18. Dijt JC, Cohen Stuart MA and Fleer GJ (1994) Adv Colloid Interface Sci 50:79–101
19. Dabros T, van de Ven TGM (1983) Colloid Polym Sci 261:694–707
20. Buijs JAG, White DD and Norde W (1995) J Imm Meth: submitted
21. Kyte J, Doolittle RF (1990) J Mol Biol 157:105–132
22. Elgersma AV, Zsom RLJ, Norde W and Lyklema J (1991) Colloids Surf 54:89–101
23. Deisenhofer J (1981) Biochemistry 20:2361–2370
24. Poljak RJ, Amzel LM, Chen BJ, Phizackerly RP and Saul F (1974) Proc Natl Acad Sci USA 71:3440–3444

Progr Colloid Polym Sci (1996) 101:9–17

25. Van der Wal A, Minor M and Lyklema J (1995) NATO-ASI on Nanoparticles: in press
26. Van der Wal (1996) PhD thesis Agricultural University Wageningen NL
27. Plette ACC, van Riemsdijk WH, Benedetti MF and van der Wal A (1995) J Colloid Interface Sci 173:354–363
28. O'Brien RW, White LR (1978) J Chem Soc Faraday Trans (II) 74:1607–1626
29. Dukhin SS, Shilov VN (1974) Dielectric Phenomena and the Double layer in Disperse systems and Polyelectrolytes. Wiley, Jerusalem
30. Fixman M (1983) J Chem Phys 78:1483–1491
31. Kijlstra J, van Leeuwen HP and Lyklema J (1992) J Chem Soc Faraday Trans 88:3441–3449
32. Kijlstra J, van Leeuwen HP and Lyklema J (1993) Langmuir 9:1625–1633
33. Lyklema J (1995) Fundamentals of Interface and Colloid Science, Volume II ch 4. Academic Press (London)
34. Maste MCL (1996) PhD thesis Agricultural University Wageningen NL

Progr Colloid Polym Sci (1996) 101:18–22

S.E. Friberg
T. Huang
L. Fei
S.A. Vona, Jr.
P.A. Aikens

Vapor pressure of phenethyl alcohol in the system with water, and polyoxyethylene, 4, lauryl ether (Brij® 30)

Prof. Dr. S.E. Friberg (✉) · T. Huang
L. Fei · S.A. Vona, Jr.
Center for Advanced Materials Processing
Clarkson University
Potsdam, New York 13669-5814, USA

P.A. Aikens
ICI Surfactants
3411 Silverside Road
Wilmington, Delaware 19850-5391, USA

Abstract The phase diagram was determined for the system water, phenethyl alcohol and a commercial surfactant, Laureth 4 (Brij® 30), and the vapor pressure of phenethyl alcohol (PEA) measured in the entire system. The phase diagram showed complete solubility of the surfactant and the phenethyl alcohol with limited solubilization of water. The water was dissolved into the surfactant forming a surfactant solution and a lamellar liquid crystal.

The variation of vapor pressure for the solutions was related to the surfactant self-association and to the formation of inverse micelles. The Henry's constant was 1.5 for PEA in the surfactant and increased systematically for the different phases to a value of approximately 10^3 for water.

Introduction

The vapor pressure, reflecting the chemical potential of the compounds, is an essential phenomenon to characterize a colloidal system and a number of contributions have treated this aspect of microemulsion systems (1–10). These investigations were concerned with the thermodynamic stability of microemulsions and the results were interpreted against models involving microemulsion droplets. However, many "microemulsions" are in fact molecular solutions (11) or a large part of the "microemulsion" region does not contain microemulsion droplets (12–14).

In this publication, we present vapor pressure determinations in a system of water, an alkylethyleneglycol adduct and a phenethyl alcohol, a fragrance. The vapor pressure of fragrances is of decisive importance for consumer acceptance of personal care products and the relation between colloidal phenomena in the formulated product and the fragrance vapor pressure is of pronounced interest. In addition, colloidal phenomena are an essential component not only in the formulation of consumer products, but in the entire fragrance perception process from the solubilization/transport through the olfactory mucus (15, 16) over the olfactory membrane (17) and its structural changes (18–20) due to the confirmational variation of receptor protein to the final ion gate transport across the membrane (21).

The publications in the area of formulations of fragrance products have been limited to early contributions on the influence by fragrance on emulsion stability and the recent approach (22, 23) focussing on solubilization in aqueous solutions by Abe and collaborators. These latter contributions are essential against the trend towards replacing the traditional solvent based formulations by ones with water as the solvent (24). The vapor pressure of perfumes over aqueous solutions of sodium dodecyl sulfate was determined by Behan (25) who, lacking information about the colloidal state, interpreted the values using a relative component volume approach.

With the present article, we extend investigations on fragrance vapor pressures to include an entire system of water, a nonionic surfactant and a fragrance, phenethyl alcohol. The vapor pressure of phenethyl alcohol in the different association structures provides information about the relative influence on the presence of the molecular interactions with water and with surfactant and also, more importantly, enables an evaluation of the fragrance vapor pressure as function of time during the evaporation process of a consumer product.

Experimental

Materials

Laureth 4 (C12E04) (Brij® 30), ICI Surfactants, Wilmington, DE, and phenethyl alcohol (PEA), 99%, Aldrich Chemical Co., Milwaukee, WI, used as is; water, doubly distilled, deionized.

Phase diagrams

The phase diagram was determined by visual observation of samples during addition of one liquid component. Liquid crystalline phases were identified by the microscopy pattern when viewed between crossed polarizers and the region was determined by the low angle x-ray diffraction (nick points in the plot of interlayer spacing versus water volume ratio).

Low angle x-ray diffraction

A small amount of sample mixture was drawn into a thin glass capillary tube with 0.5 mm diameter. Low angle x-ray diffraction data was obtained by using a Kiessig low angle camera from Richard Siefert. Ni filtered Cu-radiation was used ($\gamma = 1.542$ Å) and the reflections was determined by a Tennelec position sensitive detection system (model PSD–1100).

Vapor pressure

Fifty gram liquid samples were prepared in 200 mL, clear, glass bottles, by first weighing an appropriate amount of Laureth 4 (Brij® 30), adding phenethyl alcohol and water. The samples were mixed by magnetic stirrer until the samples were visibly monophasic and allowed to equilibrate overnight. Multiphase samples were mixed continuously overnight and headspace vials were prepared immediately after removal from the stirrer.

Static headspace vials were prepared by filling 40 mL nominal volume vials (Supelco, cat. #2-3278) then sealing with Teflon faced silicone septa (Supelco, cat. #2-3285) and hole screw caps (Supelco, cat. #2-3202).

The headspace was generated by drawing 10 mL of liquid from the vial. This was done by adapting a technique described by V.D. Roe, et al. (26) A 5 mL Becton/Dickinson diposable luer lock syringe was used, with a 22 gauge, 1.5 cm needle as a vent. One headspace vial of each sample was stored, upside down, for a minimum of two days in the laboratory atmosphere (22 °C ± 2 °) prior to analysis. The samples were turned upright prior to sampling. A 250 μL Hamilton, #1725, gas/liquid, minimum septa coring syringe needle was inserted through the septa, into the headspace, a distance of ca. 1.5 cm. The syringe was plunged six times and on the seventh, 200 μL of gas were drawn and immediately injected into the gas chromatograph.

Analysis was performed with a capillary gas chromatograph (Hewlett-Packard 5890 Series II) with a split/splitless injector in the splitless mode. A Hewlett-Packard capillar column, (0.32 mm id., 30 m in length, with 0.25 uM 5% phenyl substituted methylpolysiloxane and a flame ionization detector were used. The chromatograph was monitored by G.C. terminal (Hewlett-Packard Vectra, VL2) with following column temperature programming initial temperature, 40 °C, 1 mixture; temperature program rate, 8 °C/min; final temperature, 180 °C, 2 min. The volumetric flow rate was kept between 5 and 7 mL/min.

The vapor in each vial was measured twice, and if the values were within good agreement, an average was taken. If agreement was poor, samples were allowed to reequilibrate and repeat. This gave the quantity P. P^0 was measured by recording the peak area of a vial containing the standard state.

The retention time of phenethyl alcohol was determined by preparing a 1×10^{-3} M phenethyl alcohol in ethanol solution, and measuring the response to 2 μL of the solution in the same temperature programming in split mode.

Results

The results describe the vapor pressure of phenethyl alcohol from the different phases in the system and the phase diagram, Figure 1, is consequently described first. The solubility of phenethyl alcohol in water is small, 2% by weight and the solubility of the surfactant is even smaller < 1%. Water is soluble in phenethyl alcohol in 7.5%, while the phenethyl alcohol and the surfactant are mutually completely soluble. The water initially is dissolved in the fragrance without surfactant; with increased amount of surfactant it becomes solubilized into inverse micelles at higher surfactant amounts reaching a maximum value of 46% water by weight at a phenethyl alcohol/surfactant ratio of 3/7. The surfactant dissolved water to a maximum of 12%; higher water amounts gave a lamellar liquid crystal between 27 and 50% water. The liquid crystal solubilized phenethyl alcohol reaching a modest maximum of 2% by weight.

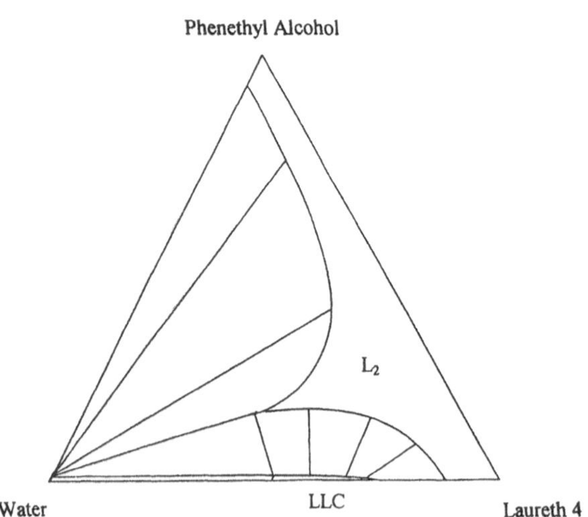

Fig. 1 Phase diagram for the system PEA/Laureth 4/H_2O. L_2 = Laureth 4/phenethyl alcohol W/O microemulsions. LLC = Lamellar liquid crystal

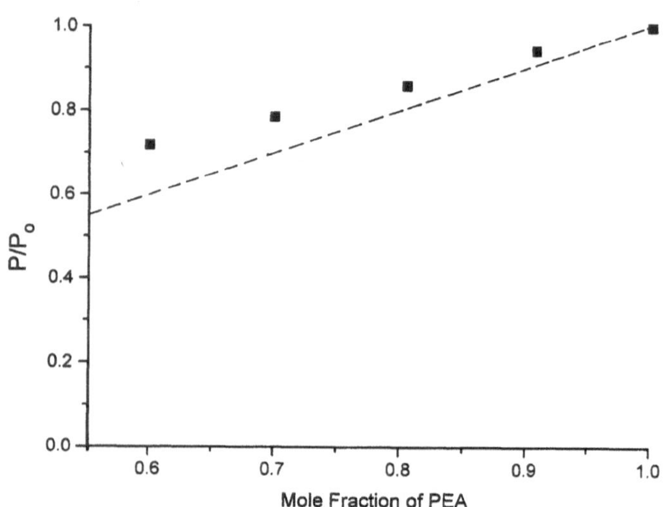

Fig. 2 Phenethyl alcohol vapor pressure with water dissolved into the alcohol. p_0 is the vapor pressure of pure phenethyl alcohol

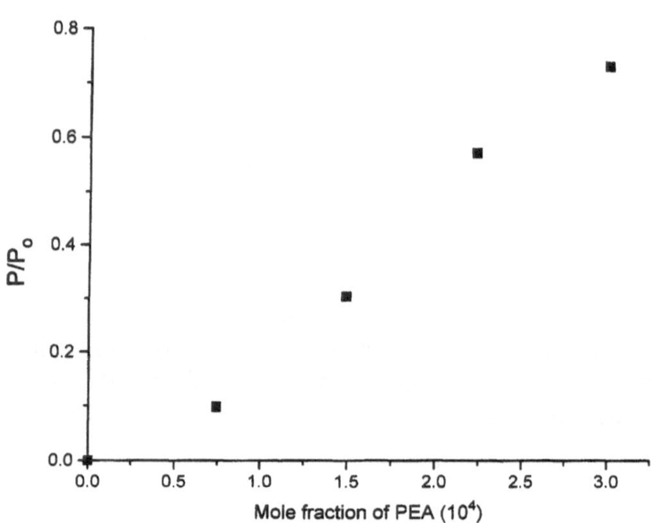

Fig. 3 Phenethyl alcohol vapor pressure of its solution in water. p_0 is the vapor pressure of pure phenethyl alcohol

The resulting two-phase region between the PEA/surfactant solution with solubilized water and the lamellar liquid crystal covered the lower part of the phenethyl alcohol/surfactant solution from zero phenethyl alcohol to a phenethyl alcohol/surfactant ratio of 3/7 at which point maximum water was solubilized. This last point was in equilibrium with an aqueous solution with 1.0% phenethyl alcohol and with the lamellar liquid crystal of high water content 49.5% and small amounts of phenethyl alcohol (1.0%) leaving a small two-phase region along the water/surfactant axis between the water and the liquid crystal.

The second two-phase region consists of the aqueous solution with phenethyl alcohol dissolved in the range 1 to 2% and the part of the phenethyl alcohol/surfactant solution with phenethyl alcohol/surfactant ratios in excess of 3/7 with water solubilized to the limit. It should be observed that the tie-lines in this two-phase region are determined exactly from the vapor pressure data.

The vapor pressure of phenethyl alcohol in its interaction with water is described in Figures 2 and 3. The first mentioned figure shows the reduction of phenethyl alcohol vapor pressure with added water; The values are in excess of those for an ideal solution. Figure 3 displays the increase of vapor pressure, when phenethyl alcohol was added to water. In this case values are extremely high, of a different magnitude than these in an ideal solution.

The phenethyl alcohol pressure in the phenethyl alcohol/surfactant solutions is found in Figure 4. The pressure variation when surfactant is added to phenethyl alcohol is characterized by an initial part of slightly lower values than these for an ideal solution, while in the surfactant rich part vapor pressures slightly higher than these in an ideal solution were found. The pressures for ideal solutions are added (hatched curve) for comparison.

The vapor pressures for compositions along the limit of water solubility in the phenethyl alcohol/surfactant region are given in Figure 5 showing the regular variation with phenethyl alcohol mole fraction calculated on the two amphiphiles only. The values follow those for an ideal solution except the range of high water content in which they are significantly higher. This trend is even more accentuated if the mole fraction of PEA is calculated as fraction of the number of moles also including water. Now the increase of vapor pressure with high water content is even more conspicuous.

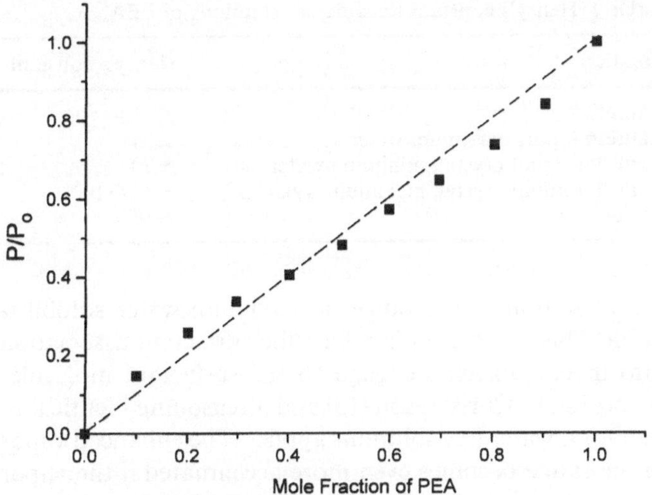

Fig. 4 Phenethyl alcohol vapor pressure in the surfactant/phenethyl alcohol solution with no water present. p_0 is the vapor pressure of pure phenethyl alcohol

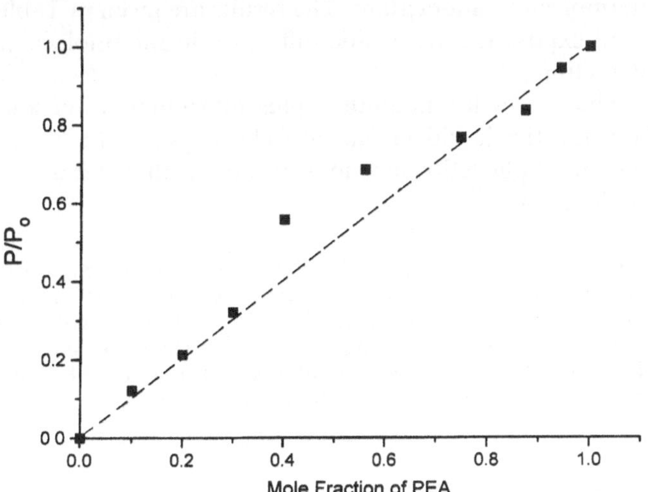

Fig. 5 Phenethyl alcohol vapor pressure in the surfactant/phenethyl alcohol solution with maximally dissolved and solubilized water. p_0 is the vapor pressure of phenethyl alcohol saturated with water. The phenethyl alcohol mol fraction is counted on the alcohol and surfactant only

Discussion

The phase diagram, Figure 1, illustrates the conditions for a temperature in excess of the HLB-level (27–30). This is expected because the phenethyl alcohol is not only a polar compound, but in addition an aromatic one. An aromatic nucleus shows a strong interaction with the polar chain of a nonionic surfactant (12, 31) and, hence, a low value for the HLB-temperature. The short polar chain of the surfactant also means that the critical association concentration of surfactant at high water content gives rise to a separate lamellar liquid crystal but not to spherical micelles (30). Such micelles would be water soluble and a significant solubility range of the surfactant in an aqueous isotropic solution would be found.

The solubilization of water into the PEA/surfactant solution follows a pattern typical of nonionic surfactant systems at temperatures in excess of the HLB-value. One finds a pronounced increase of water solubilization at high concentrations of the surfactants and the maximum water solubilization composition closely connected to these of the lamellar liquid crystal.

With the phase diagram understood, the vapor pressures may be related to the structure and content of the different phases. At first the relative interaction is evaluated between water and phenethyl alcohol and between surfactant and phenethyl alcohol.

The high vapor pressure of PEA with added water is not at a magnitude to make possible a detailed analysis, but the association between water and the alcohol hydroxy group could create some less favorable orientation for the interaction between the aromatic nuclei. The tremendous PEA pressures from opposite solution, that of PEA in water, are easily rationalized from the difference in solubility parameters, the polar and hydrogen bond forces in the alcohol are but a small part contrary to the case for water. A reasonable value for the solubility parameter for phenethyl alcohol would be equal to $23(\mathrm{MPa})^{1/2}$ while that for water is $48.0(\mathrm{MPa})^{1/2}$ (31). The ratio between the cohesive energy densities becomes 4.4, a reasonable basis for a Henry's constant of the order of 10^3.

The vapor pressure of PEA in the PEA/surfactant solution, Figure 4, is close to that of an ideal solution, but in this case the deviations may be explained from what is known about association structures between hydrocarbon and nonionic surfactants. An alkylpolyethyleneglycol surfactant, $n \simeq 4$, exists as a monomer to weight fractions of the surfactant at the level of 0.5 when added to an aromatic compound with no water present as demonstrated by Christenson (13). Surfactant contents in excess of this value leads to the formation of surfactant dimers and, consecutively, trimers, etc. For surfactant contents less than this value, added water is dissolved and becomes solubilized to inverse micelles first *after* the primary association of the surfactant has taken place.

Comparing Christenson's results to the present it is apparent that the initial reduction of the vapor pressure by addition of the surfactant, Figure 5, is related to the effect of the surfactant in monomeric form. After initiation of the self-association of the surfactant the vapor pressure curve becomes more horizontal, Figure 4, and at high surfactant contents the vapor pressure is in excess of that of an ideal solution. This higher vapor pressure is referred to the fact that the polar groups of the surfactant are interassociating,

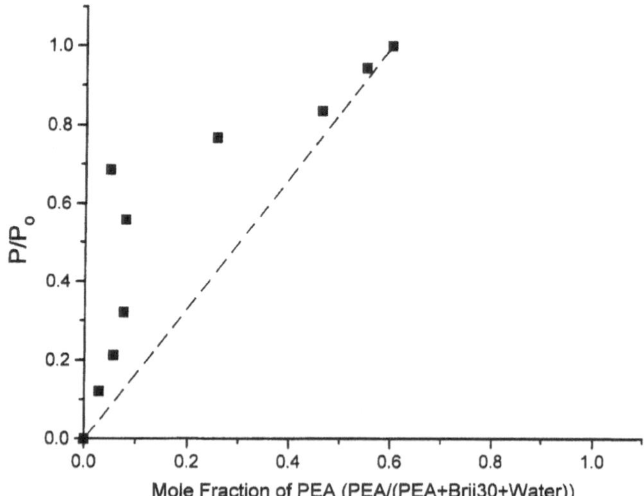

Fig. 6 The vapor pressure of phenethyl alcohol of the surfactant/phenethyl alcohol solution with maximally dissolved and solubilized water. p_0 is the vapor pressure of phenethyl alcohol saturated with water. The phenethyl alcohol mol fraction is counted on the alcohol, surfactant and water

leaving the interaction mainly with the aliphatic chains of the surfactant. As a corollary it should be noticed that solutions of phenethylacetate in decane show Henry's constants at the level of 6–8. Hence the present variation of vapor pressure of the surfactant fragrance solutions are satisfactorily explained taking the surfactant association into consideration.

Identical reasoning applies to the vapor pressure variation along the border for maximum water content in the surfactant/fragrance isotropic liquid solution Figure 5 shows a vapor pressure significantly in excess of that for an ideal solution in the range of maximum water solubilization. This is a region in which the surfactant association into inverse micelles engage all the surfactant molecules according to Christenson (13) and a reasoning identical to that for a water free solutions applies. The enhanced vapor in the range becomes even more accentuated if the vapor pressure is plotted against the PEA mol fraction including also the water, Figure 6. A comparison of Henry's constants for the different structures along the surfactant/water axis is very illustrative of the difference in intermolecular interactions. The results are given in Table 1 and expose the systematic difference in intermolecular interaction.

Table 1 Henry's constant for different solutions of PEA

Solution	Henry's constant
Laureth 4	1.5
Laureth 4 plus maximum water	$\simeq$ 4
Lamellar liquid crystal, minimum water	$\simeq$70
Lamellar liquid crystal, maximum water	$\simeq$ $2.4 \cdot 10^2$
Water	$\simeq 10^3$

The values for Laureth 4 plus maximum water and these for the liquid crystal are chosen with the vapor pressure of phenethyl alcohol saturated with water as the standard state.

Acknowledgment This paper is based upon work supported in part by the New York State Science and Technology Foundation as part of the research program of the Clarkson University Center for Advanced Materials Processing and ICI Surfactants Wilmington, DE. The constructive discussions with Professor Rusanov, St. Petersburg University, Russia are gratefully acknowledged.

References

1. Chew CH, Wong MK (1991) J Disp Sci Techn 12:495
2. Ueda M, Schelly ZA (1988) J Colloid Interface Sci 124:673
3. Cavallo JL, Rosano HL (1986) J Phys Chem 90:6817
4. Zulauf M, Eicke HF (1979) J Phys Chem 83:486
5. Weatherford WD, Jr., Naegeli DW (1984) J Disp Sci Techn 5:159
6. Weatherford WD, (1985) J Disp Sci Techn 6:467
7. Biais J, Bortherel P, Clin B, Lalanne P (1981) J Colloid Interface Sci 80:136
8. Biais J, Odberg L, Stenius P (1982) J Colloid Interface Sci 86:350
9. Sjöblom E, Johnsson A, Stenius P, Saris P, Oedberg L (1986) J Phys Chem 90:119
10. Damaszewski L, Mackay RA (1984) J Colloid Interface Sci 97:166
11. Das KP, Celglie A, Lindman B, Friberg SE (1987) J Colloid Interface Sci 116:390
12. Christenson H, Friberg SE, Larsen DW (1980) J Phys Chem 84:3633
13. Christenson H, Friberg SE (1980) J Colloid Interface Sci 75:276
14. Sjöblom E, Friberg SE (1978) J Colloid Interface Sci 67:16
15. Pelosi P (1994) Crit Rev Biochem Mol Biol 29:199
16. Getchell TV, Sec Zh, Getchell ML (1993) Proc Ciba Foundation Symp p. 27
17. Russel Y, Evans P, Dodd GH (1989) J Lipid Res 30:877
18. Nef P, (1993) Receptors and Channels 1:259
19. Jordi W, Nibbeling R, deKruijff B (1990) FEBS Ltrs 261:55
20. Fadool DA, Ache BW (1994) Proc Natl Acad Sci USA 91:947
21. Hatt H, Ache BW (1994) Proc Natl Acad Sci USA 91:6264
22. Tokuoka Y, Uchiyama H, Abe M, Christian SD (1995) Langmuir 11:725
23. Tokuoka Y, Uchiyama H, Abe M (1993) Colloid Polym Sci 272:317
24. US Patent 5,283,056 Feb 1, 1994
25. Behan JM, Perring KD (1987) Int J Cosm Sci 9:261
26. Roe VD, Lacy MJ, Stuart JD (1989) Anal Chem 61:2584
27. Shinoda K, Sagitani H (1978) J Colloid Interface Sci 64:68
28. Olsson U, Shinoda K, Lindman B (1986) J Phys Chem 90:4083
29. Kahlweit M (1982) J Colloid Interface Sci. 90:197
30. Friberg SE (1987) Eicke HF, Parfitt GD (eds), In: Interfacial Phenomena in Apolar Media, Marcel Dekker, NY, Ch 3, pp. 93.
31. Barton AFM (1983) Handbook of Solubility Parameters and Other Cohesion Parameters. CRS Press, Boca Raton, FL

Progr Colloid Polym Sci (1996) 101:23–29

F. Blockhaus
J.-M. Séquaris
H.D. Narres
M.J. Schwuger

Interactions of a water-soluble polymeric detergent additive (polycarboxylate) with clay minerals from soil

F. Blockhaus · Dr. J.-M. Séquaris (✉)
H.D. Narres · M.J. Schwuger
Institute of Applied Physical Chemistry
Research Center (KFA)
52425 Jülich, FRG

Abstract Acrylic/maleic acid (PAA-PMLA) copolymer is a low biodegradable dispersing agent. Its interfacial behaviour is investigated at clay minerals which control the transport of organic compounds in the soil compartment. Adsorption affinities and surface coverage results indicate that binding sites are located on edge faces of clay minerals. Contributions of long-range electrostatic (anion exchange reaction) and of short-range chemical interactions (ligand exchange reaction) to the adsorption/desorption process are discussed. Thus, the roles played by protonated aluminol binding sites at kaolinite surface and by the ionization degree of the copolymer are studied from variations of pH and ionic strength and from binding competitions of phosphate compounds. A dependence of the PAA-PMLA copolymer desorption on the adsorbed layer structure can be concluded. Furthermore, investigations concerning the interactions with kaolinite components, Al and Si, show that only the non-crystalline form of the adsorbed aluminium on the kaolinite can be solubilized.

Key words Polycarboxylate – kaolinite – adsorption – desorption – phosphate – aluminol

Introduction

As with other water-soluble polymers [1], the use of low-biodegradable polycarboxylates (PCA) in detergents gave rise to a discussion about their environmental behaviour with respect to input and possible pathways [2]. About 90% of PCAs entering sewage treatment plants may be eliminated by adsorption at sludges or by precipitation of insoluble calcium polymers. However, in case of an inefficient waste water treatment or of sewage sludges used as fertilizers, a possible residue of non-biodegradable PCAs in soils is to be taken into account.

Thus the role played by clay mineral surfaces, a major part of aquifers, in controlling the mobility of PCA must be elucidated. It is generally admitted that the presence of a permanent negative charge at the basal plane of clay minerals is unfavourable from electrostatic considerations for the adsorption of partially ionized PCA [3]. However, the presence of pH-dependent positively charged aluminol groups at edge faces must also be considered for electrostatic adsorption and ligand exchange mechanism of a more chemical nature [4].

In this report, an overview of the results obtained with acrylic-maleic acid copolymer (PAA-PMLA copolymer) [5–7], an ingredient of modern detergents, will be presented.

Regarding the binding behaviour of PCA carboxyl groups with metal such as aluminium [8], a major component of clay mineral, the solubility of total Al species in clay mineral suspensions with added PCA has also been studied.

Materials and methods

Materials

Clay minerals are characterized by procedures as described elsewhere [4–7]. The unlabelled and ^{14}C-labelled sodium form of acrylic-maleic acid copolymers (PAA-PMLA) of Mw 70000–92000 g $*$ mol^{-1}, based on weight percentage of acrylic acid (70%) and maleic acid (30%), was obtained from BASF. Other chemicals are Merck products.

Water purified through Millipore filters was used in all experiments.

Methods

Adsorption isotherms

Clay mineral suspensions (5 gl^{-1}) with added polymer were equilibrated for 65 h on a horizontal shaker at room temperature. The adsorbed amounts are calculated from the difference between initial and equilibrium concentrations after centrifugation (20 000 rpm; 30 min.).

Desorption measurements

Half of the supernatant obtained after the first centrifugation was replaced by fixed concentrations of displacer solutions. The mixtures were further shaken for about 80 h.

Analytical

Unlabelled polymers were quantitatively determined by direct polyelectrolyte phototitration [9] and recently improved potentiometric colloidal titration [10]. The ^{14}C labelled copolymers were measured by a radiotracer method using a liquid scintillation counter Beckman LS 5000TA.

Al and Si concentrations in supernatant of kaolinite suspension were measured by graphite furnace atomic absorption spectrometry (GF-AAS).

Results and discussion

Binding sites of clay minerals for PAA-PMLA copolymers

In order to localize the binding sites of clay minerals for PAA-PMLA copolymers, the adsorption isotherms of PAA-PMLA copolymer on three defined clay mineral microstructures were compared in Fig. 1, under identical solution conditions: pH 4.5–5 and 0.01 M NaCl. The adsorbed amount is expressed in moles of carboxyl units or in polymer weight per gram of clay minerals. Apparent "high affinity" isotherms can be observed in a range of

Fig. 1 Adsorption isotherms of PAA-PMLA copolymer on different clay minerals at pH 4.5–5 in 0.01 M NaCl. A: linear representation; B: semi-logarithmic representation. ∗–∗, montmorillonite; △–△, illite; ○–○, kaolinite; clay minerals 5 gl^{-1}. Drawn curves are calculated by the Langmuir adsorption isotherm equation

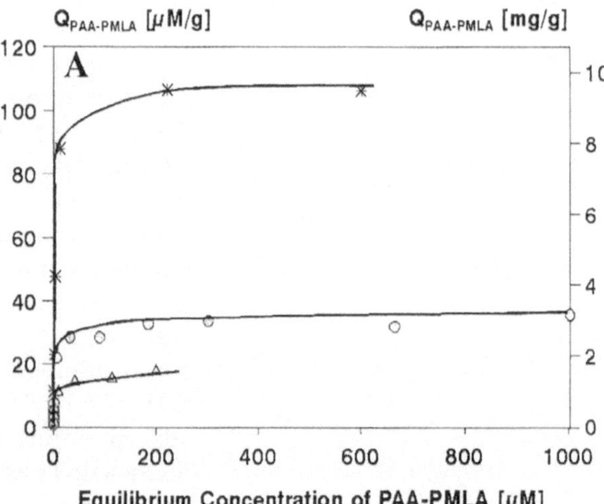

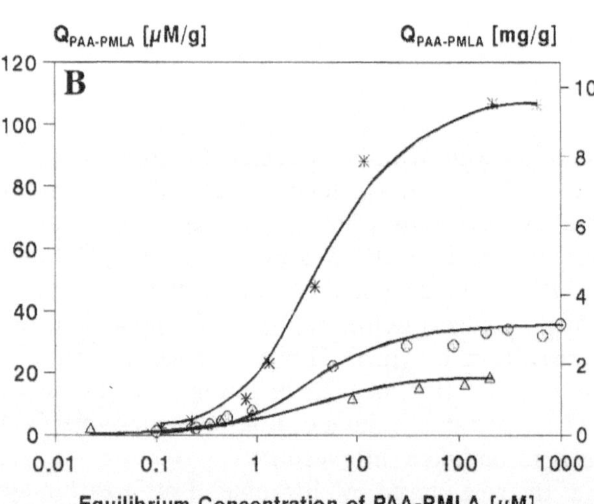

equilibrium concentrations from 0.02 μM to 1 mM, (Fig. 1A). In the case of montmorillonite, kaolinite and illite, it can be shown [6–7], by taking an average molecular surface area of 0.22 $nm^2 \pm 0.03$ nm^2 for $-CH_2-COOH$ in the copolymer, that the maximum surface area occupied by PAA-PMLA copolymer molecules matches the estimated area of the edge surface of the different clay minerals. It must be noted that the use of ^{14}C-labelled polymers has allowed a large domain of free concentration for PAA-PMLA copolymer to be covered down to 0.02 μM of carboxyl units. Thus, a semi-logarithmic representation in Fig. 1B reveals that the adsorption isotherm can be fitted by an isotherm equation of the Langmuir type. A derived adsorption constant $K^{PCA}_{ads.}$ of about $3*10^5$ M^{-1} for all three types of clay minerals can be found. This confirms the same nature of the binding site attributed to the aluminol groups of the edge surface.

Adsorption/desorption processes of PAA-PMLA copolymer at kaolinite surface

Adsorption behaviour of PAA-PMLA copolymer at clay minerals is governed by forces which can be roughly divided into long-range electrostatic interactions and short-range interactions of a more chemical nature.

Electrostatic interaction

The ionic interactions depend on the electrical properties of both polyelectrolyte and clay mineral surfaces, which are controlled by a pH-dependent surface ionization variably screened by the effect of the ionic strength. The influence of these parameters has been investigated in the adsorption study of PAA-PMLA copolymer on kaolinite. In Fig. 2, the pH dependences of the maximum adsorption values of PAA-PMLA copolymer are shown for three NaCl salt concentrations corresponding to ionic strengths of 0.01, 0.1 and 1 M. A rapid overview of the experimental results indicates that the adsorption of PAA-PMLA copolymer is decreased by increasing the pH or by lowering the salt concentration.

At acidic and neutral pH, the presence of protonated aluminol groups on clay mineral faces can be established by potentiometric titration [6–7]. It can be thus asserted that a charge compensation (anion exchange reaction with chloride anions) between partially negative ionized PAA-PMLA copolymer (pK_a 6.3 in 0.01 M NaCl [6–7]) and a positively charged edge face (pH_{pzc} of edge face 7.5 [11]) is a driving force.

Taking into account a total negative charge of the kaolinite particles due to a predominant effect of permanent negative charge, as well as the growing importance of the ionization of the carboxyl groups up to an alkaline pH range, the results also show that long-range electrostatic repulsion forces, according to the DLVO theory, operate at low ionic strength. Charge screening for higher ionic strengths decreases the electrostatic repulsion between PAA-PMLA copolymer and kaolinite surface and thus promotes adsorption through other non-electrostatic forces. Moreover, the increase of the ionic strength decreases the internal repulsion between adsorbed PAA-PMLA copolymer segments and thus favours more coiled structures of higher segment density.

Ligand exchange reaction

As already noted above, short-range interactions of a more chemical nature are involved in the adsorption processes. In regard to the weak acid properties of PAA-PMLA

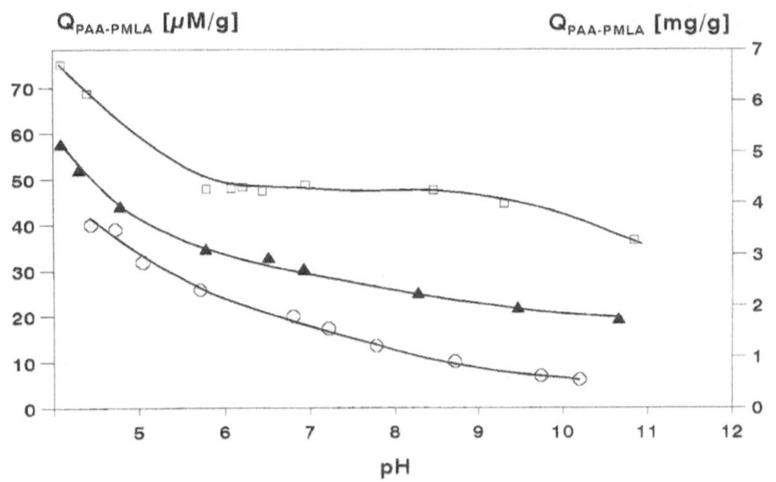

Fig. 2 Effects of pH and NaCl concentration on the PAA-PMLA copolymer adsorption on kaolinite. ○–○, 0.01 M; ▲–▲, 0.1 M; □–□, 1 M NaCl; kaolinite 5 gl^{-1}; PAA-PMLA copolymer 58 $mg\,l^{-1}$

copolymer and the aluminol nature of the binding site, a ligand exchange mechanism [12] can be generally considered in the investigated pH range from pH 4 to pH 11 where a hydroxyl group from aluminol site $>$ AlOH is exchanged with a carboxylate group from PAA-PMLA copolymer. It follows that this reaction is favoured at neutral and acidic pH, i.e. under conditions of low competing OH^- concentration. This adsorption process involves an inner-sphere complex with crystalline surface aluminiums as has been described in the case of the specific adsorption of phosphate compounds [13]. Indeed, the adsorption competition of simple organic acids and phosphates on soil components has been widely investigated. Thus an adsorbed maximum concentration of monophosphate (8–10 μM/g kaolinite) equivalent to the aluminol group concentration at edge faces of kaolinite can be put forward as evidence [6, 7]. In Fig. 3, a decrease of the adsorption of PAA-PMLA copolymer on modified kaolinite surface by the preadsorption of phosphate compounds also confirms the identical nature of the binding sites. It results in a competitive ligand exchange reaction between phosphate and carboxyl units. Thus, in the case of the strongly preadsorbed sodium tripolyphosphate (STP) ($K_{ads.}$ $3.5*10^4$ M^{-1} [6–7]), it can be roughly shown that a 55–60% occupation of binding sites by STP corresponds to a 60–65% decrease in adsorbed PAA-PMLA copolymer. For the monophosphate compound with a lower affinity ($K_{ads.}$ $1*10^{+4}$ M^{-1} [6–7]) to kaolinite surface, a retardation effect can be observed with respect to reaching the saturation of the edge faces with PAA-PMLA copolymer.

In a similar way, it can be shown that the desorption of PAA-PMLA copolymer from kaolinite also depends on the relative displacer efficiency of phosphate compounds where oligomeric STP is more efficient than the monophospate compound in Fig. 4.

For comparison, the effects of other anions such as SO_4^{2-} and NO_3^- on the displacement of preadsorbed PAA-PMLA copolymer are also reported. Independently of an ionic strength effect by raising the salt concentration (see Fig. 2), it can be shown that these less specifically or only electrostatically adsorbed anions have no effect on polymer desorption.

The configuration in the adsorbed polymer layer also seems to be of crucial importance in the displacement mechanism [14]. This observation is demonstrated by desorbing the PAA-PMLA copolymer in the entire region of an adsorption isotherm with a fixed STP concentration of 150 mM phosphate monomer in Fig. 5. The observed percentages of desorption are plotted against the amount of PAA-PMLA copolymer initially adsorbed at the kaolinite surface. The percentage of desorption is found to be lowest for surface concentrations lower than about 0.7 mg/g kaolinite (or 8 μM/g) of adsorbed copolymer. This is followed by a rise in the percentage of desorption up to a constant value of about 70% under surface saturation conditions as already shown in Fig. 1.

The initial almost nondesorbing region indicates that the PAA-PMLA copolymer is strongly bound to the edge surfaces and is only desorbable under very drastic conditions. It can be assumed that the adsorbed PAA-PMLA copolymer adopts a totally flat configuration at low equilibrium (≤ 1 μM) exclusively formed by train segments equivalent to the aluminol groups concentration (8–10 μM/g kaolinite). The observed desorption in the intermediate and final regions of the adsorption isotherm

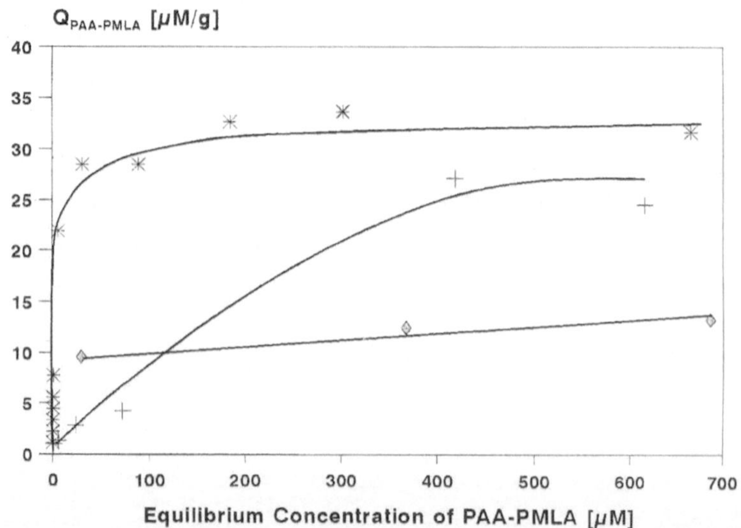

Fig. 3 Adsorption isotherms of PAA-PMLA copolymer onto kaolinite with preadsorbed phosphate compounds (Θ: surface coverage ratio and $c_{eq.}$ P : equilibrium concentration of phosphate compounds before addition of PAA-PMLA copolymer). *–*, in absence of phosphate compounds; + – +, Θ = 100% and $c_{eq.}$ P = 1 mM of monophosphate; ◇–◇, Θ = 55% and $c_{eq.}$ P = 0.08 mM of triphosphate

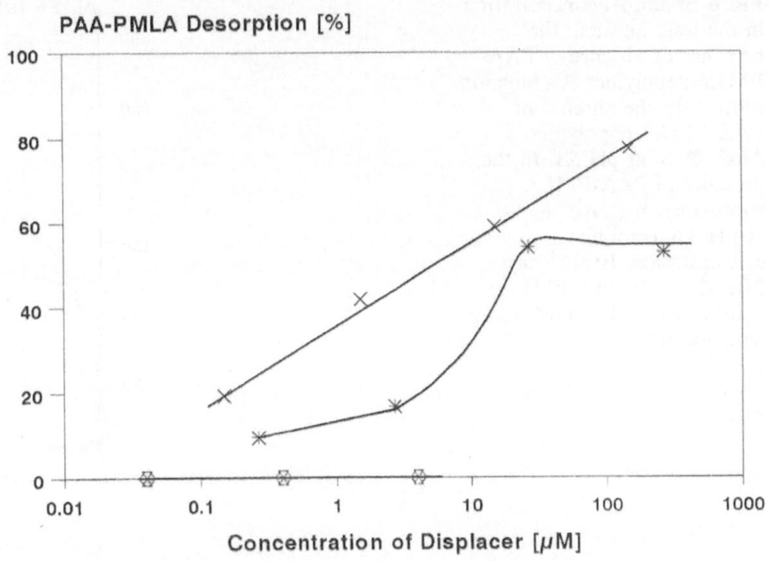

Fig. 4 Percentage of PAA-PMLA copolymer desorption from a presaturated kaolinite as a function of displacer concentration (semi-logarithmic representation). *–*, monophosphate; ×–×, STP; O–O, nitrate; ⧖–⧖ sulfate. Initial adsorbed PAA-PMLA copolymer amount 3 mg g^{-1}. See also Materials and Methods and other conditions in Fig. 1

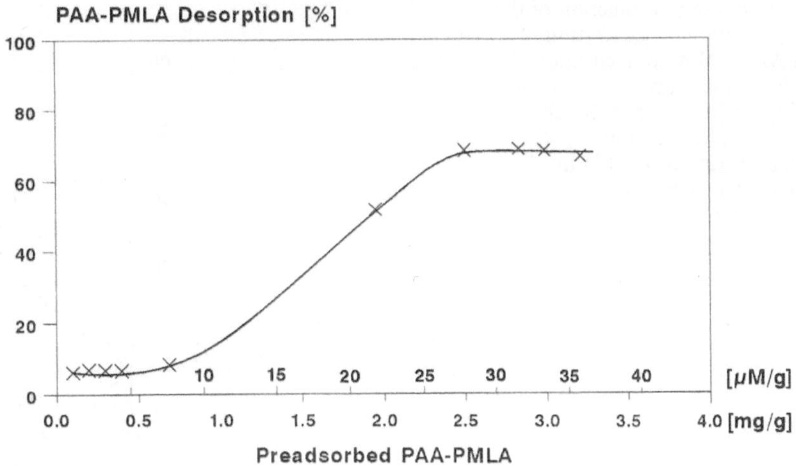

Fig. 5 Desorption percentage of PAA-PMLA copolymer in the presence of STP (150 mM phosphate monomer) as a function of initial preadsorbed PAA-PMLA copolymer amount on kaolinite. See also Materials and Methods and other conditions in Fig. 1

confirms that adsorbed polymer layers containing more loops and tails are formed at higher equilibrium concentrations (≤ 1 mM). The relatively fewer adsorbed segments per macromolecule in this region could thus be displaced more easily by strong competing displacers than in the flat configuration state.

In regard to the polymer concentration traces found in treated sewage [2], this finding shows that a negligible desorption of PAA-PMLA copolymer from desorbing clay mineral surface can be expected.

Interactions of PAA-PMLA copolymer with clay mineral components

Clay minerals like kaolinite show a pH-dependent low solubility in aqueous solution [15]. In a pH range down to pH about 3, the concentration ratio of the main dissolved kaolinite components Al/Si in the bulk phase is markedly lower than in the solid phase so that the dissolution seems to be non-stoichiometric. A cation exchange reaction with positively charged aluminium (cationic Al) species, Al^{3+} and $Al(OH)^{2+}$ at the kaolinite surface in acidic pH as well as the formation of poorly soluble hydrolysed aluminium solid phases in a more neutral pH region are responsible for this incongruent dissolution. However, it has been shown that the complexing properties of organic acids like di-, tricarboxylic acids [16] or humic materials [17] can influence the dissolution process by shifting the equilibrium concentration of aluminium species towards the solution or, more actively, by changing the dissolution rate through a surface interaction [11]. These two potential mechanisms have been examined in the case of PAA-PMLA copolymer. In Fig. 6, the release of Si and Al from

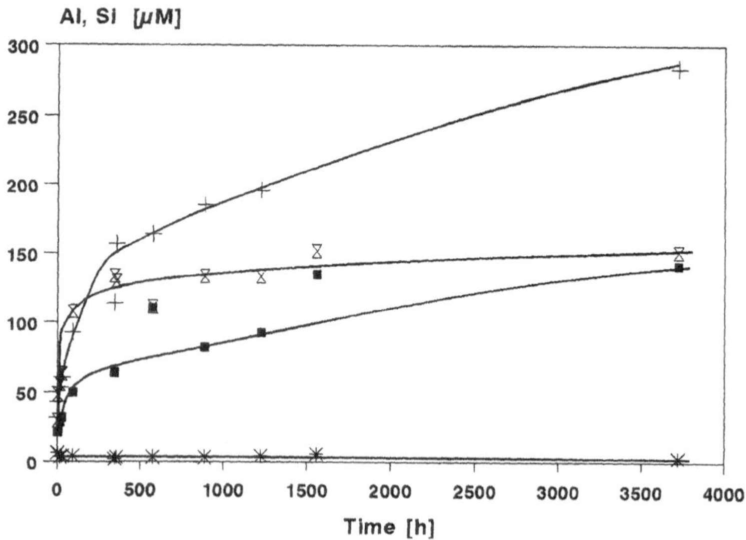

Fig. 6 Si and Al concentration in the bulk phase in the presence or absence of PAA-PMLA copolymer as a function of time. In the absence of PAA-PMLA copolymer: *–*, Al ■–■, Si at pH 5.1. In the presence of PAA-PMLA copolymer: ⧖–⧖, Al; + – + Si at pH 4.6. Kaolinite concentration: 10 gl^{-1}, in 0.01 M NaCl; PAA-PMLA concentration: 1.12 mM; room temperature

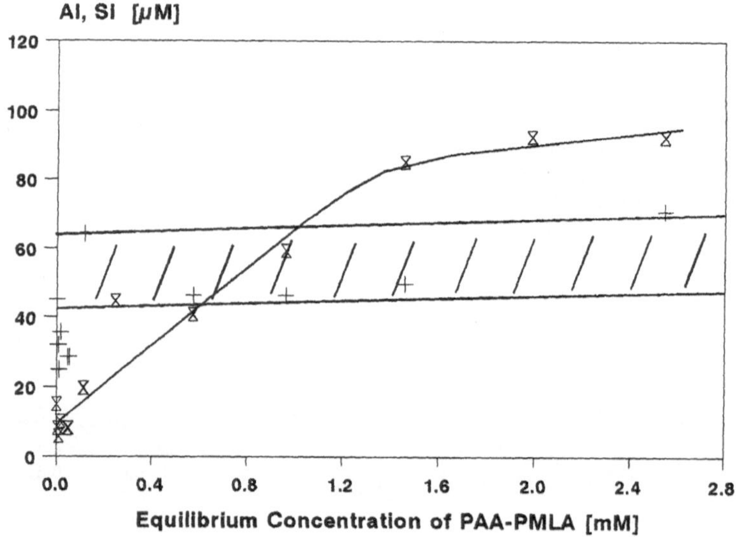

Fig. 7 Dependence of the Si and Al concentration in the bulk phase as a function of the equilibrium concentration of PAA-PMLA concentration after an equilibration time of 65 h at pH 4–4.5. ⧖–⧖, Al, + – +, Si. Kaolinite concentration: 10 gl^{-1} in 0.01 M NaCl; room temperature

the kaolinite particles at acidic pH is reported as a function of time in the presence or absence of added PAA-PMLA copolymer. Considering the highly soluble Si as a conservative tracer of the dissolution process, the observed dissolution kinetics apparently follow, as already noted, a parabolic curve [18]. However, investigations have demonstrated that this parabola-like initial short-term time reaction ($t < 100$ h) has to be cautiously examined [19]. Therefore, the long-term or steady-state dissolution period ($t > 100$ h) only has to be considered in the dissolution rate calculation from the curves slopes in Fig. 6. After taking into account pH variations, a constant proton-promoted release rate for Si of about $0.3 * 10^{-9}$ mol m^{-2} h^{-1} at pH 4.6 can be derived independently of the presence of PAA-PMLA copolymer in solution. This result with the PAA-PMLA copolymer thus gives no proof of an enhancement of the rate-limiting detachment process in the dissolution of the kaolinite crystalline structure by a surface complexation as in the case of oxalate- and salicylate-promoted dissolution of kaolinite [11]. The observed drastic increase of Al solubility can only be related to the equilibrium shift of cationic non-crystalline Al species from a bound state at the kaolinite surface into a bound state with carboxylate groups of PAA-PMLA copolymer in solution. The levelling of the Al concentration in solution for a longer period of time depends on the binding strength of the corresponding PAA-PMLA copolymer concentration at acidic pH [6]. In Fig. 7, the concentration effect of the PAA-PMLA copolymer on the release of cationic Al species and Si is shown. In accordance with the kinetics results, the dissolved concentration of Si is almost independent of the PAA-PMLA copolymer equilibrium concentration. In the case of cationic Al species, the concentrations are enhanced up to a plateau value of about 90–100 μM, which is equivalent to the amount of easily exchangeable non-crystalline cationic Al species as has been determined by a short-time extraction in a same concentration range of pyrophosphate or STP [6]. However, it must be considered that under soil conditions the widest distribution of these phosphate compounds as well as the high concentration up to 1 mM of active weak organic acids [20] predominate the aluminium mobilization and speciation.

It must be remarked that the PAA-PMLA copolymer in solution can be precipitated in the presence of an Al^{3+} concentration excess at pH 4 which also leads to an immobilization [6].

Conclusion

The results show that the adsorption behaviors of PAA-PMLA copolymer traces at clay minerals involves electrostatic (anion exchange reaction) and chemical (ligand exchange reaction) contributions. Under environmental conditions, it can be concluded that the carboxyl groups from PAA-PMLA copolymer in the aqueous phase or at the surface of clay minerals have a great tendency to interact with exchangeable non-crystalline Al species and crystalline surface Al which promotes an immobilization through precipitation and adsorption respectively.

References

1. Swift G (1993) Acc Chem Res 26:105–110
2. Opgenorth H-J (1992) In: Hutzinger O (eds) The Handbook of Environmental Chemistry, Vol 3, Part F. Springer, Berlin, pp 337–350
3. Theng BKG (1982) Clays Clay Miner 30:1–10
4. Sastry NV, Séquaris J-M, Schwuger MJ (1995) J Colloid Interface Sci 171: 224–233 and references cited therein
5. Blockhaus F, Séquaris J-M, Schwuger MJ (1991) Tenside Surf Det 28:447–451
6. Blockhaus F (1996) PhD thesis, Univ. Düsseldorf
7. Blockhaus F, Séquaris J-M, in preparation
8. Young SD, Bache BW (1985) J Soil Sci 36:261–269
9. Wassmer K-H, Schroeder U, Horn D (1991) Makromol Chem 192:553–565
10. Séquaris J-M, Kalabokas P (1993) Anal Chim Acta 281:341–346
11. Wieland E, Stumm W (1992) Geochim Cosmochim Acta 56:3339–3355
12. Sposito G (1989) The Chemistry of Soils. Oxford University Press, New York, pp 127–147
13. Muljadi D, Posner AM, Quirk JP (1966) J Soil Sci 17:212–229
14. Dodson PJ, Somasundaran J (1984) J Colloid Interface Sci 97:481–487
15. Stumm W, Wollast R (1990) Rev Geophys 28:53–69
16. Chin P-KF, Mills GL (1991) Chem Geol 90:307–317
17. Tan KH (1980) Soil Sci 129:5–11
18. Carroll-Webb S, Walther JV (1988) Geochim. Cosmochim. Acta 52:2609–2623
19. Holdren GR Jr, Berner RA (1979) Geochim. Cosmochim. Acta 43:161–171
20. Drever JI, Vance GF (1994) In: Pittsman ED, Lewan MD (eds) Organic acids in Geological Processes, Springer, Berlin, pp 138–160

Progr Colloid Polym Sci (1996) 101:30–37

K. Tauer
I. Kühn
H. Kaspar

Some colloid-chemical features of emulsion polymerisation

Dr. K. Tauer (✉) · I. Kühn · H. Kaspar
Max-Planck-Institut für Kolloid- und Grenzflächenforschung
Kantstraße 55
14513 Teltow-Seehof, FRG

Abstract Emulsion polymerisation is of great technical and economical importance as well as a process with a lot of colloid chemical features. Two of these colloid features, namely particle nucleation and swelling of polymer particles with monomer, are considered in more detail. It is shown that the particle nucleation can be described with a model based on the classical nucleation theory. This consideration is able to predict the chain length of the nucleating oligomers which is mainly influenced by the water solubility of the oligomers. With increasing water solubility the chain length of the nucleating oligomers becomes longer in good accordance with experimental findings. The activation energy of nucleation turned out to be a crucial parameter for further theoretical developments on particle nucleation in emulsion polymerisation. In this sense the development of a set-up for experimental investigations of the nucleation process is of importance. With a new developed experimental set-up based on a combination of on-line transmission and conductivity measurement with off-line particle size analytic it is possible to investigate the nucleation period. The results obtained so far indicate a strong influence of the emulsifier concentration on the particle concentration time curves in the very early stages of an emulsion polymerisation. Swelling experiments have been performed with toluene and latexes carrying chemically different stabilising groups. The latexes have been cleaned very carefully by ultrafiltration before they have been used. The results prove the enormous influence of the nature of the particle water interface on the swelling capability of the particles. It is to conclude that the Morton-Kaizerman-Altier equation cannot be applied for a complete description of latex particle swelling.

Key words Emulsion polymerisation – particle nucleation – classical nucleation theory – latex particle swelling

Introduction

Polymer synthesis by radical polymerisation in nm-sized particles dispersed in water is of great technical as well as economical importance [1]. Water-based paints and synthetic rubbers underline the importance of polymer dispersions as well as of their preparation techniques. Figure 1 shows as an example actual production amounts

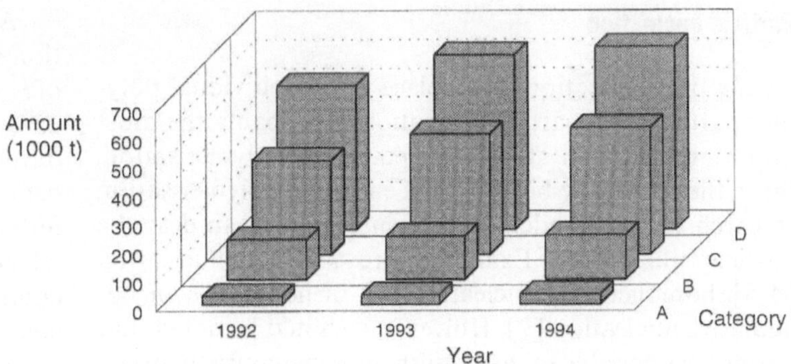

Fig. 1 Production amounts of several water-based dispersion paints in Germany. Category A – for concrete B – outdoor C – indoor D – sum A–C

of selected water-based paints for indoor and outdoor building protection in Germany [2]. The most important technical heterophase polymerisation techniques are suspension and emulsion polymerisation. The paper is confined to emulsion polymerisation which is defined as the polymerisation of monomers in aqueous medium whereby the resulting polymers are water-insoluble and form a polymer dispersion or latex. This definition implies that in the course of the polymerisation the polymer molecules usually precipitate in shape of spherical particles. After the particles have been formed the polymerisation (monomer consumption) takes place predominantly within the swollen latex particles. Especially, in technical processes stabilisers are employed during the polymerisation to ensure the latex stability up to a polymer content higher than 50%. All known kinds of colloid stabilisers can be used, whereas the nature of the stabiliser particularly used is determined by requirements of the polymerisation process and/or final application of the latex. So, in a "curriculum vitae" of an average particle of a polymer dispersion, one can distinguish the following stages: nucleation, swelling with monomer, growth by particle coalescence, growth by monomer consumption, and finally the application as dispersion (paints, glues) or as bulk polymer after separation from the dispersion (rubber, poly(vinyl chloride) for pastes). With this definition of an emulsion polymerisation, the colloid-chemical features as for instance particle nucleation as phase formation, swelling of latex particles with monomer, particle stability against coagulation or coalescence, and stabiliser adsorption or desorption become obvious. However it is necessary to point out that all of these colloid-chemical events take place in "living" systems which means that most of the latex parameters like particle size and number as well as the particle water interfacial composition are changing with polymerisation time. So, it cannot be excluded that non-equilibrium situations may be important

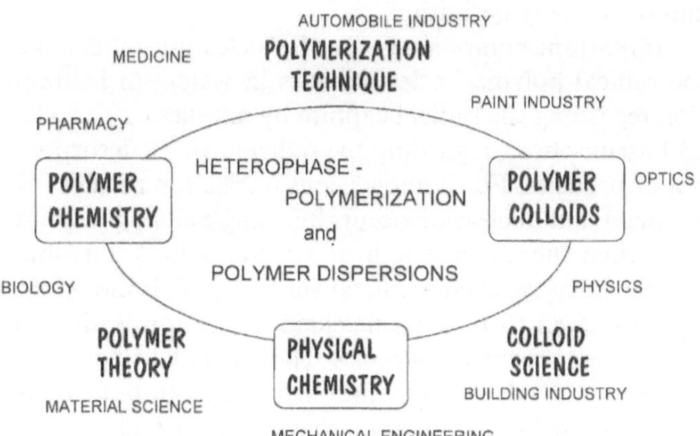

Fig. 2 Relation of "Heterophase polymerisation/polymer dispersions" to other scientific areas

especially with respect to the emulsifier adsorption/desorption.

With regard to the complexity of the topic, it is obvious that studies on heterophase polymerisations as well as on polymer dispersions require knowledge of several scientific fields. Figure 2 summarises the relations of the subject "polymer dispersion/heterophase polymerisation" to other scientific areas as well as to major application fields for polymer dispersions. Of course, the upper description is a simple figure of an emulsion polymerisation. For a more profound picture of the state of the art the recent monograph of R.G. Gilbert [3] is recommended.

In the present paper particle nucleation and particle swelling will be considered in more detail as both processes are important already in the very early stages for the success of an emulsion polymerisation. As there is no experimental part in this paper a reference to the corresponding original contribution is given for experimental details.

Particle nucleation

As the polymerisation takes place inside individual polymer particles the particle formation is probably the most important reaction step of an emulsion polymerisation. Since the very early beginning of emulsion polymerisation research, several models have been developed to describe particle nucleation. Examples are micellar nucleation [4, 5], homogeneous nucleation [6], or homogeneous coagulative nucleation [7]. However, it should be noted that it is not yet possible to distinguish unambiguously between these models using the current experimental data. From an experiment point of view it is crucial that no experimental data for complete particle number time curves are available, e.g., from the first appearance of particles up to the end of the polymerisation.

Important components for all nucleation models are the radical polymerisation kinetics in water, an assumption regarding the radical capture by micelles or particles, and assumptions regarding the radical exit or desorption out of particles. For homogeneous nucleation models it is assumed that nucleation occurs by a single chain precipitation when the chain length of a water soluble oligomer radical, j, has reached a critical value, j_{crit}. This model can be considered to be a single-chain nucleation leading to particles or precursor particles. The fate of these particles strongly depends on the emulsifier concentration in the system. If the free surfactant concentration is high enough, all of the particles generated may be stabilised and therefore no decrease in the particle number takes place just after nucleation.

According to the micellar nucleation mechanism a particle is formed when a radical from the water phase enters a micelle and continues to grow inside.

With the help of experimental particle number-time curves it should be possible to distinguish between homogeneous and micellar nucleation since it should take more time to form the first particles by homogeneous nucleation than by micellar nucleation.

As the particle formation is the formation of a new phase, several attempts have been published to explain the particle nucleation in heterophase polymerisation with the classical nucleation theory (CNT) [8, 9, 10]. However, all of these attempts resulted only in qualitative treatments of the nucleation process. Surprisingly, one can easily use the classical nucleation theory together with the Flory-Huggins theory of polymer solutions and the radical polymerisation kinetics to develop an applicable quantitative model of the nucleation process in emulsion polymerisation [11]. Table 1 summarises the basic equations of this model. For the derivation as well as for the model parameters is referred to ref. [11].

According to these ideas based on the CNT the nucleation starts with the formation of small clusters consisting of m water soluble oligomers with a chain length j produced by radical polymerisation in water. Figure 3 shows the development of the free energy (ΔG) of the cluster in dependence on the number of oligomers forming the cluster. ΔG passes with increasing m through a maximum which corresponds to the critical nucleus with the critical composition m_C. ΔG_{max} can be considered as the activation energy of nucleation. A nucleus that consists of a number of oligomers greater than m_C is stable, whereas a nucleus with $m < m_C$ is unstable and will dissolve again. Furthermore, the dependence shown in Fig. 3 makes clear that a cluster with a value of m greater than m_C will grow infinitely. So, this really simple model on the basis of the CNT can be valid only for the very first nucleation step. It has to be modified for nucleation in the presence of particles, as oligomers can be captured by already existing particles. The solutions of the mathematical model may be considered as matrices with the variables t and j. To get a plastic view for which values of j and t nucleation occurs the following algorithm was applied:

a) the nucleation condition is $\Delta G_{max} \gg v \cdot kT$ whereby v is a simple parameter to modify activation energy,

b) physically meaningful in the sense of nucleation are only solutions where $M_C(j,t)$, $D_C(j,t)$, $N_C(j,t)$, and $R_n(j,t)$, are positive,

c) if $M_C(j,t)$, $D_C(j,t)$, $N_C(j,t)$, and $R_n(j,t)$ were negative their values have been set equal to zero.

d) by multiplying the matrices $M_C(j,t)$, $D_C(j,t)$, $N_C(j,t)$, and $R_n(j,t)$ element by element a nucleation plane was constructed whereby values greater than zero indicate nucleation for the corresponding j and t values (in the j-t plane spikes occur for nucleation).

Figure 4 shows as a typical result of model calculations the so-called nucleation plane. It is obvious that beside the activation energy of nucleation represented here by v, the water solubility of the monomers has a strong influence on the nucleation behaviour. It is interesting to note that the higher ΔG_{max} (or v), the longer the time until nucleation occurs. The predictive power of this model is documented by a comparison of the nucleation chain length with experimental data (Table 2). The good agreement between experimental and calculated values of the critical nucleation chain length, j_{crit}, is even more remarkable as there were no special fits of constants or model parameters necessary [11]. The experimental data of j_{crit} for styrene (STY) and methyl methacrylate (MMA) were taken from [12] and that for vinyl acetate (VAC) from [13].

Another result of these model calculations is the recognition of ΔG_{max} as a crucial parameter also for the nucleation process in emulsion polymerisation. For further

Table 1 Basic equations for modelling particle nucleation

Classical nucleation theory	
DG_{max} – activation energy of nucleation	$\Delta G_{max} = \frac{4}{27} \cdot \frac{c_2^3}{c_1^2} \cdot \frac{j^2 \cdot \sigma^3}{(\ln S)^2}$; $M_C = \left(\frac{2}{3} \cdot \frac{c_2}{c_1} \cdot \frac{j^{2/3} \cdot \sigma}{\ln S}\right)$
M_C – number of chains per nucleus N_C – Number of nuclei formed D_C – diameter of a critical nucleus j – chain length of the oligomers S – supersaturation of the oligomers	$N_C = \frac{1}{v_w} \cdot \exp(-\Delta G_{max}/kT)$
σ – interfacial tension nucleus to water c_1, c_2 – constants MG_{max} – molecular weight of the monomer	$D_C = 10^7 \cdot \frac{2}{3} \cdot \left(\frac{6}{\pi}\right)^{1/3} \cdot \frac{C_1}{C_2} \cdot \left(\frac{MG_{mon}}{d_p \cdot N_A}\right)^{1/3} \cdot \frac{j \cdot \sigma}{\ln S}$
Radical polymerisation kinetics	
$C(t,j)$ – current cumulative concentration of oligomers	$C(t,j) = \frac{1}{2} \cdot \frac{\beta}{(1+\beta)^j} \cdot I_0 \cdot (1 - \exp(-k_d \cdot t))$
I_0, I – initial and actual initiator concentration, resp. $2fk_d$ – radical flux k_p – propagation rate constant	$\beta = \frac{(2 \cdot f \cdot k_d \cdot I \cdot k_t)^{1/2}}{k_p \cdot M_w}$; $I = I_0 \cdot \exp(-k_d \cdot t)$
k_t – termination rate constant M_w – monomer concentration in water	
Flory Huggins theory	
$C_0(j)$ – solubility of oligomers in water	$\ln \phi_2 = j \cdot (1 - \chi - 1/j)$; $C_0(j) = \frac{\phi_2}{j \cdot MG_{mon}/d_p \cdot (1 - \phi_2)}$
f_2 – volume fraction of oligomers in water χ – FLORY-HUGGINS interaction parameter d_p – polymer density	

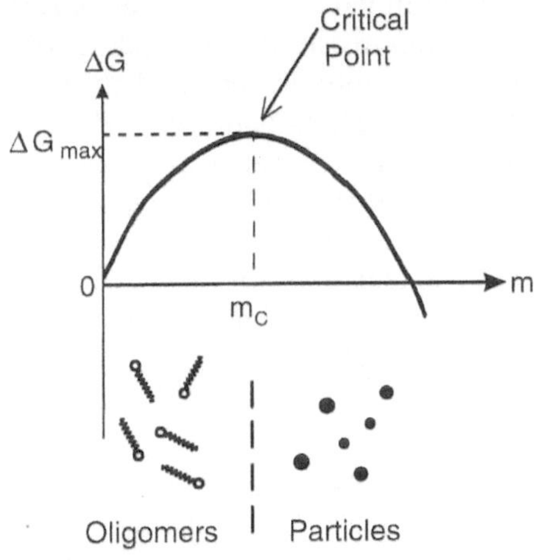

Fig. 3 Free energy of nucleus formation dependence on the number of oligomers forming a nucleus

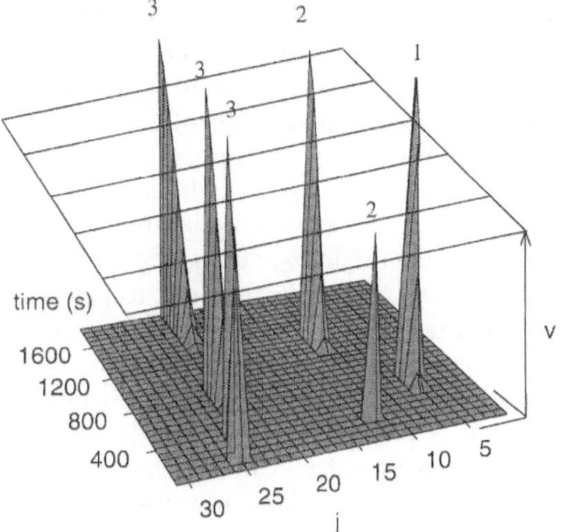

Fig. 4 Calculated nucleation plane for styrene (1), methyl methacrylate (2), and vinyl acetate (3) for v = 1; the upper grid indicates the influence of an increasing v on the calculated nucleation plane

improvements experimental values of ΔG_{max} have to be estimated for different monomers. However, this requires a suitable set-up for experimental investigations during the nucleation step. From several investigations of particle nucleation in heterophase polymerisation it is known that the reproducibility of experimental data is remarkably poor [14]. A combination of on-line measurements of optical transmission and conductivity of the latex with

Table 2 Numerical results of model calculations characterising particle nucleation; parameters for calculations see [11]

Monomer	$j_{crit,exp}$	j_{crit}	t_C (min)	D_C (nm)	M_C
$v = 10$; STY		6	6	11.8	887
$v = 1$; STY	5	6	7	12	890
		7	1	4	22
$v = 10$; MMA		11	16	8.7	223
$v = 1$; MMA	10	12	4	7	114
		13	1	6	77
$v = 10$; VAC		22	20	8.0	110
$v = 1$; VAC	18–20	23	9	21	1.640
		24	7	11	217
		25	2	8	70
		26	1	6	33

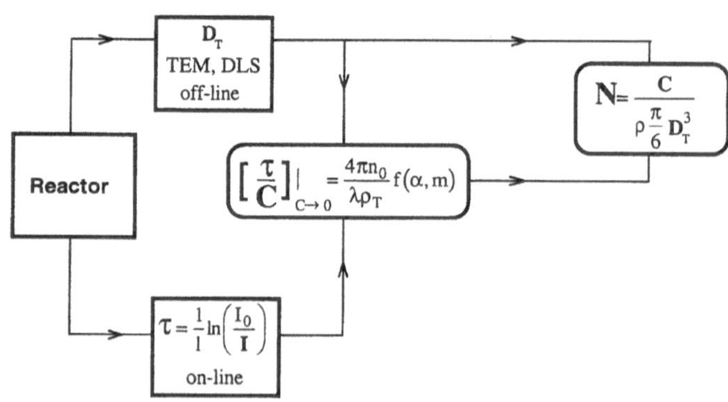

Fig. 5 Block diagram of the experimental set-up for the investigation of particle nucleation

off-line particle size analysis by dynamic light scattering or electron microscopy turned out to be very well suited for experimental investigations of the nucleation period at the very beginning of an emulsion polymerisation [15]. The reproducibility of the data is drastically enhanced up to a factor of eight when first, all solutions are carefully degassed before use (even no purging with an inert gas) and second, the water is filled into the reactor with a temperature slightly above the polymerisation temperature. A block diagram of the experimental set-up which makes the basic ideas of the investigations clear is shown in Fig. 5. The basic idea is that if turbidity (τ) and particle diameter (D_T) are known the tables of scattering functions for spherical particles (a relation between $f(\alpha, m)$, τ/C and D_T) [16] can be used in the reverse sense to calculate the polymer concentration (C) and subsequently the particle concentration (N). In Fig. 5, l is the optical path length, I_0 is the incident light intensity, I is the light intensity after passing through the polymerisation mixture, λ is the wavelength of the incident light, n_D and n_0 are the refractive indices of the spheres and of the medium, respectively, and ρ is the polymer density. For a detailed description of the experimental conditions as well as for the utilisation of the primary experimental results see [15]. With this experimental set-up it is possible to recognise the onset of particle nucleation by both a change in the conductivity as well as a change in the transmission. However both measured quantities respond to nucleation at different times. First, a decrease in the slope of the conductivity curve is observed followed by a decrease in the transmission (increase in turbidity). This behaviour can be explained by an aggregation process of water soluble oligomers that leads to a decrease in mobility.

Figure 6 shows particle concentration-time curves for two extreme cases with respect to particle nucleation: a surfactant-free emulsion polymerisation and another one with an initial surfactant concentration well above the critical micelle concentration (CMC). In both cases a completely different behaviour is observed. During the emulsifier-free polymerisation a decrease of the particle concentration takes place over the whole time period that could be investigated. Contrary to this behaviour, in the case of the polymerisation in the presence of emulsifier, a nearly steady increase of the particle concentration is observed. These experimental results indicate not only a quantitative but even a qualitative difference between both cases. It is obvious that the duration of the nucleation period and the particle concentration as well as the stability of the particles are quite different. Further differences between both polymerisations exist with respect to the final mean particle diameter (volume weighted average diameter measured with dynamic light scattering) which is 81.9 nm after $1.943 \cdot 10^3$ s for the emulsifier-free case and 19.4 nm after $1.1924 \cdot 10^4$ s for that in the presence of emulsifier. In the latter case the polymer content of the latex is 1.91 $g\,l^{-1}$ and the transmission still 74.9% whereas in the emulsifier-free latex the polymer content is 0.15 $g\,l^{-1}$ and the transmission is only 20.2%.

Swelling of the particles with monomer

Especially, from the colloidal point of view the swelling process in itself is a very interesting event. Swelling means the transfer of monomer into the polymer particles as long as free monomer is present. However, this does not lead to a dissolution of the particles even if the bulk polymer is completely soluble in the monomer as is, for instance, the case in the system poly(styrene)/styrene. Furthermore, it was shown that there are no polymer molecules dissolved in the coexisting free monomer phase and also not in water [17]. Swelling stops when the thermodynamical

Fig. 6 Comparison of the time development of the particle concentration for an emulsion polymerisation of styrene with sodium laurylsulfate (20 mM) and without a surfactant; 60°C polymerisation temperature and an initiator concentration of 2.5 mM potassium peroxidisulfate

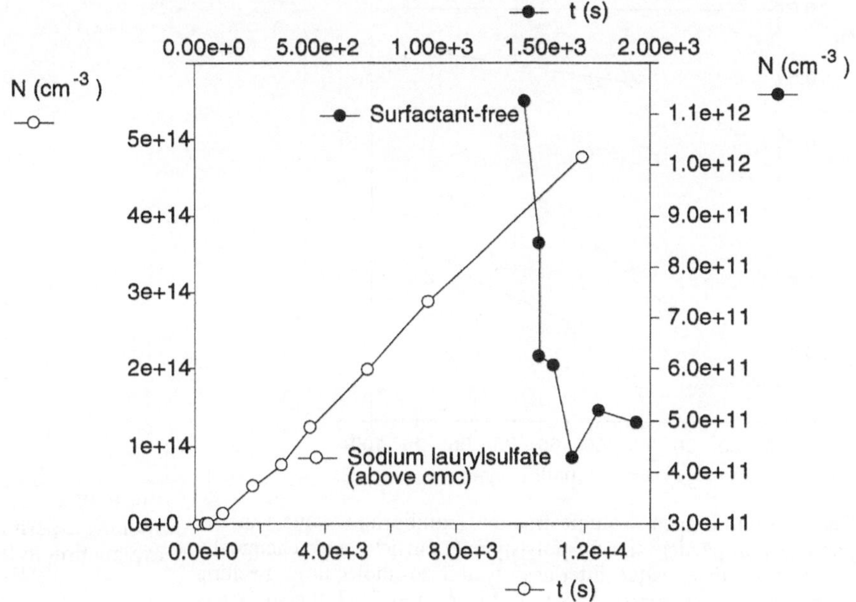

equilibrium is reached at a well defined size of the swollen particles. From another point of view swelling means the extension of an existing surface or the creation of a new surface. This is a well known process in colloid science however, with an important difference compared to the case of a pure one component system. In the latter case the new surface has the same properties as the old one and also the volume properties remain completely unchanged. The surface area enlargement of a latex particle during swelling with a second component means an increase in the average distance between the surface molecules (charges, stabilisers, etc.) as well as between the polymer chains in the particle volume. This is schematically shown in Fig. 7.

Morton, Kaizerman, and Altier published in 1952 the first theoretical approach to describe latex particle swelling [18]. This approach is still accepted today [3]. They considered the swollen particle in equilibrium with the free solvent and described the equilibrium condition for the partial molar free energy of the solvent ($\Delta\bar{F}_1$) with the osmotic contribution ($\Delta\bar{F}_{ml}$) and the interfacial free energy contribution ($\Delta\bar{F}_t$). In equilibrium $\Delta\bar{F}_1$ is zero and the result is Eq. (1), known as Morton-Kaizerman-Altier (MKA) equation.

$$\ln(1-\phi_2)+\left(\phi_2-\frac{1}{DP}\right)+\chi\cdot\phi_2^2=-\frac{2\cdot V_1\cdot\gamma}{r\cdot RT} \quad (1)$$

The left-hand side of Eq. (1), the osmotic contribution, ($\Delta\bar{F}_{ml}$), follows from the Flory-Huggins theory of polymer solutions where ϕ_2 is the polymer volume fraction in the

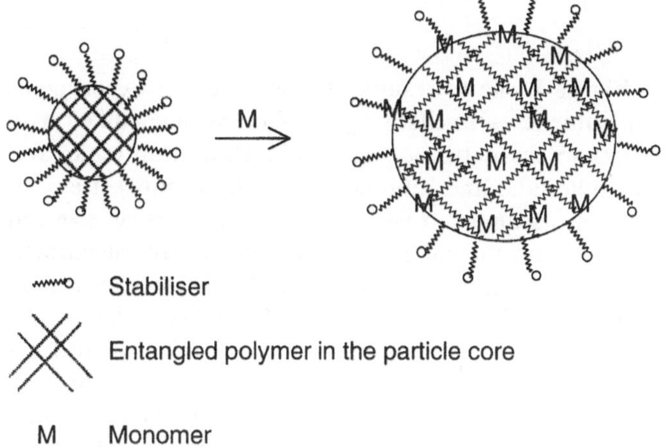

Fig. 7 Schematic illustration of the changes during particle swelling

swollen particle, DP is the number average degree of polymerisation of the polymer molecules within the particles, and χ is the Flory-Huggins interaction parameter. Furthermore, V_1 is the molar volume of the monomer (swelling agent), γ is the interfacial tension particle to water, RT is the thermal energy, and r is the particle radius.

Equation (1) predicts that in case of a high molecular weight polymer ($1/DP\approx 0$) for given values of χ and γ the swelling depends only on the particle size in a way that the greater the size, the higher the swelling agent volume content ϕ_1, whereby $\phi_1 = 1-\phi_2$ as there is only polymer and swelling agent in a particle.

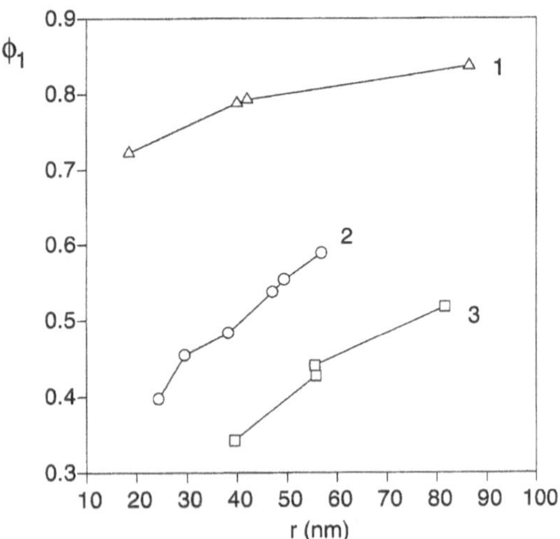

Fig. 8 Swelling agent volume fraction (equilibrium values) in dependence on particle size for poly(styrene) particles with chemically different particle water interfaces (stabiliser molecules); swelling agent toluene; temperature 25°C 1) MKA latex [18]; the surface is fully covered with potassium laurate in swelling equilibrium 2) Latex stabilised only by covalently bound sulfonate groups after purification by ultrafiltration [19] 3) Latex stabilised only by covalently bound non-ionic siloxane surfactants after purification by ultrafiltration [19]

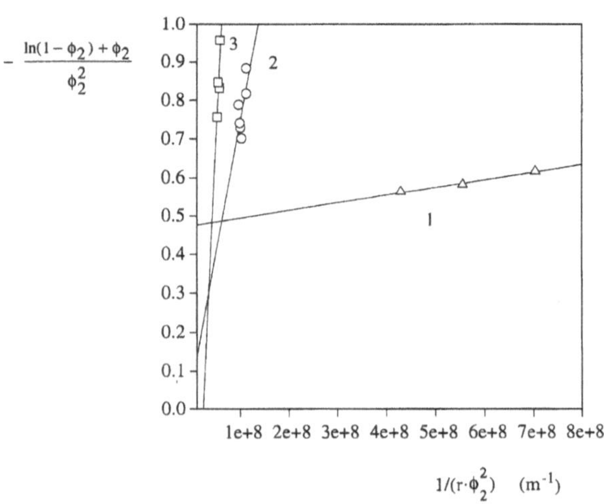

Fig. 9 Plot according to Eq. (2) to estimate χ and γ from latex swelling experiments; latex identification and swelling conditions see explanation in legend of Fig. 8

Table 3 χ and γ for the latexes 1, 2 and 3 estimated from the linear regression according to equation (2); $V_1 = 1,15 \cdot 10^{-4}\,m^3\,mol^{-1}$; $RT = 2479\,Nm\,mol^{-1}$

Latex	χ	$\gamma\,(Nm^{-1})$
1	0.474	0.00216
2	0.065	0.0729
3	− 0.58	0.2758

Figure 8 shows results of latex swelling experiments with toluene as swelling agent. The particles differ with respect to the nature of the stabilisation (non-ionic, anionic sulfonate, and anionic carboxylic groups). Reasonably, γ is assumed to be independent of particle size but dependent on the chemical nature of the particle surface, which is qualitatively confirmed by these experiment (cf. Fig. 8). The latexes 2 and 3 have been cleaned by ultrafiltration as long as all water-soluble by-products have been washed out. The particles are stabilised only by the covalently bound sulfonate groups (latex 2) and siloxane-glucono groups (3), respectively. Contrary to these latexes, latex 1 is saturated with such an amount of potassium laurate at a pH of 10 that the latex surface is still completely covered at swelling equilibrium. This means that micelles are present during swelling and consequently a lot of toluene is solubilised. The surfactant is in that case the reason for the observed enhanced swelling. The surfaces of samples 2 and 3 can be considered as naked compared to that of sample 1. Morton, Kaizerman and Altier describe a similar effect, as they reported that if no soap was added to the latex before swelling the uptake of solvent was only half as high as in the case in which the surface was saturated with soap [18].

As Eq. (1) seems to be applicable to describe latex particle swelling, it should be possible to estimate γ (slope) as well as χ (intercept) as swelling equilibrium from a plot $-(\ln(1-\phi_2)+\phi_2)/\phi_2^2$ versus $1/(\phi_2^2 \cdot r)$ (cf. Eq. (2)).

$$-\frac{-\ln(1-\phi_2)+\phi_2}{\phi_2^2} = \chi + \frac{2\gamma}{r\cdot\phi_2^2}\cdot\frac{V_1}{RT} \quad (2)$$

The swelling data for the three different latexes show the expected linear dependence (Fig. 9). Table 3 summarises the results of a linear regression. These results with respect to χ seem to be fairly reasonable. But, with respect to the interfacial tension this estimation leads to very unlikely results especially for the latexes 2 and 3 as these γ-values are higher than that for pure water.

A linear regression analysis according to Eq. (1) with the χ-values given in Table 3, i.e. a plot $(\ln(1-\phi_2) + \phi_2 + \chi \cdot \phi_2^2)$ versus $1/r$ should give straight lines passing through the origin. Indeed, Fig. 10 shows that the data result in straight lines, but they do not pass through the origin as required by the MKA-relation (Eq. (1). The results of the regression analysis are summarised in Table 4. It is obvious that the values for both γ and the intercept depend on the different chemical composition of the particle water interface.

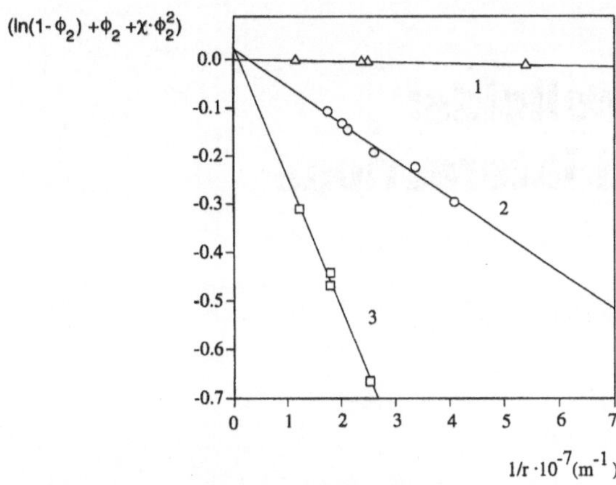

Fig. 10 Plot according to Eq. (1) data from latex swelling experiments; latex identification and swelling conditions see explanation in legend of Fig. 8

Table 4 Intercept and γ for the latexes 1, 2 and 3 estimated from the linear regression according to equation (1); $V_1 = 1,$ $15\cdot10^{-4}\,m^3\,mol^{-1}$; $RT = 2479\,Nm\,mol^{-1}$

Latex	γ $(N\,m^{-1})$	Intercept
1	0.0034	1.04 10^{-4}
2	0.127	0.0216
3	0.449	0.0277

As all three curves do not pass through the origin there must be something not yet completely understood in latex particle swelling. The data should be reliable as they were obtained with different techniques in different groups. At this time, the first thing that one might be able to conclude is that the intercept according to Eq. (2) cannot be equal to χ as in Eq. (1) an additional additive term is missing. Obviously, a χ-value that fits the experimental data must be greater than the intercept according to Eq. (2).

Conclusion

Emulsion polymerisation is of great technical and economical importance. Moreover, it is also a process with a lot of colloid-chemical features. Two of these colloid features are particle nucleation and swelling of polymer particles. It is shown that particle nucleation can be described with a model based on the classical nucleation theory. This consideration is able to predict the chain length of the nucleating oligomers which is mainly influenced by the water solubility of the oligomers. With increasing water solubility the chain length of the nucleating oligomers becomes longer. This is in good accordance with experimental findings. The activation energy of nucleation turned out to be a crucial parameter for further theoretical developments on particle nucleation in emulsion polymerisation. In this sense the development of a set-up for experimental investigations of the nucleation process is of importance. With a new developed experimental set-up based on a combination of on-line transmission and conductivity measurement with off-line particle size analytic it is possible to investigate the nucleation period. The results obtained so far indicate a strong influence of the emulsifier concentration on the particle concentration time curves in the very early stages of an emulsion polymerisation. Swelling experiments have been performed with toluene and latexes carrying chemically different stabilising groups. The latexes have been cleaned very carefully by ultrafiltration before they have been allowed to swell with toluene. The results prove the enormous influence of the nature of the particle water interface on the swelling capability of the particles. It is concluded that the Morton-Kaizerman-Altier equation cannot be applied for a complete description of latex particles swelling. Improvements of the swelling theory of latex particle are necessary in the future. New approaches have to consider the chemical nature of the particle water interface.

References

1. Markert G (1984) Angew Makromol Chem 123/124:285–306
2. Angaben des Statistischen Bundesamt der Bundesrepublik Deutschland, Berlin
3. Gilbert RG (1995) Emulsion Polymerization, Academic Press, London
4. Harkins WD (1947) J Am Chem Soc 69:1428–1444
5. Smith WV, Ewart RH (1948) J Chem Phys 16:592–599
6. Fitch RM, Tsai CH (1971) In: Fitch RM (ed) Polymer Colloids. Plenum, New York, pp 73–102
7. Lichti G, Gilbert RG, Napper DH (1983) J Pol Sci-Polym Chem 21:269–291
8. Adamson AW (1990) Physical Chemistry of Surfaces. John Wiley & Sons, Inc, New York, pp 364–378
9. Barrett KEJ (1975) Dispersion Polymerization in Organic media. John Wiley & Sons, London, pp 131–177
10. Carra S, Morbidelli M, Storti G (1985) Physics of Amphiphiles: Micelles, Vesicles and Microemulsions, XV Corso. Soc. Italiana di Fisica, Bologna, pp 483–512
11. Tauer K, Kühn I (1995) Macromolecules 28:2236–2239
12. Morrison BR, Gilbert RG (1995) Macromol Symp 92:13–30
13. Schmutzler K, Kakuschke R, Hergeth W-D (1989) Acta Polymerica 40:238–242
14. Shen S, Sudol ED, El-Aasser MS (1994) J Pol Sci Part A: Pol Chem 32:1087–1100
15. Kühn I, Tauer K (1995) Macromolecules (in press)
16. Verner B, Bárta M, Sedlácek B (1976) Table of Scattering Functions for Spherical Particles. Edice Macro, Prague
17. Gardon JL (1968) J Polym Sci 6:2859–2879
18. Morton M, Kaizerman S, Altier MW (1954) J Coll Sci 9:300–312
19. Antonietti M, Kaspar H, Tauer K (to be published)

Progr Colloid Polym Sci (1996) 101:38–44

E. Matijević

Monodispersed colloids: preparations and interactions

Supported by the NSF grant CHE-9423163

Prof. E. Matijević (✉)
Center for Advanced Materials Processing
Clarkson University
Potsdam, New York 13699-5814, USA

Abstract The progress in the preparation of uniform colloids, consisting of simple and composite particles of different shapes, by precipitation from electrolyte solutions is described. The use of such dispersions in studies of interactions with various solutions and solids is discussed. As an example, the incorporation of dyes into monodispersed particles to produce pigments of precise and reproducible optical properties is illustrated. It is shown that the phenomena in mixed dispersions of well defined particles (heterocoagulation) and of their adhesion (deposition on and detachment from solid surface) can be quantitatively evaluated and interpreted.

Key words Adhesion – colloids monodispersed – coated particles – heterocoagulation – particle adhesion

Introduction

Systematic research in well defined colloidal dispersions was initiated approximately a quarter of a century ago. The first such widely studied and used systems were various polymer latexes, produced as exceedingly uniform spheres. Since then, methods for the preparation of a large variety of monodispersed simple and composite inorganic colloids of different morphologies have been described in the literature [1–8]. Most common syntheses of these dispersions involve precipitation in solutions of electrolytes or organometallic compounds, which may also include intermediate stages, such as in the sol/gel or gel/sol processes.

Originally, the research in this field was mostly appreciated by the academic community, because the availability of monodispersed systems made it possible to quantitatively evaluate the properties of finely divided matter as a function of particles size and shape. Presently, the importance of such systems is recognized in applications as diverse as catalysis, ceramics, cosmetics, medical diagnostics, pigments, magnetic recordings, color display filters, etc. [9]. In view of these interests, it is quite important that the scientific progress be followed by engineering solutions for scaling up processes, or for continuous production of uniform dispersions based on precise specifications for given uses.

In this article some of the well defined colloids are illustrated and their interactions with dyes, to produce high quality pigments, are described. Furthermore, the results of studies of the stability of dispersions containing more than one kind of uniform particles are shown. As a special case are considered systems of two kinds of particles of greatly different size, which are employed to elucidate particle adhesion phenomena, i.e. particle deposition on and detachment from solid surfaces. In principle, the same theoretical treatment is applicable to the latter processes (adhesion) as to the stability of mixed dispersions of comparable particle sizes (heterocoagulation).

Preparation of monodispersed colloids

Most monodispersed colloids have been prepared by precipitation from homogeneous solutions by controlling the rate of the formation of solute complex precursors, which lead to the final solid products. Thus, metal (hydrous) oxides can be synthesized by "forced hydrolysis", i.e., by aging metal salt solutions at elevated temperatures, which greatly accelerate the deprotonation of hydrated cations, yielding intermediate species in the precipitation process. Depending on the conditions, either spheres (Fig. 1) or particles of different morphologies (Fig. 2) are so obtained.

Other kinds of colloidal metal compounds can be generated by the controlled release of anions into metal salt solutions during the aging process. Several such dispersions are exemplified in Fig. 3.

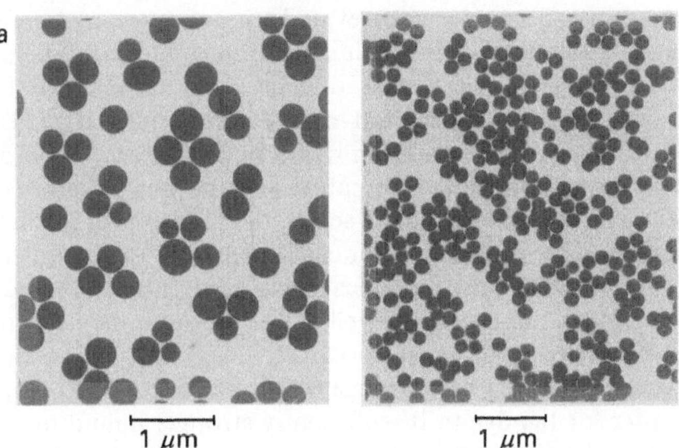

Fig. 1 Transmission electron micrographs (TEM) (A) of amorphous chromium hydroxide particle obtained by aging at 75 °C for 24 h a 4×10^{-4} mol dm^{-3} chrome alum solution [10], and (B) of cerium oxide particles obtained by aging at 90 °C for 48 h a 1.2×10^{-3} mol dm^{-3} $Ce(SO_4)_2$ solution [11]

Fig. 2 TEM (B) of hematite (α-Fe_2O_3) particles obtained by aging at 100 °C for 48 h a solution 2.0×10^{-2} mol dm^{-3} $FeCl_3$ and 3.0×10^{-4} mol dm^{-3} NaH_2PO_4 [12], and (A) of hematite particles obtained by aging at 100 °C for 16 h a solution 1.9×10^{-2} mol dm^{-3} $FeCl_3$, 1.2×10^{-3} mol dm^{-3} HCl and 40% (vol) ethanol [13]

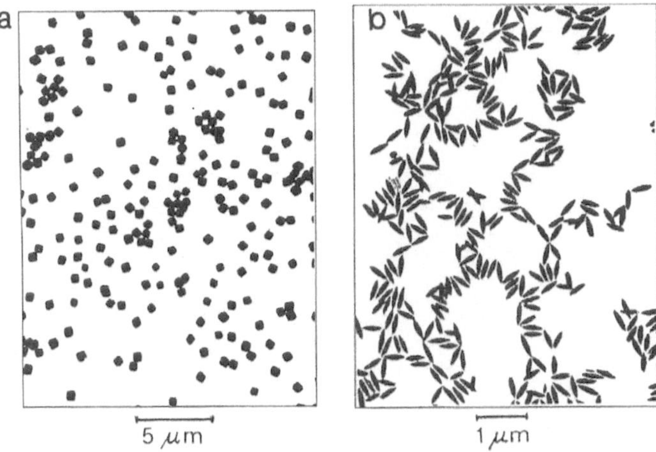

Fig. 3 (A) TEM of zinc sulfide particles obtained by aging at 26 °C for 5 h a solution of 2×10^{-2} mol dm^{-3} $Zn(NO_3)_2$, 6.2×10^{-2} mol dm^{-1} HNO_3 and 1.1×10^{-1} thioacetamide (TAA), and then continued aging it at 60 °C for 6 h [4]. (B) Scanning electron micrograph (SEM) of lead sulfide particles obtained by first preparing a seed sol of PbS to which was then added 1.25×10^{-3} mol dm^{-3} TAA and continued aging for 1 h [15]. (C) TEM of manganese (II) phosphate particles obtained by aging at 80 °C for 3 h a solution of 5.0×10^{-3} mol dm^{-3} $MnSO_4$, 5.0×10^{-3} mol dm^{-3} NaH_2PO_4, 1 mol dm^{-3} urea, and 1.0×10^{-2} mol dm^{-3} sodium dodecyl sulfate (SDS) [16]. (D) SEM of $CdCO_3$ particles obtained by mixing equal volumes of a 10 mol dm^{-3} urea, solution preheated at 80 °C for 24 h, and a solution 2.0×10^{-3} mol dm^{-3} $CdCl_2$ [17]

There is much ongoing discussion regarding the mechanisms of the formation of uniform particles. The original concept of a short lived nucleation stage with subsequent growth, due to the incorporation of the constituent species by a diffusion process, has been shown inapplicable in most cases. A more involved sequence of events takes place, whereby the nuclei grow to nanosize primary particles, which then aggregate to form the final uniform dispersions by a mechanism still not fully understood [2].

The precipitation can also yield particles of mixed composition, if solutions containing two or more different

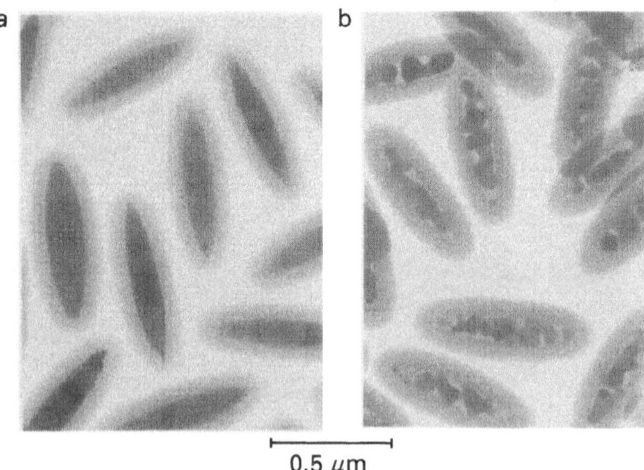

Fig. 4 TEM (A) of hematite (α-Fe_2-O_3) particles coated with silica by aging at 40°C aqueous dispersions of hematite particles in the presence of 2-propanol, NH_3, and tetraethyl orthosilicate (TEOS) [20]. (B) The same particles after the cores were reduced to metallic iron in hydrogen at 450°C [21]

electrolytes are aged at an appropriate temperature [8, 9]. However, the so prepared solids are – as a rule – internally inhomogeneous.

Of special interest are the syntheses of coated particles, which may consist of inorganic cores and shells of different chemical compositions, of organic cores and inorganic shells, or vice versa. In doing so, one can alter the surface properties of the dispersed matter, or produce particles of a given shape, when the coating material cannot be directly obtained in the desired morphology, while the core of such a shape is available.

Figure 4a illustrates hematite particles coated with a continuous layer of silica [20]. This dispersion behaves as pure silica in terms of interface reactions and surface charge characteristics. On treating the same powder at elevated temperatures in the hydrogen atmosphere, the cores are reduced to pure iron, while the silica layer is sealed, yielding magnetic particles fully protected by the coating from the possible effects of the environment [21].

Interactions of monodispersed colloids with solutes

Uniform dispersions are essential for the quantitative evaluation of the adsorption of solute species onto a solid, or of their incorporation into a solid. Using different particles it is possible to study solid/solute interactions as a function of the size, shape, surface charge, and other characteristics of an adsorbent, in addition to its chemical composition.

A special case to be used as an example of such processes deals with the formation of pigments by the reaction of dyes with monodispersed particles. The physical characteristics of color, i.e. the dominant wave length and purity, can be exactly calculated for spheres of a given radius and refractive index by the well known Mie theory [22]. Indeed, the particle size affects the color more strongly than often recognized, as illustrated by the calculated spectra of spherical hematite particles (Fig. 5) [23].

Commercial pigments are generally rather polydisperse and, consequently, their color properties are difficult to reproduce. It is also unfortunate that by far most of these pigments are essentially insoluble in any solvent and that they decompose at elevated temperatures, making it impossible to convert such solids into uniform particles, preferably spherical. This difficulty can be rectified by combining monodispersed particles with pure dyes. By selecting carriers of a given radius and refractive index, as well as by controlling the nature and the concentration of dyes incorporated in these solids, the pigments of predictable and reproducible color characteristics can be created.

For this purpose spherical silica particles of low refractive index are especially well suited, because they can be synthesized as dispersions of exceedingly narrow size distribution over a broad range of modal diameters [24]. In order for the dye to be sufficiently strongly bound to the core, chemical interactions with surface sites must take place. Thus, silica will preferably combine with basic dyes. The uptake of other kinds of dyes can be affected by appropriate surface modifications of the carriers, such as by coating SiO_2 particles with aluminum (hydrous) oxide to enhance the binding of acidic dyes.

Fig. 5 Extinction efficiency, Q_{ext}, as a function of the diameter and the wavelength, calculated according to the Mie theory for spheres having refractive index of hematite [23]

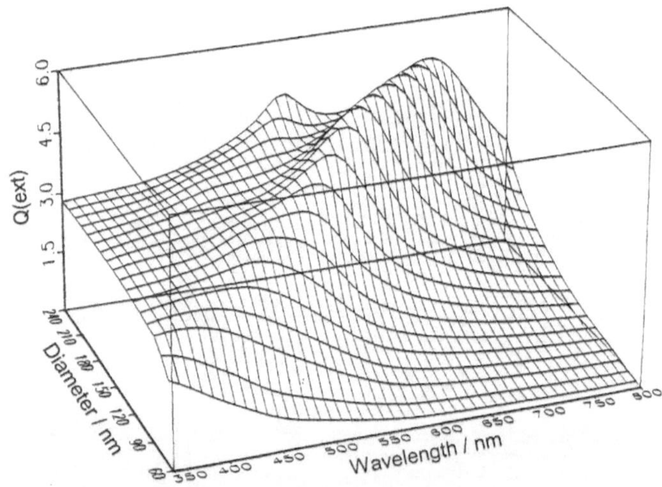

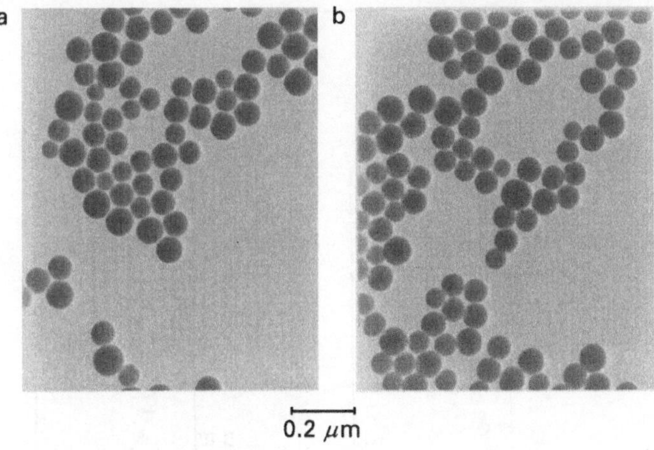

Fig. 6 TEM of (a) Violamin R and (b) Acid Red 183, coupled to amino-modified silica [25]

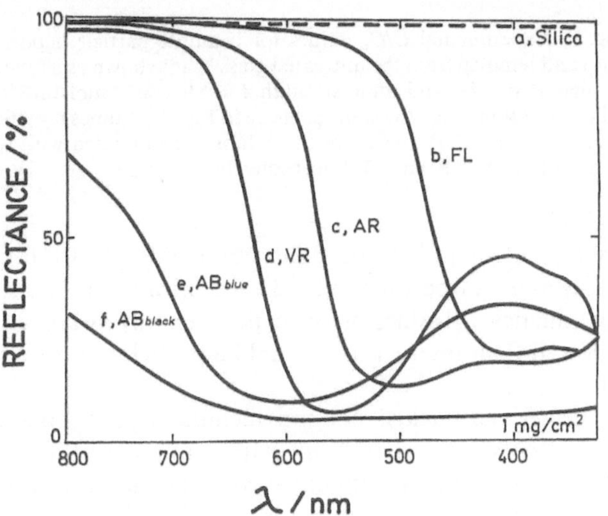

Fig. 7 Reflectance spectra of powder samples of pure amino-modified silica particles (a), and of pigments prepared by incorporating different dyes into these particles as follows: Flavazin L (b), C.I. Acid Red 183 (c), Violamin R (d), C.I. Acid Blue 45 (e) and (f). The spectra refer to a constant amount of 1 mg sample per cm^2 [25]

Figure 6 illustrates two so prepared pigments [25], and Fig. 7 shows a series of reflectance spectra obtained by reacting different dyes with amino-modified SiO_2 particles. It is noteworthy that silica cores had no effect on the spectral characteristics of the final products [25]. By determining the chromaticity parameters, it is possible to exactly define the primary color and the purity of each pigment.

Interactions in composite systems

The use of monodispersed colloids in the studies of composite systems will be discussed in terms of heterocoagulation and particle adhesion phenomena. Spherical inorganic particles are especially convenient for such investigations, because the necessary parameters are sufficiently well known in order to compare the rate of heterocoagulation or particle adhesion with theoretically calculated values, using either the approximate or exact expressions for colloid stability. The latter are based on the double layer models extended to dispersions consisting of more than one kind of particles.

Heterocoagulation

Systematic studies of the stability of mixed dispersions have been carried out with different inorganic and polymer latex colloids. By changing the pH, the sign and the magnitude of the surface potential of these particles can be altered, making it possible to evaluate the effect of this most influential parameter on the interactions of unlike particles. Figure 8 illustrates such an investigated system [28].

The rate of heterocoagulation was followed in dispersions of two kinds of particles of different sizes, present in various number concentration ratios, over a broad pH

Fig. 8 ζ-potential as a function of the pH of $Cr(OH)_3 \cdot xH_2O$ (●, ▲, ◆, ⬢) and of polyvinyl chloride (PVC) latex (○, △, ◇, ⬡) at four different ionic strengths (κ). All systems were prepared with $NaNO_3$

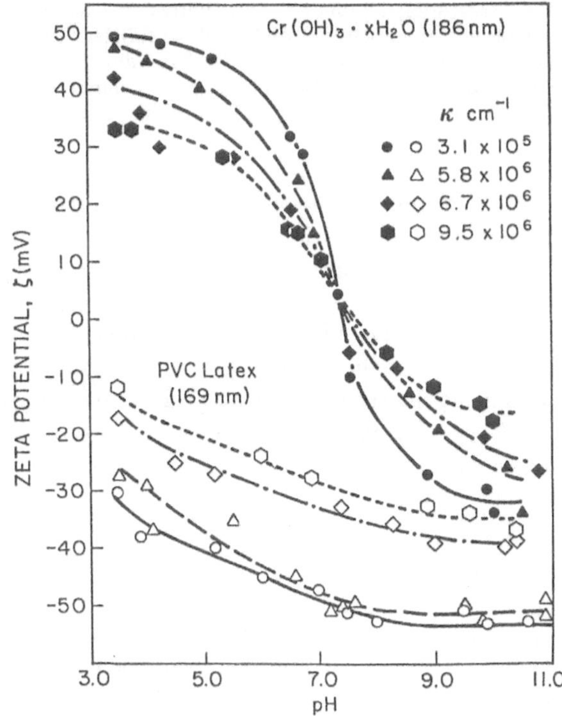

Table 1 Theoretical and experimental (in parentheses) values of log W_T at three concentrations of $NaNO_3$ (C_e) for the systems shown in Fig. 8

$C_e(M)$	log W_T at pH 8.0	log W_T at pH 9.0
0.016	5.75 (0.70)	30.7 (1.60)
0.032	4.60 (0.38)	28.6 (1.00)
0.063	3.32 (0.27)	26.2 (0.95)

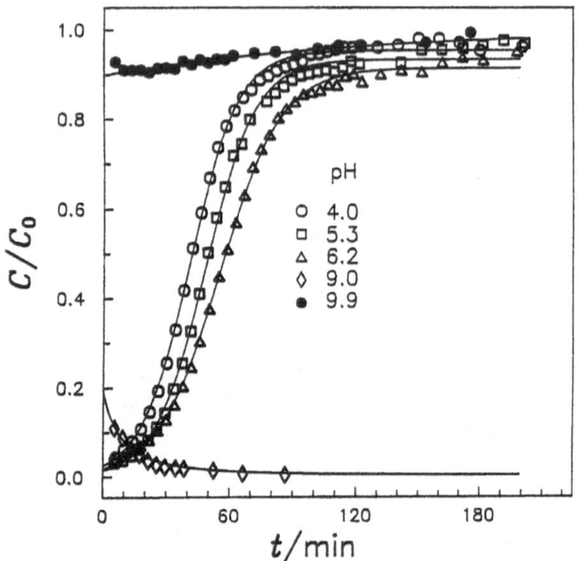

Fig. 9 Experimental C/C_0 values for the deposition of hematite particles (C_0, the initial concentration of 2.1×10^{10} particles cm^{-3}, C their concentration at times t) onto untreated glass beads as a function of time at suspension pH 4.0 (○), 5.3 (□), 6.2 (△), 9.0 (◇), and 9.9 (●) and ionic strength $I = 1.0 \times 10^{-4}$ mol dm^{-3}, at the fluid superficial velocity $u = 1.0 \times 10^{-3}$ $m s^{-1}$ in a column of height $z_0 = 1.5 \times 10^{-2}$ m. The solid lines are theoretically calculated [30]

range. While the mixed dispersions qualitatively behaved as anticipated from the double layer theory, dramatic differences were noted in all cases, when theoretical and experimental rate data were compared. As an example, Table 1 lists the values of the overall stability coefficient, W_T, for the system shown in Fig. 8, as determined and calculated using the Hogg, Healy and Fuerstenau approximation [26]. A possible explanation for such discrepancies is offered below.

Particle adhesion

The deposition of particles on solid surfaces from dispersions in liquids, and their removal from such substrates, represent phenomena of great practical and theoretical interest. In principle, particle attachment and detachment

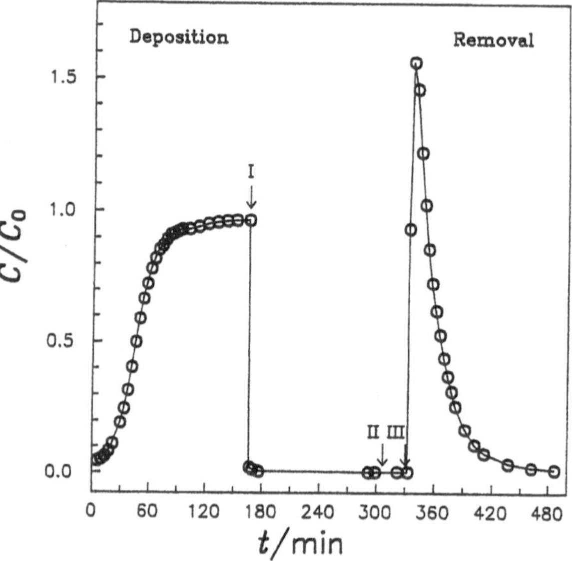

Fig. 10 Experimental C/C_0 values for heamtite particle deposition onto and removal from the untreated glass beads shown as a function of time at pH 4.0 and ionic strength $I = 1.0 \times 10^{-4}$ mol dm^{-3}. All other conditions were the same as those in Fig. 9. At times t = 165(I), 305(II), and 331(III), rinsing of the column was initiated with solutions of pH 4.0, 9.8, and 11.7, respectively

are analogous to heterocoagulation and peptization processes and can be thus treated theoretically. In addition, the kinetics of adhesion is experimentally more easily followed than interactions of unlike particles in bulk systems.

Using the packed column technique [20], adhesion experiments were carried out with a number of aqueous dispersions of uniform spherical colloidal particles passing through beds of different beads, as a function of varying surface potentials, ionic strengths and other parameters. Figure 9 illustrates the effect of the pH on the deposition of spherical hematite particles on glass beads as a function of time. It should be noted that the collector is negatively charged over the entire pH range, whereas hematite particle have an isoelectric point (i.e.p.) at pH = 8.5 [30]. No retention is observed in the most basic solutions used, while multilayer deposition occurs near the i.e.p. of Fe_2O_3. Single layer attachment takes place at other investigated pH values.

Particles deposited at pH 4.0 could be completely detached, when the column was rinsed with a solution of pH 11.7 (Fig. 10) [30], indicating that no chemical bonds were formed between hematite and glass during the attachment process. In contrast, if the collector beads are coated with gelatin, the same adhered particles remain permanently bound, regardless of the pH of the rinsing solution.

A comparison of experimental and theoretically calculated stability coefficients for the coagulation of

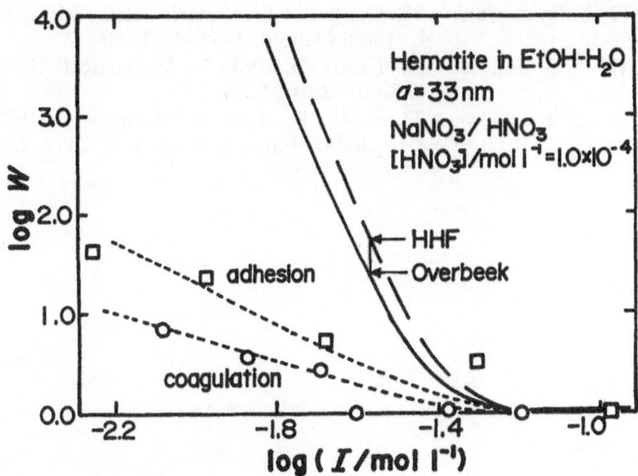

Fig. 11 The stability coefficients for homocoagulation, W, of spherical hematite particles (a = 33 nm) (○) and for their multilayer adhesion on glass beads (□) as a function of the ionic strength at $[HNO_3] = 1.10^{-4}$ mol dm^{-3} [31]. The Overbeek and the HHF expressions [26] are given by the solid and the broken line, respectively [27]

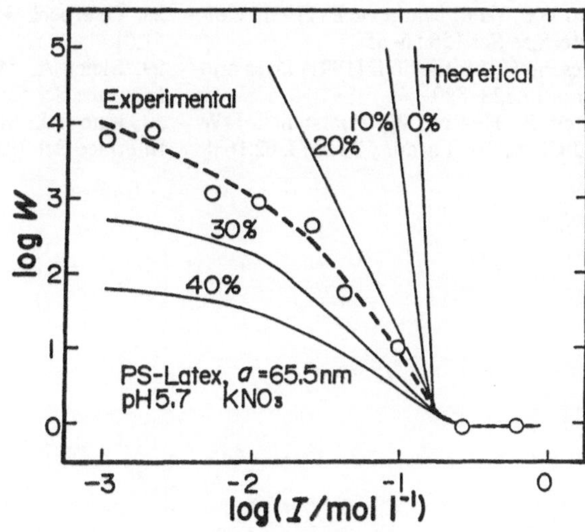

Fig. 12 The effect of the surface charge segregation on the stability of the polystyrene latex (PS) plotted as the function of the ionic strength. The solid lines are calculated, while the experimental values are given by open circles

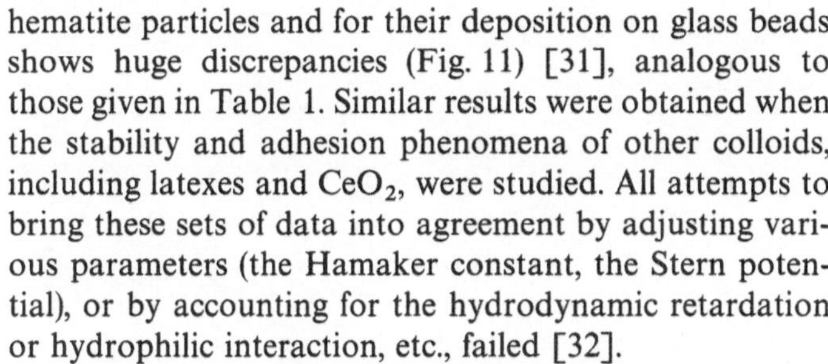

hematite particles and for their deposition on glass beads shows huge discrepancies (Fig. 11) [31], analogous to those given in Table 1. Similar results were obtained when the stability and adhesion phenomena of other colloids, including latexes and CeO_2, were studied. All attempts to bring these sets of data into agreement by adjusting various parameters (the Hamaker constant, the Stern potential), or by accounting for the hydrodynamic retardation or hydrophilic interaction, etc., failed [32].

Finally, it was established that taking into consideration discreetness of the surface charge, rather than a smoothed out charge on the particles, assumed by the theoretical model, the experimental and calculated curves could be brought into agreement [32]. Figure 12 offers, as an example such a correction for a polystyrene latex dispersion, with 25% surface charge segregation. Again, analogous results were obtained with a variety of systems, by treating the kinetics both of heterocoagulation and adhesion processes.

The foregoing examples are offered to illustrate the usefulness of monodispersed colloids in studies that address some fundamental aspects of interactions in systems consisting of fine particles.

References

1. Matijević, E (1994) In: Wedlock DJ (ed) Controlled Particle, Droplet and Bubble Formation. Butterworth–Heinemann, London pp 39–59
2. Matijević E (1994) Langmuir 10:8–16
3. Matijević E (1993) Chem Mater 5:412–426
4. Matijević E (1992) Discuss Faraday Soc 92:229–239
5. Matijević E (1985) Annu Rev Materials Sci 15:483–516
6. Overbeek J Th G (1982) Adv Colloid Interface Sci 12:251–277
7. Haruta M, Delmon B (1986) J Chim Phys 83:859–868
8. Sugimoto T (1987) Adv Colloid Interface Sci 28:65–108
9. Fine Particles, Parts I and II. A series of articles published in the MRS Bull (1989) 14: No. 12; (1990) 15:No 1
10. Matijević E (1989) MRS Bull 14:18–20
11. Demchak R, Matijević E (1969) J Colloid Interface Sci 31:257–262
12. Hsu WP, Rönnquist L, Matijević E (1988) Langmuir 4:31–37
13. Hamada S, Matijević E (1981) J Colloid Interface Sci 84:274–277
14. Ozaki M, Kratohvil S, Matijević E (1984) J Colloid Interface Sci 102:146–151
15. Murphy Wilhelmy D, Matijević E (1984) J Chem Soc Faraday Trans I 80:563–570
16. Murphy Wilhelmy D, Matijević E (1985) Colloids Surf 16:1–8
17. Springsteen L, Matijević E (1989) Colloid Polymer Sci 267:1007–1015
18. Janeković A, Matijević E (1985) J Colloid Interface Sci 103:436–447
19. Ribot F, Kratohvil S, Matijević E (1989) J Mater Res 4:1123–1131
20. Quibén J, Matijević E (1994) Colloids Surf 82:237–246
21. Ohmori M, Matijević E (1992) J Colloid Interface Sci 150:594–598
22. Ohmori M, Matijević E (1993) J Colloid Interface Sci 160:288–292
23. Kerker M (1969) The Scattering of Light and other Electromagnetic Radiation. Academic Press New York, NY
24. Ryde N, Matijević E (1994) Appl Optics 33:7275–7281

25. Hsu WP, Yu R, Matijević E (1993) J Colloid Interface Sci 156:56–65
26. Giesche H, Matijević E (1991) Dyes and Pigments 17:323–340
27. Hogg R, Healy TW, Fuerstenau DW (1966) J Chem Soc Faraday Trans I, 62:1638
28. Overbeek J Th G (1990) Colloids Surf 51:61
29. Bleier A, Matijević E (1976) J Colloid Interface Sci 55:510–524
30. Kuo RJ, Matijević E (1980) J Colloid Interface Sci 78:407–421
31. Gangolli S, Partch RE, Matijević E (1989) Colloids Surf 41:339–344
32. Kihira H, Ryde N, Matijević E (1992) Colloids Surf 64:317–324
33. Kihira H, Ryde N, Matijević E (1992) J Chem Soc Faraday Trans I 88: 2379–2386

Progr Colloid Polym Sci (1996) 101:45–50

N.V. Churaev

Surface forces in wetting, flotation and capillary phenomena

Dr. N.V. Churaev (✉)
Institute of Physical Chemistry of Russian Academy of Sciences
Laboratory of Thin Liquid Layes
Leninsky Prospect 31
117915 Moscow, Russia

Abstract Surface forces play a crucial role in disperse systems, containing colloidal particles or thin films. Theory of wetting, based on the Frumkin-Derjaguin approach, relates the contact angles with isotherms of disjoining pressure $\Pi(h)$ of wetting films. On the basis of the isotherms obtained, contact angles were calculated and compared with experimental data. It was shown that in the region of contact angles from 10 to 50 degrees it is sufficient in some cases to take into account the molecular and electrostatic forces only. Complete wetting is guaranteed by the action of hydrophilic repulsion forces, whereas large contact angles arise under the action of hydrophobic attraction forces. The effects of surfactants on contact angles' formation, wetting films' stability and flotation are discussed. Capillary phenomena, influencing kinetics of penetration and mutual displacement of fluids were investigated using model systems. The kinetics is crucially influenced by addition of surfactants. Capillary pressure of a moving meniscus changes as a result of a mass exchange of surfactant molecules between the meniscus, solid wall and forming film interfaces. The arising effects were investigated in dependence on flow rates, electrolyte and surfactant concentration.

Key words Surface forces – contact angles – wetting films – capillary penetration of surfactant solutions

Introduction

Surface forces play a crucial role in disperse systems containing small particles and in thin films. The behavior of disperse systems and porous bodies, their stability, aggregative state and strength, processes of mass transfer depend on the forces acting between particles and interfaces. Consideration of the phenomena in such systems is based on the DLVO theory that takes into account joint action of molecular and electrostatic forces [1, 2]. The DLVO theory is applicable to the systems of a moderate hydrophilicity. At high degree of hydrophilicity or hydrophobicity the third component of surface forces, namely the structural forces, comes into play. The latter originate from the overlapping of boundary layers with a structure modified as compared with bulk liquid [1, 2].

Wetting

The phenomenon of wetting may be considered also in the framework of the surface forces theory [3–5]. Thin wetting films, like the liquid interlayers between particles, are subjected to action of the surface forces. However, wetting films, as distinct from colloidal systems, represent a non-symmetrical case. The film is bounded by two different phases: a solid and a gas one. Due to this, surface forces manifest themselves in wetting films and symmetrical liquid interlayers in very different manner.

Molecular forces in colloidal systems containing particles of the same nature are always attractive and responsible for the aggregation of the particles. Molecular forces in wetting films are attractive only when the liquid is more polarizable than a substrate. This occurs, for instance, in the case of low-energy solids. Molecular forces in wetting films on dielectrics or metals are repulsive and stabilize the films. The isotherm of molecular component of disjoining pressure has the following form [1, 2]:

$$\Pi_m = A/6\pi h^3, \tag{1}$$

where A is the Hamaker constant and h is the film thickness.

Positive values of the Hamaker constant correspond to the repulsion forces between film interfaces, and are negative to the attraction ones. The disjoining pressure is determined as a difference between the pressure P acting on interlayer or film surfaces and the pressure P_0 in the bulk phase: $\Pi = P - P_0$ [1].

Electrostatic repulsion takes place between identically charged colloidal particles immersed into an aqueous solution. The strength of electrostatic interaction depends on the magnitude of the particles surface potential ψ. Electrostatic forces in wetting films depend on electrical potentials of both solid-liquid ψ_1 and liquid-gas ψ_2 interfaces which are seldom equal to each other. When the ψ_1 and ψ_2 potentials have the same sign but are different in magnitude, electrostatic forces change from repulsion to attraction at a film thickness $h > h_* = (1/\kappa)\ \ln(\psi_1/\psi_2)$, where $1/\kappa$ is the Debye length. At $h < h_*$ and $\kappa h \ll 1$ the electrostatic forces of attraction may be expressed by the following equation applicable at arbitrary ψ_1 and ψ_2 values [1]:

$$\Pi_e = -\varepsilon(\psi_1 - \psi_2)^2/8\pi h^2, \tag{2}$$

where ε is the dielectric constant.

Structural forces are now being intensively studied both experimentally and theoretically [1, 2]. The isotherm of the short range structural forces may be approximately represented in an exponential form:

$$\Pi_s = K \exp(-h/\lambda), \tag{3}$$

where the parameter K is responsible for the strength of the forces and the decay length λ is on the order of correlation length λ_0 in bulk liquid. For aqueous systems $\lambda \simeq 1$ nm. The structural forces of the repulsion ($K > 0$) operate between two hydrophilic surfaces. The attraction ($K < 0$) takes place between two hydrophobic surfaces in water [6–10].

The Frumkin-Derjaguin approach [11–13] gives a relation between the equilibrium contact angel θ_0 and the disjoining pressure isotherm $\Pi(h)$ of plane wetting films:

$$\cos\theta_0 = 1 + (1/\gamma_{lv}) \int_{h_0}^{\infty} \Pi(h)\,\mathrm{d}h = 1 + [G(h_0)/\gamma_{lv}], \tag{4}$$

where γ_{lv} is the surface tension and $G(h_0)$ is the excess free energy of a plane wetting film being in equilibrium with bulk meniscus or droplet. The equilibrium film thickness h_0 may be found as a smallest magnitude of the root of an equation that represents the complete $\Pi(h)$ isotherm as algebraic sum of the above-mentioned components

$$\Pi(h) = \Pi_m(h) + \Pi_e(h) + \Pi_s(h) = 0 \tag{5}$$

Computer calculation of equilibrium contact angles was performed using Eq. (4) and Eqs. (1–3) for three components of surface forces:

$$\cos\theta_0 = 1 + (1/\gamma_{lv})[(A/12\pi h_0^2) - [\varepsilon(\psi_1 - \psi_2)^2/8\pi h_0] + K\lambda \exp(-h_0/\lambda)], \tag{6}$$

where h_0 is the equilibrium film thickness calculated from the solution of Eq. (5).

The results of the calculation of contact angles for aqueous solutions on solids in dependence on the potential difference $\Delta\psi = \psi_1 - \psi_2$ are shown in Fig. 1. Here are used: the Hamaker constant for water $A = 7\times10^{-7}$ J, $\gamma_{lv} = 72$ mN/m, $\varepsilon = 80$ and $\kappa = 10^5$ cm^{-1}. When the structural forces are absent (curve 1), the contact angles grow with increasing forces of electrostatic attraction that depend quadratically on the $\Delta\psi$ value. Hydrophilization of the substrate, characterized by the $K\lambda$ values, decreases the contact angles. Thus, the known phenomenon of improved wetting by surface hydrophilization may be explained as a manifestation of structural forces of repulsion between wetting film interfaces. At $K\lambda > 1.5\times10^{-2}$ J/m^2, complete wetting is attained, in spite of a counteraction of the electrostatic attraction forces.

Figure 2 shows the results of calculation of the contact angles that include the effect of hydrophobic attraction forces, when $K\lambda < 0$. At the degree of hydrophobicity corresponding to the negative $K\lambda$ values higher than 2×10^{-2} J/m^2, the influence of the electrostatic forces becomes negligible. The formation of contact angles θ_0 larger than 50 degrees is determined in this case by the predominant action of the hydrophobic attraction forces. This effect manifests itself even at smaller values of the contact angles, and the larger the extent, the smaller are the $\Delta\psi$ values that characterize the forces of electrostatic attraction.

Filled points in Fig. 2 represent the $K\lambda$ values obtained from the direct measurements of hydrophobic attraction forces acting between two hydrophobic surfaces in water [6–10]. The degree of hydrophobicity of the surfaces was characterized by advancing contact angles. Comparison of

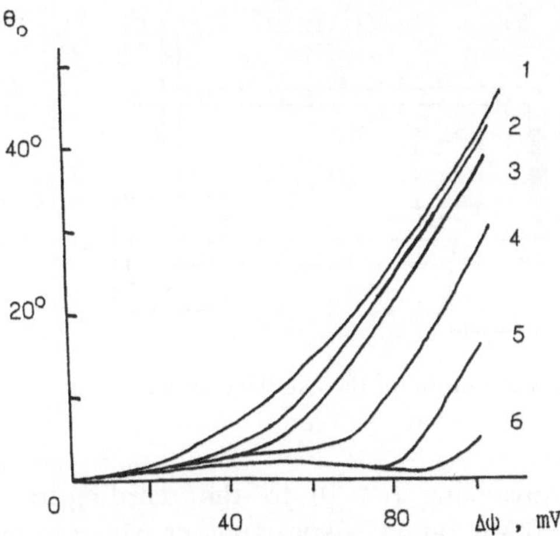

Fig. 1 Calculated dependences of contact angle θ_0 on the potential difference $\Delta\psi$ at various values of the parameter $K\lambda$ that characterizes the structural forces of hydrophilic repulsion: $K\lambda = 0$ (1); 10^{-3} (2); 2×10^{-3} (3); 10^{-2} (4); 1.5×10^{-2} (5) and 2×10^{-2} J/m^2 (6)

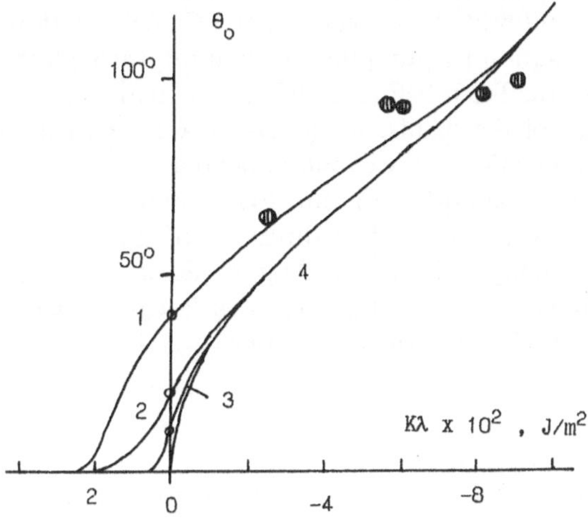

Fig. 2 Calculated dependences of contact angle θ_0 on the $K\lambda$ values that characterizes the hydrophobic attraction forces ($K\lambda < 0$) and hydrophilic repulsion forces ($K\lambda > 0$) at various values of the potential difference $\Delta\psi = 0$ (1); 0.05 (2); 0.07 (3) and 0.1 mV (4)

the points with curves 1 and 4 shows that the constants $K\lambda$ are nearly the same both in the cases of water interlayers between two hydrophobic surfaces and in aqueous wetting films on a hydrophobic substrate.

Therefore, inclusion of structural forces into consideration becomes necessary not only in the case of colloids but also in the case of wetting phenomena. The DLVO theory is applicable at a moderate degree of surface hydrophilicity in the range of the contact angles from 10 to 40–50 degrees.

Flotation

Addition of cationic surfactants results in a substantial increase in the contact angles of quartz particles, improving the efficiency of froth flotation. The adsorption of the cationic surfactants influences the surface charge of both solid-liquid and film-air interfaces of an aqueous wetting film.

In Fig. 3 are shown, as an example, the dependences of the potentials ψ_1 of quartz surface (curve 1), paraffin surface (curve 2) and the ψ_2 potential of the film-air interface (curve 3) on the concentration C_s of cetyltrimethylammonium bromide (CTAB) in the background 10^{-4} M KCl electrolyte. The values of the ψ_1 potentials were obtained from the measurements of streaming potentials in single quartz capillaries [14, 15]. The values of the ψ_2 potential were obtained from the experiments with foam films performed in D. Exerowa's laboratory.

Figure 4 shows the results of measurements of the contact angles on quartz in dependence on the concentration C_s of the CTAB solution. Curve 1 refers to the data obtained at the background electrolyte 10^{-4} M KCl [16], and curve 2 refers to 10^{-6} M KCl [17]. The contact angles for the CTAB solution on quartz surface amount of about 40°. Theoretical assessment of the contact angle based on Eq. (6) while using $\Delta\psi \simeq 100$ mV (compare curves 1 and 3) and neglecting structural forces gives nearly the same result. This means that in this case the main contribution is given by the electrostatic attraction forces. At the CTAB concentration $C_s = 10^{-3}$ M, near to CMC, both film interfaces acquire positive potentials of nearly the same magnitude (see Fig. 3). The electrostatic forces become repulsive providing the formation of thick wetting films and complete wetting (see Fig. 4).

Fig. 3 Experimental dependences of the potential ψ_1 of quartz-solution (curve 1), paraffin-solution (curve 2) and ψ_2 of solution-air (curve 3) interfaces on concentration C_0 of CTAB solution

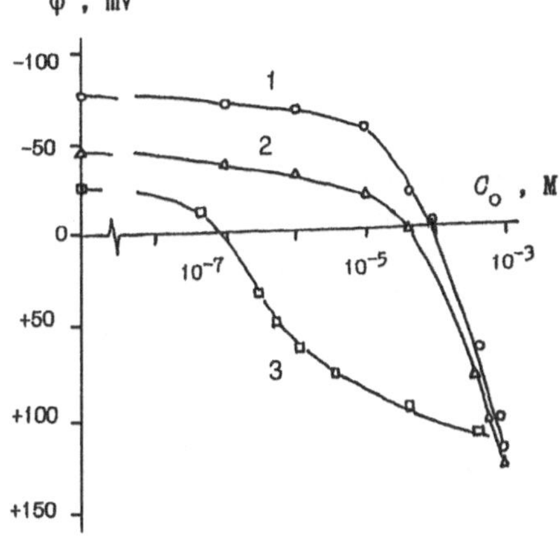

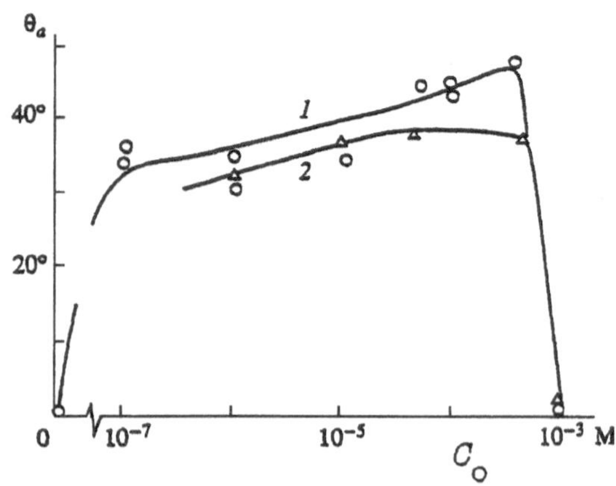

Fig. 4 Results of measurements of contact angles of CTAB solutions on quartz in dependence on the solution concentration C_0. Background electrolyte 10^{-4} (curve 1) and 10^{-6} M KCl (curve 2)

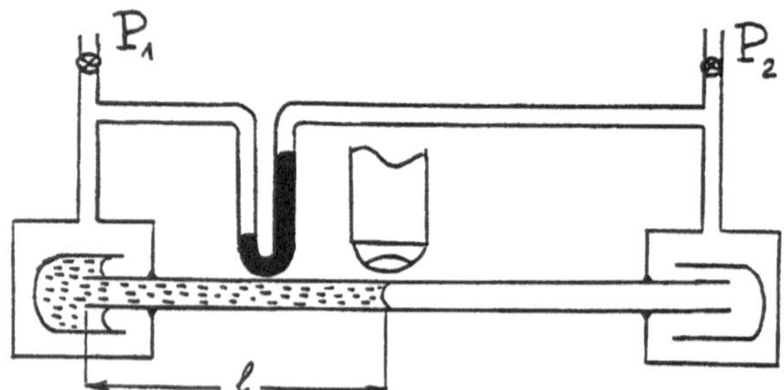

Fig. 5 Schematic representation of the capillary set-up

For thin aqueous films of the CTAB solutions placed between quartz and oil surfaces the difference between the ψ_1 and ψ_2 potentials is far less (compare curves 1 and 2 in Fig. 3) and the effect of hydrophobic attraction forces prevails. Equation (6) gives in this case (at $\Delta\psi = 40$ mV, $K\lambda = 2 \times 10^2$ J/m^2, $\lambda = 1$ nm) the theoretically estimated value of the contact angles $\theta_0 \simeq 50°$ for tetradecane ($\gamma_{lv} = 52$ mN/m) and $\theta_0 \simeq 60°$ for silicon oil ($\gamma_{lv} = 35$ mN/m) which fits the experimental data [18]. When the hydrophobic forces are ignored the calculation gives much smaller values of the contact angle ($\theta_0 \simeq 18°$).

Thus, the flotation efficiency may be enhanced while using two different methods: i) imparting the potentials ψ_1 and ψ_2 that are noticeably different in magnitudes to wetting film interfaces: ii) hydrophobization of particles using, for instance, dehydration of their surface or adsorption of non-ionic surfactants. An addition of the cationic surfactants (in the case of negatively charged particles) results both in enhanced electrostatic and hydrophobic attraction forces. This explains the widespread application of the cationic surfactants in practice.

Capillary phenomena

The effect of the surface forces on a capillary penetration of surfactant solutions was investigated using thin single quartz capillaries having radii 1 to 10 μm [18–21].

Schematic representation of the experimental set-up is shown in Fig. 5. The left end of the capillary is placed into a vessel filled with the solution under investigation. The gas pressures (nitrogen) P_1 and P_2 in the chambers could be regulated separately. The pressure difference ΔP was measured using a mercury manometer. Depending on the ΔP values, the direction of the meniscus motion can be changed from advancing ($v > 0$) to the receding one ($v < 0$). At $\Delta P = 0$, the rate of a spontaneous penetration can be measured.

This method was used for investigations of liquid-air and liquid-liquid immiscible displacement of pure liquids and surfactants solutions [18–22]. As an example, here the results obtained for the CTAB solutions in quartz capillary, $r = 6.6$ μm, would be considered. To prevent the effect of the CTAB adsorption, the capillary surface was previously equilibrated with each of the CTAB solutions under investigation by pumping the solution through the capillary during 2–3 h before performing experiments.

The rates of the meniscus movement were measured within a small section of the capillary of about 0.1 cm long. The length l of the liquid column (Fig. 5) remains in this case practically constant. The direction of motion was successively changed from receding to advancing. The capillary pressure of a moving meniscus P_c can be determined from the Washburn equation [23]:

$$v = r^2(\Delta P + P_c)/8\eta l, \tag{7}$$

where η is the viscosity of the solution.

When the capillary pressure P_c remains constant in the course of displacement it follows from Eq. (7) that the flow rates v must linearly depend on ΔP. Figure 6 shows the results obtained for the CTAB solution, $C_0 = 5 \times 10^{-4}$ M, at pH 3.4 in the background KCl electrolyte, 10^{-4} M. At high flow rates, $v > 10^{-3}$ cm/s (curves 1 and 2), the $v(\Delta P)$ dependences are linear. Their intersections with the ΔP axis give the capillary pressures P_c of an advancing (curve 1) and of a receding (curve 2) meniscus, the pressures that do not depend on the flow rates.

In the case of receding meniscus, the tension $\gamma \cos \theta_\gamma = P_c r/2$ equals 70 mN/m, and is much higher than the surface tension of the bulk solution $\gamma_0 = 40$ mN/m approaching the surface tension of water. This means that the concentration of CTAB near the receding meniscus is substantially lowered and $\theta_r \simeq 0$. The effect was explained considering an exchange of surfactant molecules between a moving meniscus and a wetting film formed on capillary

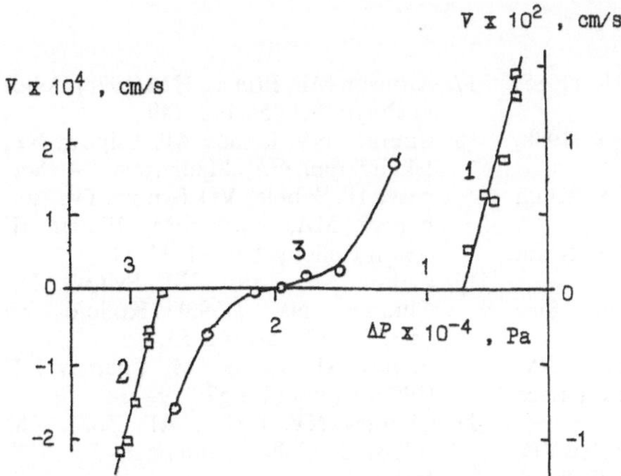

Fig. 6 Experimental $v(\Delta P)$ dependences for CTAB solution, $C_0 = 5 \times 10^{-4}$ M, at high (curves 1 and 2, right scale of v) and low (curve 3, left scale) flow rates v. Advancing motion: $v > 0$. Receding motion: $v\langle 0.\rangle$

walls after retreating meniscus [18–21]. The film stability is secured by the electrostatic repulsion forces between both positively charged solid-liquid and film-gas interfaces (see Fig. 3). In the steady state at the given flow rate, the flux of surfactant molecules from the mensicus to the film surface is compensated by diffusion of the CTAB molecules to the meniscus from the solution inside the capillary. An analysis of the mass transfer results in the following equation for the surfactant concentration C_m near meniscus [18–21]:

$$C_m/C_0 = \{1 + [k\,\mathrm{Pe}/r(1 + \mathrm{Pe}_s)]\}^{-1} \tag{8}$$

Here $k = \Gamma_m/C_m$ is the Henry constant, where Γ_m is the adsorption on the meniscus; $\mathrm{Pe} = v\delta/D$ and $\mathrm{Pe}_s = v\delta_s/D_s$ are the bulk and surface Peclet numbers; D and D_s, δ and δ_s are the coefficients of diffusion and diffusion lengths respectively.

Figure 7 shows the dependence of C_m/C_0 on the flow rate v, calculated using Eq. (8) with $k = 10^{-3}$ cm; $D = 10^{-6}$ and $D_s = 5 \times 10^{-7}$ cm^2/s; $\delta = 5 \times 10^{-3}$ and $\delta_s = 5 \times 10^{-4}$ cm. At $v \Rightarrow 0$, the surface tension of receding meniski remains equal to C_0. At high flow rates (when $\mathrm{Pe}_s \gg 1$), the relation C_m/C_0 tends to a constant value

$$C_m/C_0 \simeq (r\delta_s D/k\delta d_s) \tag{9}$$

and does not depend on v, in accordance with the experimental data (curves 1 and 2, Fig. 6).

Comparing the dynamic tension of the receding meniscus $\gamma_m = 70$ mN/m and the bulk tension of the solution, $\gamma_0 = 40$ mN/m, we obtain using the $\gamma(C_0)$ isotherm for CTAB solution that $C_m/C_0 \simeq 10^{-2}$. Substitution of this value into Eq. (9) gives $k \simeq 10^{-3}$ cm, as was decided above.

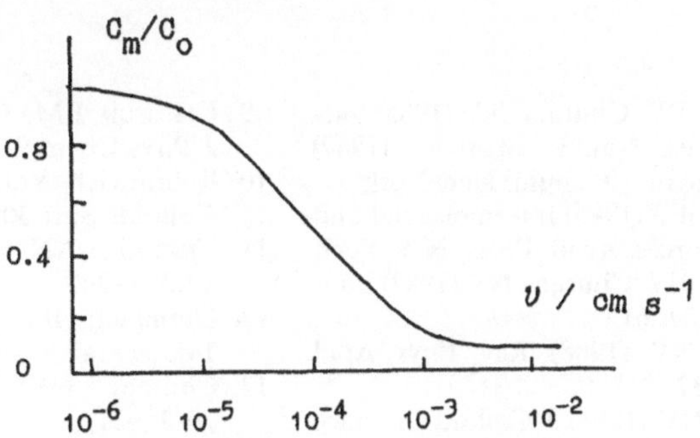

Fig. 7 Calculated dependence of the solution concentration C_m near to receding meniscus on the flow rate v

The advancing contact angle θ_a may be estimated from the intersection of line 2 in Fig. 6 with the ΔP axis assuming that in this case $\gamma = \gamma_0$, and $\cos\,\theta_a = P_c r/2\gamma_0$. The calculations have given $\theta_a \simeq 50°$. In the same way the dependences of θ_a on the CTAB concentration have been obtained including ones for solution-oil systems [18, 21].

Besides, the $v(\Delta P)$ dependences at low flow rates, $v < 10^{-4}$ cm/s, in the same capillary were obtained (curve 3, Fig. 6). For every point of curve 3, capillary pressures $P_c(v)$ were calculated on the basis of Eq. (7) using the corresponding pairs of the local v and ΔP values:

$$P_c(v) = -\Delta P + (8\eta l v/r^2) \tag{10}$$

In this case the capillary pressure $P_c = 2\gamma_m/r$ of a slowly moving meniscus depends on v in accordance with Eq. (8). As was shown in Fig. 7, the dynamic surface tension of the meniscus γ_m is at $v < 10^{-3}$ cm/s rate dependent. Similar effect is also observed with advancing meniscus (curve 3, Fig. 6, $v > 0$) [21].

The effects of the dynamic surface tension of the surfactant solutions observed at $v < 10^{-3}$ cm/s are more pronounced in thin capillaries. As follows from Eq. (8), $C_S/C_0 \Rightarrow 0$ at $r \Rightarrow 0$.

Therefore, it may be concluded that formation of wetting films of surfactant solutions changes the mechanism of displacement in thin capillaries. Mass exchange of surfactant molecules with forming wetting films results in the dynamic surface tension of the moving menisci that influences the kinetics of the solution displacement. At flow rates $v < 10^{-3}$, the capillary pressure of the moving menisci is rate dependent. In the region of the flow rates v from 10^{-3} to 2×10^{-2} cm/s both the capillary pressure and contact angles of the receding and advancing meniscus remain constant.

References

1. Derjaguin BV, Churaev NV (1985) Surface Forces, Nauka, Moscow; (1987) Cons. Bureau – Plenum, New York
2. Israelachvili J (1985) Intermolecular and Surface Forces, Acad. Press, New York
3. Derjaguin BV, Churaev NV (1987) Langmuir 3:607–612
4. Churaev NV (1988) Rev Phys Appl 23:975–987
5. Churaev NV (1995) J Colloid Interface Sci 172:479–484
6. Israelachvili JN, Pashley RM (1982) Nature 300:341–342
7. Pashley RM, McGuiggan PM, Ninham BV, Evans DF (1985) Science 229:1088–1089
8. Claesson PM, Blom CE, Herder PC, Ninham BV (1986) J Colloid Interface Sci 114:234–242
9. Claesson PM, Christenson HK (1988) J Phys Chem 92:1650–1655
10. Rabinovich YaI, Derjaguin BV (1988) Colloids Surf 30:243–249
11. Frumkin AN (1938) Zh Fiz Khim 12:337–345
12. Derjaguin BV (1940) Zh Fiz Khim 14:137–147
13. Churaev NV (1993) Colloids Surf 79:25–31
14. Sergeeva IP, Sobolev VD, Churaev NV, Derjaguin BV (1981) J Colloid Interface Sci 84:451–460
15. Churaev NV, Sergeeva IP, Sobolev VD, Zorin ZM, Gasanov EK (1993) Colloids Surf 76:23–32
16. Churaev NV (1994) Kolloid Zh 56:707–723
17. Aronson MP, Princen HM (1978) Colloid Polym Sci 256:140–149
18. Churaev NV, Ershov AP, Esipova NE, Iskandarjan GA, Madjarova EA, Sergeeva IP, Sobolev VD, Svitova TF, Zakharova MA, Zorin ZM, Poirier JE (1994) Colloids Surf 91:97–112
19. Ershov AP, Zorin ZM, Svitova TF, Churaev NV (1993) Kolloid Zh 55(3):39–47; 55(4):45–53
20. Ershov AP, Zorin ZM, Churaev NV (1995) Kolloid Zh 57:329–334
21. Churaev NV, Ershov AP, Zorin ZM (1996) J Colloid Interface Sci 177: 589–601
22. Zorin ZM, Churaev NV (1992) Adv Colloid Interface Sci 40:85–108
23. Washburn EW (1921) Phys Rev 17:273–283

Progr Colloid Polym Sci (1996) 101:51–57

M.L. Hair
C.P. Tripp

Adsorption of block polymers on well-defined silica surfaces

Dr. M.L. Hair (✉) · C.P. Tripp
Xeror Research Centre of Canada
2600 Speakman Drive
Mississauga, Ontario L5LK 2LI, Canada

Abstract Block copolymers of polyethylene oxide and polystyrene have been examined in the Surface Force Apparatus and their adsorption on mica is well understood. The results of the measurements suggest that these block copolymers can be subdivided into three compositional groups according to Marques/Joanny theoretical relationships. The main focus of the SFA investigations was to determine the extended length of the polymer brush as this in turn is known to affect the flocculation-deflocculation behavior of particles. This paper describes the use of a "colloidal cell" which can be used to prepare dispersions of silica particles in carbon tetrachloride using a totally evacuated system. The surface of this silica has been defined by heat treatment in vacuo. The cell also enables infrared spectroscopy to be applied to the dispersion. The adsorption of the polymer, the specific interactions of the polyethylene oxide and polystyrene segments with the surface and the rate of settling of the particles can be quantified. Block copolymers with three different asymmetries have been examined and the results compared with data obtained for polyethylene oxide and polystyrene homopolymers.

Key words Adsorption – silica surfaces – block copolymers – flocculation

Introduction

The adsorption of polymers onto surfaces is a critical step in the protection of colloidal particles and their stabilization against flocculation. The use of polymer stabilized particles in industry, in detergency, and in science is well established. The supporting solvent is known to be critical to the stabilization process and the role of polymer-solvent interactions has been the subject of many studies. Over the past ten years, since the widespread introduction of scaling concepts, the area has received considerable stimulus. Theoretical work in particular has provided a great deal of understanding and has emphasized polymers which are "tethered" and adsorbed to particle surfaces via one end of the polymer chain. When the adsorbed polymer segments are in semi-dilute solution and adsorbed in such a way that the space occupied by each molecule is less than the normal radius of gyration (Rg), then the polymers are extended from the surface and form "brushes". Such brushes have the potential for providing unusual viscoelastic properties to the resultant dispersions. The original theory of Alexander [1] and DeGennes [2] suggested that the height (i.e. length) of the adsorbed brush can be predicted simply from a knowledge of the coverage of the polymer molecule on the surface and its solution properties (Eq. (1)). In this formulation

$$L \sim aN\sigma^{1/3} \quad (1)$$

σ is a reduced surface density and the equation implies that as the surface density of the polymer is increased, then the brush extends further out from the surface. The surface force apparatus originally developed by Israelchvili [3] is unique in its ability to directly measure the extended length of an adsorbed polymer in solution. If the polymer is adsorbed onto the mica surface then, when the two surfaces are brought together, a repulsive force is measured at the first contact of the tails of the opposing polymers. The extension of the polymer from the surface (L) is then taken to be half the distance of separation of the two surfaces ($\frac{1}{2}$ D). There is some question regarding the accuracy in the extended length which is measured as the point of initial contact is an osmotic force and thus represents the total extension of the polymer without any consideration of the segment density distribution. In the simple Alexander-DeGennes model it is assumed that the segment density is uniform until, at the termination of the polymer, there is a precipitous fall off in segment density (i.e. a short tail). More recent calculations on segment density distributions (both mean field and self consistent field) suggest that the segment density is not uniform throughout the length of the brush but undergoes a more gradual decay. However, these results still suggest the scaling of Eq. (1) though the prefactor is obviously different.

From a compositional point of view one of the interesting extensions of the scaling model has derived from the work of Joanny and his co-workers [5, 6]. In a series of papers, these authors have examined the adsorption of di-block copolymers onto a surface as a function of the molecular weights of the two blocks. In these models the block which extends into solution (the buoy block) is highly soluble and does not interact with the surface. The anchor block which adsorbs to the surface may either be soluble or insoluble in the surrounding fluid. If it is insoluble it will adsorb onto the surface and the strength of the bond with the surface will be determined by its wetting properties and the normal adhesive forces. Alternatively, the anchor polymer may be soluble in the solvent but able to adsorb chemically onto the surface. In both cases, differences in the solvencies of the two blocks will force the anchor block towards the surface and increase the adsorption. Examples of both types of di-block copolymers adsorbed on mica have been examined in the surface force apparatus. The buoy block has normally been polystyrene, the insoluble anchor block has been polyvinyl pyridine (PS-b-PVP), the soluble anchor block has been polyethylene oxide (PS-b-PEO) and the solvent has been toluene. Good agreement with theory has been observed.

The inclusion of varying molecular weights of the two blocks into calculations based on Eq. (1) has led to some interesting predictions. Basically, Joanny et al. have shown on the basis of geometric and energetic arguments that the scaling properties of the adsorbed layer vary with the asymmetry of the block polymer. Specifically, for the case of a diblock adsorbed from a non-selective solvent, they have shown that as the composition changes from moderately asymmetric to highly asymmetric the length of the absorbed polymer brush changes according to the following equations:

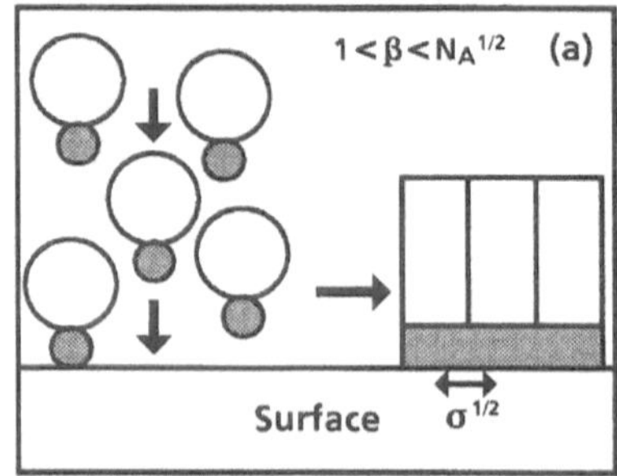

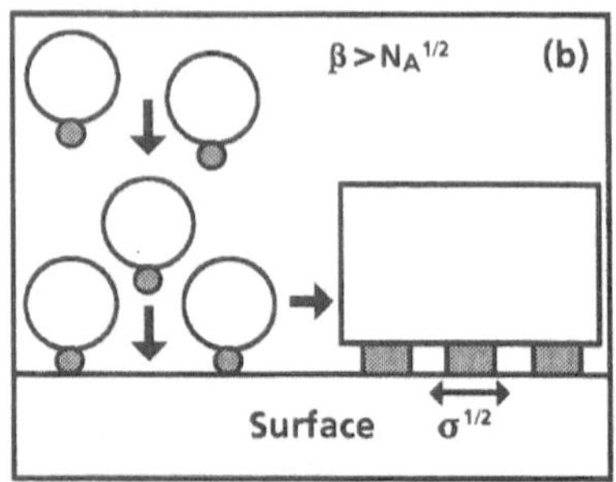

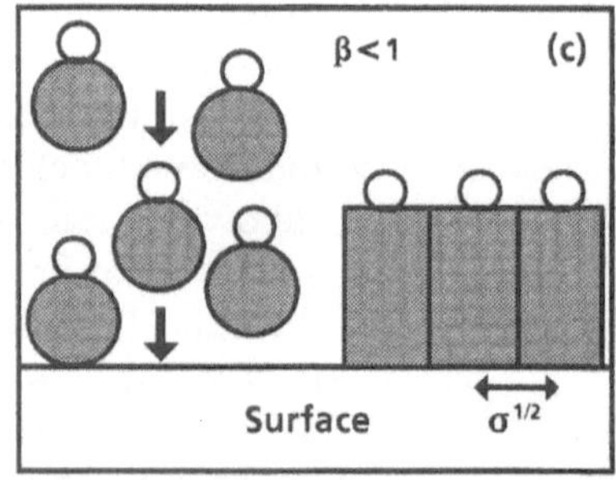

Fig. 1 Schematic representation of block copolymer adsorption on surface

$$\text{When } 1 < \beta < N_A^{\frac{1}{2}} \quad L \sim a\, N_B\, N_A^{-3/5} \tag{2}$$

$$\text{When } 1 < \beta < N_A^{\frac{1}{2}} \quad L \sim a\, N_B^{3/5}\, N_A^{2/5} \tag{3}$$

$$\text{Where } \beta = (N_{ps}/N_{peo})^{3/5} \tag{4}$$

This has led to the concept that the compositional space of block copolymers can be separated into three distinct areas, each with different scaling laws. In this work, we will report on the absorption of three polymers from these three different regions onto a silica surface and compare the results with adsorption data obtained with the same polymers adsorbed on mica.

Adsorption onto silica is probably more relevant than adsorption onto mica! However, the assumptions of the

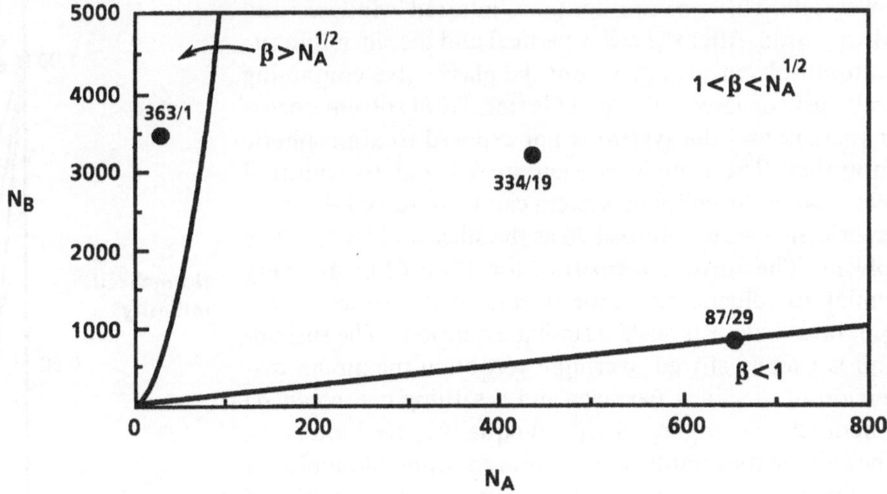

Fig. 2 Compositional space diagram

Marques-Joanny model become stressed as the polystyrene block is able to adsorb onto a silica surface (albeit much less strongly than the polyethylene oxide). The ability of both components to adsorb on the surface has certain implications in the bridging of particles and also on the effect of small additive molecules to the flocculation.

In order to study these systems, we have developed a liquid colloidal cell which enables us to measure transmission infrared spectra of the dispersed silica particles [8]. The spectroscopy enables us to directly determine the specific interactions of both the polyethylene oxide and polystyrene components parts of the block copolymer with the hydroxyl groups which exist on the silica surface [9]. The total amount of polymer absorbed onto the silica surface can be determined in this cell using a "settling" experiment and thus the effect of the mass and conformation of the adsorbed block copolymer on the particle stability is directly observed. A major factor in designing this cell was to enable the total use of evacuated conditions throughout the experiment. We have previously shown the importance of using a well defined silica which has not been exposed to water vapor. If silica is exposed to atmospheric conditions the amount of polystyrene which is adsorbed on the silica surface is dependent upon the number of hydroxyl groups which exist on the surface and specifically on the number of hydroxyl groups which are not covered by adsorbed water. In other words, the effect of adsorbed water on the surface is to reduce the amount of polymer adsorption. Consequently, the present cell has been designed so that the silica can be thermally pretreated in a vacuum, suspended in solvent so as to prepare the dispersed system and exposed to the block copolymer and purified solvents without exposure to atmosphere. The cell, which has been described fully elsewhere [8], is shown in Fig. 3. The adsorbed solvent is contained in a graduated

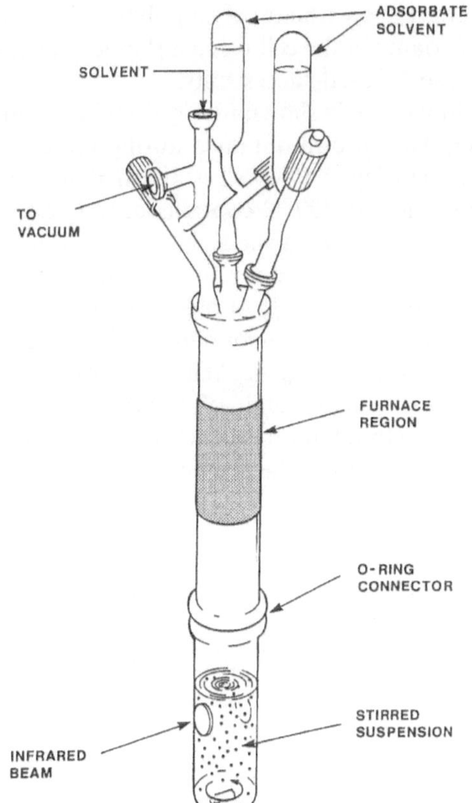

Fig. 3 The colloidal cell

valve and is degased using standard vacuum techniques before being attached to the upper portion of the cell. The cell is then mounted horizontally so that the silica sample can be inserted into the furnace region which is then evacuated. After heating (in these experiments to 400 °C) the cell is turned upright and the silica dropped into the lower cell. This portion contains infrared windows and stirring bean. After the cell is vertical and the silica is in the bottom, solvent is added from the glass valve containing previously degased solvent. This transfer of solvent occurs in vacuum and the system is not exposed to atmospheric impurities. The sample is then stirred and the infrared spectrum of the colloidal system can be observed. In these experiments we use Aerosil 90 as the silica and CCl_4 as the solvent. The solvent properties for PS-b-PEO are very similar to toluene and experiments in the surface force apparatus have established similar extensions. The suspension is usually stirred overnight to obtain maximum dispersion of the silica particles and a settling curve can be generated by stopping stirring at equally spaced intervals. The silica settles out from the infrared beam and a plot of the intensity of bands due to polymer and silica allows calculation of the total amount of polymer adsorbed onto the surface. Detail on the spectroscopy of the interaction of the surface silanol groups with adsorbed polymer is given elsewhere [10]. Variants of the cell using a thinner optical path length can also be used successfully.

A typical settling curve is shown in Fig. 5 and this will be discussed later. The spectra obtained during these experiments are shown in Fig. 4. It is well established that the surface of a silica dried at 400 °C contains only free hydroxyl groups which give rise to a single absorption at 3747 cm^{-1}. This absorption band is shifted to 3690 cm^{-1} in the present experiments because of physical interaction of the silanol group with the surrounding solvent. The starting spectrum is shown in Fig. 4a. When polystyrene is added to the colloid a further shift of 100 cm^{-1} in the position of hydroxyl stretching frequency is noted. This accompanies the adsorption of the polymer and is due to direct interaction of the surface hydroxyl group with the aromatic ring of the PS. When polyethylene oxide is adsorbed (Fig. 4f) the interaction is much stronger and the spectrum shows that the hydroxyl stretching frequency appears at 3300 cm^{-1} – a shift almost four times as large as for the polystyrene. These two absorption frequencies

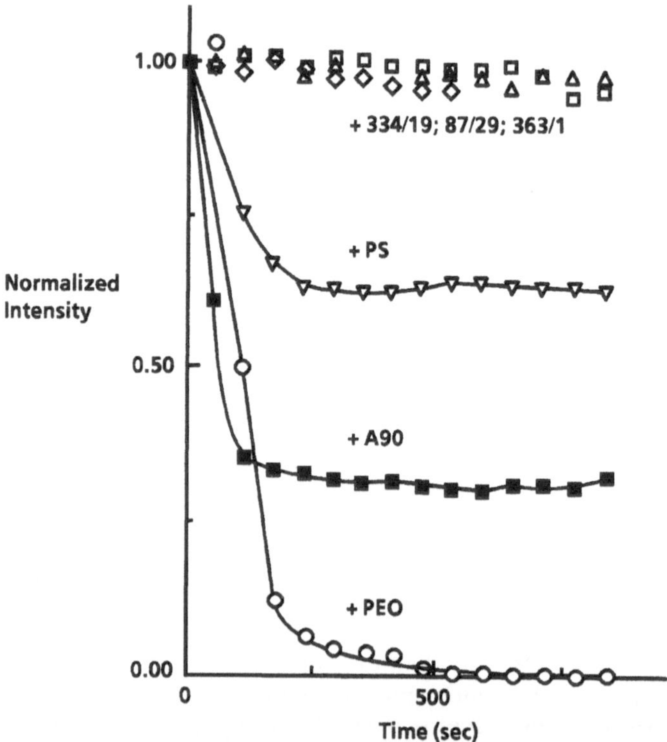

Fig. 5 Settling curves: Before and after addition of polymers

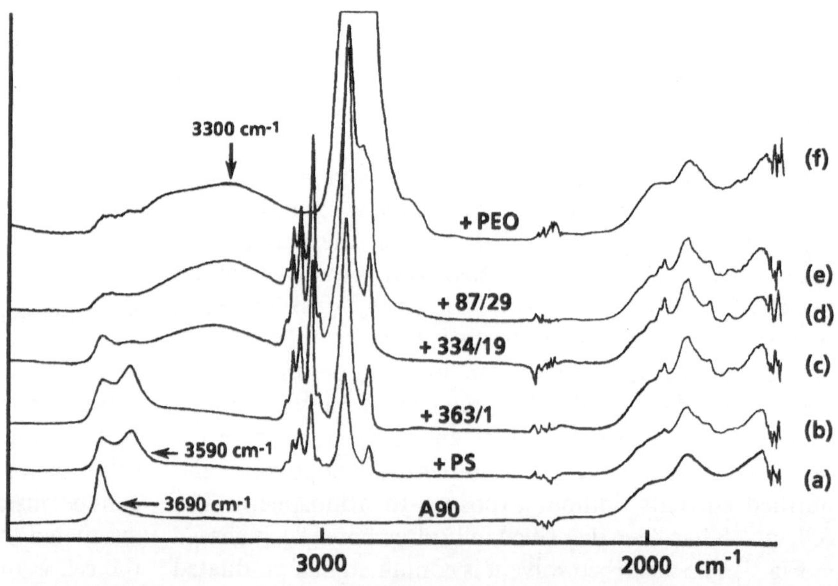

Fig. 4 Polymer adsorption on Aerosil 90. Spectra are (a) Aerosil 90 (b) + PS (c) + 363/1 (d) + 334/19 (e) + 87/29 (f) + PEO

Table 1 Characteristics of the polymer used in this study

Sample	Mw_{PS}	$(PS)_X$	Mw_{PEO}	$(PEO)_Y$	Mw/Mn	β
PS	300 000	2884	–	–	1.06	–
363/1	363 000	3490	1000	23	1.14	2.0
334/19	334 000	3211	19 000	431	1.37	3.31
87/29	87 000	836	29 000	659	1.27	1.16
PEO	–	–	1900	43	1.08	–

are sufficiently distinct and well separated so that they can be used to monitor the interaction with the different segments of the block copolymers. The blocks we have examined were selected from the three different regions of the Marques-Joanny plot and their properties are shown in Table 1. For simplicity they are referred to in this paper by their molecular weights. Thus, 363/1 refers to a block polymer of PS-b-PEO in which the molecular weights are 363 000 and 1000 respectively. They include a highly asymmetric polymer (363/1), a moderately asymmetric polymer (343/19) and a symmetric polymer (87/29). The spectra obtained after their addition to the silica are shown in Fig. 4c,d and e. In all cases, a sufficient amount of the polymer was added in order to achieve equilibrium saturation of the surface. Details on the kinetics of the adsorption and conformational changes of the adsorbed molecules will be presented elsewhere. The intensity of the bands at 3590 cm^{-1}, 3300 cm^{-1}, are used to measure the fraction of surface sites which are occupied by the polymer chains Θ_T and also the fraction of the PS or PEO segments which are attached to the surface (i.e. the bound fraction, ρ). Θ_T is computed directly from the relative decrease in the intensity of the band at 3690 cm^{-1} and specific values for ρ_{ps} are obtained from the changes in intensity of the bands at 3590 and 3300 cm^{-1} respectively. The fraction of segments attached to the surface (ρ) is calculated from Eq. (5),

$$\rho_x = \frac{\Theta_x \eta_{OH}}{D} \quad (5)$$

where x refers to PEO or PS, η_{OH} is the hydroxyl group density and D is the number of polymer segments per nm^2. For the silica which has been pretreated at 400 °C, η_{OH} is 1.4. The data obtained from these experiments is summarized in Table 2 and can be interpreted as follows:

363/1

When the asymmetric polymer is added to the silica it adsorbs readily on the surface. However, a band at 3300 cm^{-1} cannot be detected. In view of the strong absorption of PEO compared to PS it would be expected that the PEO segment adsorbs onto the surface of the silica and the non observance of the perturbed band is presumably due to the small number of interactions that could be expected. This is supported by the data given in Table 2 where it is seen that the decrease in the number of silanol groups is not fully accounted for by the formation of PS-silanol interactions. About 5% of the total coverage is unaccounted for and is attributed to adsorbed PEO segments. This value is not unreasonable since it is the number that would be anticipated if all the PEO segments of the asymmetric 363/1 were adsorbed on the surface. Further support for interaction of the PEO segments with the silanol groups comes from a comparison of the total amount of PS which is adsorbed relative to 363/1. The introduction of even a very small number of PEO groups onto the end of the PS chain increases the amount of polymer which is absorbed by almost a factor of two. This increase can only be rationalized by assuming the strong interaction of PEO with SiOH.

334/19

Within the geometric arguments of the Marques-Joanny formulations we might expect this polymer ($\beta = 3.3$) to adsorb in a form where both the PS and PEO segments are overlapping and in semi-dilute solution. For adsorption on mica the PS does not interact with the surface and so the model (Fig. 1) shows the PEO attached to the surface. With the silica surface, however, the styrene does interact with the silanol groups and this is reflected in the spectrum (Fig. 4c) where bands at both 3300 and 3590 cm^{-1} are

Table 2 Measured values for adsorbed polymers

Sample	q (mg/m^2) ± 0.02	θ_T ± 0.02	θ_{PS}	θ_{PEO}	PS Segments/nm^2	ρ_{PS}	PEO Segments/nm^2	ρ_{PEO}
PS	0.58	0.50	0.50	–	3.0	0.23	–	–
363/1	1.05	0.60	0.57	0.03	5.9	0.13	0.04	0.93
334/19	1.20	0.70	0.10	0.60	6.3	0.02	0.88	0.94
87/29	1.10	0.88	–	0.88	4.6	–	3.70	0.33
PEO	0.36	0.90	–	0.90	–	–	4.90	0.25

observed. It is noted, however, that the major proportion of surface interaction is due to PEO and approximately six times as many OH groups are covered by PEO as by PS even though the number of PS segments/nm^2 is almost the same (6.3 versus 5.9). The number of PEO segments per nm^2 increases from 0.4 to 0.88. Thus, although the theoretical model is not precise the adsorption densities show it to be a useful approximation.

37/29

This polymer is almost-symmetric ($\beta = 1.16$) and when it adsorbs on the silica surface there is no evidence of any PS adsorption on the surface silanol groups. For both the polymers discussed previously the free volume effects which prevent the overlap of the PS segments also prevent full coverage of the surface sites by the PEO segments. In the case of 87/29, however, the coverage by PEO segments is not limited by the overlap of the PS tails. The PEO segments thus access all the surface silanols and prevent any PS adsorption. This overlap supply of PEO segments for available silanol sites results in a low value for ρ (0.33 versus 0.94)

Settling curves

The settling curves obtained for the colloidal silica system after equilibration of all five polymers are shown in Fig. 5. It is noted that the adsorption of PS causes a slight increase in stability of the silica dispersion whereas PEO causes almost complete flocculation. In all three cases, the adsorbed block copolymers prevent any settling of the silica particles. All three block copolymers are adsorbed in sufficient quantity to give a PS brush. The subsequent prevention of interpenetration gives the observed stability.

Addition of polymer: multiple dose vs. single dose

It is well known in the chemical and engineering literature that the dispersion of polymer stablized dispersions is strongly dependent upon the way the polymer is added to the particulate. In many cases, it is desirable to produce flocculation in a system and this can be achieved by adding the adsorbed polymer in several small doses rather than in one single dose. In this case of multiple dose addition, the polymer cannot immediately completely coat the surface and bridging between particles is induced. As both PEO and PS interact with the silica surface this effect can occur with all three polymers used in this work. This bridging is observed in the infrared spectra (not shown) but is particularly apparent in the settling curves. Figure 6 shows the settling curve obtained after addition of multiple doses of polymer 87/29 to the silica dispersion. Direct comparison can be made with the data of Fig. 5 where single dose addition has introduced stability to the colloidal system. Obviously, the multiple dose addition of the same amount of polymer causes flocculation. Indeed, the spike which is seen in the settling curve demonstrates the presence of large aggregates.

The ability to monitor the infrared spectrum of the total system allows an interesting demonstration of effect of displacer molecules in stabilizing flocculated systems. It is well known that small molecules can displace polymer segments from the surface if the strength of the interaction

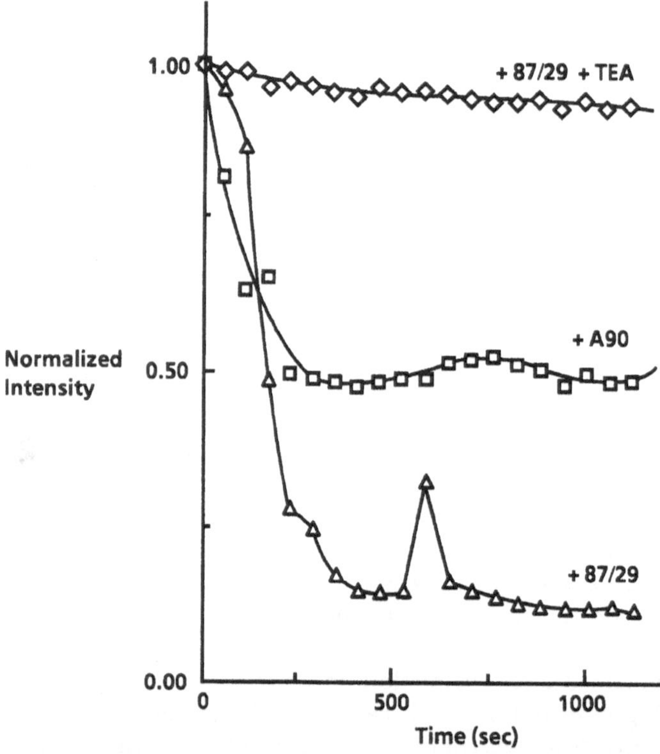

Fig. 6 Effect of TEA on settling curves obtained after multiple addition of small amounts of polymer

Table 3 Comparison: colloid cell vs. dBFTIR (on SiO_2) (on Mica)

Sample	Amount absorbed (mg/m^2)	
	Colloid cell	dBFTIR
363/1	1.05	1.0
334/1	1.2	1.4
87/29	1.1	0.43*

of the small molecule with the surface site is greater than that of the polymer segment and this effect has been used to estimate adsorption energies using displacer molecules in chromatographic-like experiments. In the present system our single dose experiments have confirmed that the adsorbed polymers exist in a brush form with the PEO adsorbed onto the surface with the PS segments forming a stabilizing brush. If, however, in the multiple dose system, the flocculation is caused by the creation of SiOH/PEO-b-PS/SiOH bridges then the addition of a small molecule such as triethylamine (TEA) or pyridine to the system should cause preferential cleavage of the PS/SiOH bond with subsequent restructuring of the adsorbed polymer to give a PS brush. Because of the different frequency shifts associated with the SiOH/PS, SiOH/PEO and SiOH/TEA species, the interactions with the surface can be observed and "titrated". The results are shown in Fig. 6. In this experiment, the 87/29 polymer has been added to the Aerosil suspension giving the flocculation which was noted earlier. On addition of small quantities of TEA the infrared spectra show cleavage of the SiOH/PS bridged bond. Figure 6 dramatically demonstrates the result: Flocculation is reversed and the system becomes very stable.

Conclusion

In many theoretical studies of block copolymers adsorption it is assumed that only one block adsorbs on the surface and that the other block remains in solution. Conformation of the adsorbed polymer is then dictated by the relative sizes and solubilities of the blocks. We report on the adsorption of a block copolymer on a silica surface where both the anchor and the buoy block can adsorb, although with differing strengths. In this case, the polymer conformation is dictated by the relative block size and also by the strenth of the interaction of the block segments with the surface silanol groups. It is confirmed that the adsorption of PEO/PS on a well defined silica surface resembles that on mica and that the adsorption gives rise to increased colloidal stability. However, if the block is added in single, small multiple doses, the fact that both parts of the block can adsorb on the silica causes a bridging between the particles and flocculation of the suspension. The molecularity of the adsorption process can be shown by titrating away the weakest polymer surface linkage with a displacer molecule such as triethylamine.

References

1. Alexander S (1977) J Phys (Paris) 38:983
2. de Gennes P-G (1982) Macromolecules 15:492
3. Israelachvili JN, Adams G (1978) J Chem Soc, Faraday Trans 1 79,975
4. Milner S (1991) Science 251:905
5. Marques CM, Joanny JF, Leibler L (1988) Macromolecules 21:1051
6. Marques CM, Joanny JF (1989) Macromolecules 22:1454
7. Guzonas DA, Boils D, Tripp CP, Hair ML (1992) Macromolecules 25:2434
8. Tripp CP, Hair ML (1992) Langmuir 8:1961
9. Tripp CP, Hair ML (1994) Langmuir 10:4031
10. Tripp CP, Hair ML (1993) Langmuir 9:3523
11. Cohen, Stuart MA, Fleer GJ, Scheutjens JMHM (1994) J Colloid Interface Sci 97:515
12. Cohen, Stuart MA, Fleer GJ, Scheutjens JMHM (1994) J Colloid Interface Sci 97:526

Progr Colloid Polym Sci (1996) 101:58–68

K. Grundke
T. Bogumil
T. Gietzelt
H.-J. Jacobasch
D.Y. Kwok
A.W. Neumann

Wetting measurements on smooth, rough and porous solid surfaces

Dr. K. Grundke (✉) · T. Bogumil
T. Gietzelt · H.-J. Jacobasch
Institut für Polymerforschung Dresden eV
Hohe Straße 6
01069 Dresden, FRG

D.Y. Kwok · A.W. Neumann
University of Toronto
Department of Mechanical Engineering
5 King's College Road
Toronto M5S3G8, Canada

Abstract The solid-vapour surface tension has been determined by contact angle measurements with polar and non-polar liquids on flat solid surfaces using Axisymmetric Drop Shape Analysis (ADSA) and by capillary penetration experiments on rough and porous solids. For smooth and inert, well prepared solid surfaces (PTFE, FC 721 on mica, FEP, PET) the plot of $\gamma_{lv}\cos\Theta$ versus γ_{lv} yields smooth curves which are consistent with the equation of state approach to calculate solid-vapour and solid-liquid interfacial tensions. Other experimental patterns of contact angle data are caused by surface roughness and non-inert solids which may result in contact angles incompatible to Young's equation. An alternative way to obtain the solid-vapour surface tension of rough and porous solids are capillary penetration experiments. The determination of the penetration velocity of liquids into rough and porous solids yields $K\gamma_{lv}\cos\Theta$ versus γ_{lv} plots, which provide γ_{sv} values for these systems; K is an unknown parameter of the constant geometry of the porous solid. The application of this concept was demonstrated for a hydrophobic PTFE powder and for hydrophilic Cellulose membranes.

Key words Wetting – contact angle – interfacial tension – capillary penetration

Introduction

Wetting phenomena are of considerable technical interest. Many processes in polymer production, processing and modification include wetting of solids by liquids. Examples of such technological processes are polymer blending, coating, and the production of polymer composites by reinforcement with fibres or inorganic fillers. Most liquids in these technical solid/fluid systems are solutions or polymer melts. The solid may have a simple surface or be finely divided (porous media such as powders or fibres). Despite their importance, the interfacial phenomena in these technical systems are still poorly understood [1]. Contact angles measured on these complicated solid/liquid systems are often used simply as empirical parameters to quantify wettability [2, 3].

For that reason another strategy is normally applied to understand wetting and adhesion phenomena in multicomponent technical systems. By measuring the contact angles of well defined model liquids on a solid surface, the solid interfacial tension can be calculated. Young's equation interrelates the measurable quantities, liquid-vapour interfacial tension γ_{lv} and the contact angle Θ, to the non-measurable interfacial tensions γ_{sv} and γ_{sl} of the solid-vapour and solid-liquid interfaces (see Fig. 1):

$$\gamma_{lv}\cos\Theta = \gamma_{sv} - \gamma_{sl} \quad (1)$$

Since only γ_{lv} and Θ are directly measurable, one requires an additional equation (information) about the interfacial

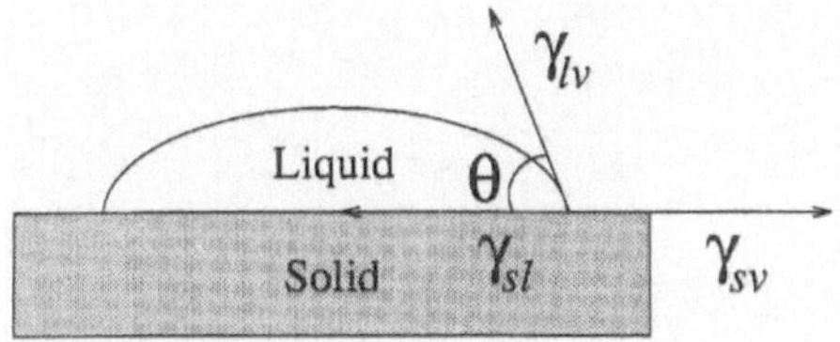

$$\gamma_{lv} \cos\theta = \gamma_{sv} - \gamma_{sl}$$

Fig. 1 A schematic of a sessile drop

tensions to determine γ_{sv} and γ_{sl}. At present, approaches [4–7] to determine solid surface tension from contact angles were largely inspired by this idea. The calculated γ_{sv} and γ_{sl} values are used to interpret and to predict the wetting and adhesion properties in technically relevant systems.

It is the objective of this paper to present contact angle studies with pure test liquids on smooth and chemically inert solid surfaces in comparison with studies on real technical solids which are more or less rough, porous, heterogeneous and chemically non-inert. A new contact angle technique (Axisymmetric Drop Shape Analysis) provides possibilities to obtain very precise and reproducible contact angles for smooth and inert, but also for rough and chemically non-inert solid surfaces. However, the results of these direct contact angle measurements cannot be used to calculate solid-vapour surface tensions of rough solid surfaces. It will be shown that the capillary penetration of liquids into porous solids can be related to the wetting tension and applied to determine the surface tension of solids with rough and porous surfaces such as powders and membranes.

Contact angle measurements on flat and smooth solid surfaces

Contact angle measurements with Axisymmetric Drop Shape Analysis (ADSA)

A very appropriate methodology to measure contact angles is Axisymmetric Drop Shape Analysis – Profile (ADSA-P). ADSA-P is a technique to determine liquid-fluid interfacial tensions and contact angles from the shape of axisymmetric menisci, i.e., from sessile as well as pendant drops. The strategy employed is to fit the shape of an experimental drop to the theoretical drop profile according to the Laplace equation of capillarity,

$$\Delta P = \gamma \left[\frac{1}{R_1} + \frac{1}{R_2} \right] \tag{2}$$

using surface (interfacial) tension γ as one of the adjustable parameters; ΔP is the pressure difference across the liquid and fluid phases, and R_1 and R_2 are the two principal radii of curvature of the drop. The best fit identifies the correct interfacial tension and, in the case of a sessile drop, contact angle. Details of the methodology can be found elsewhere [8].

Apart from local gravity and densities of liquid and fluid phases, the only information required by ADSA is several arbitrary but accurate coordinate points selected from the drop profile. An automatic digitization technique utilizing digital image acquisition and analysis has been used [9]. Computer software has been developed to implement this method and computational results provide the values of interfacial tension, drop volume, surface area and, in the case of a sessile drop, contact angle and the radius of the three-phase contact line.

The experimental set-up for ADSA-P pendant and sessile drop is shown in Fig. 2. A Cohu CCD camera is mounted on a Wild-Heerbrugg M75 microscope. The video signal of the pendant or sessile drop is transmitted to a digital video processor, which performs the frame grabbing and digitization of the image with 256 gray levels for each pixel, where 0 represents black and 255 represents white. A SPARCstation computer is used to acquire images from the image processor and to perform the image analysis and computation. To perform dynamic surface tension measurements for pendant drops or dynamic contact angle measurements for sessile drops, a motor driven syringe can be employed in the set-up shown in Fig. 2. In the case of a sessile drop, the advancing and receding dynamic contact angle measurements can be performed, respectively, by pushing or pulling the syringe plunger, leading to the increase or decrease in the drop volume; the change in drop volume forces the three-phase contact line to advance or recede.

Theory

Recently, Li and Neumann [10] and Li et al. [11] have published accurate contact angle data of various polar and non-polar liquids on three carefully prepared solid surfaces: a fluorocarbon (FC-721) dip-coated on mica, heat pressed Teflon (FEP) and poly(ethylene terephthalate) (PET). These contact angle data are shown in Fig. 3, by plotting $\gamma_{lv}\cos\Theta$ versus γ_{lv}. As can be seen in this figure, the curves for the three solid surfaces are so smooth that we have to conclude that $\gamma_{lv}\cos\Theta$ depends only on γ_{lv} and γ_{sv}, i.e. $\gamma_{lv}\cos\Theta$ is a function of γ_{lv} and γ_{sv}:

$$\gamma_{lv}\cos\Theta = F(\gamma_{lv}, \gamma_{sv}) \tag{3}$$

Combining Eq. (3) with the Young Equation (1) yields

$$\gamma_{sv} - \gamma_{sl} = F(\gamma_{lv}, \gamma_{sv}) \tag{4}$$

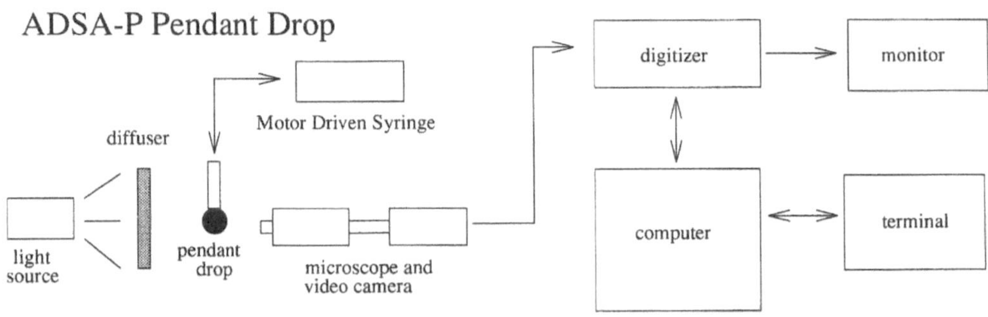

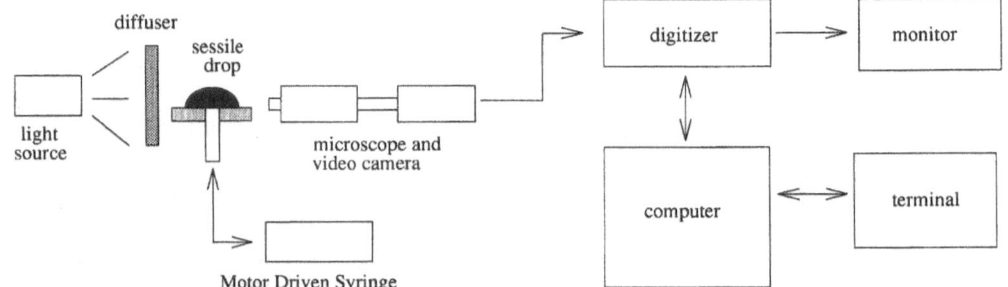

Fig. 2 A schematic of the experimental set up for ADSA-P pendant and sessile drop measurements

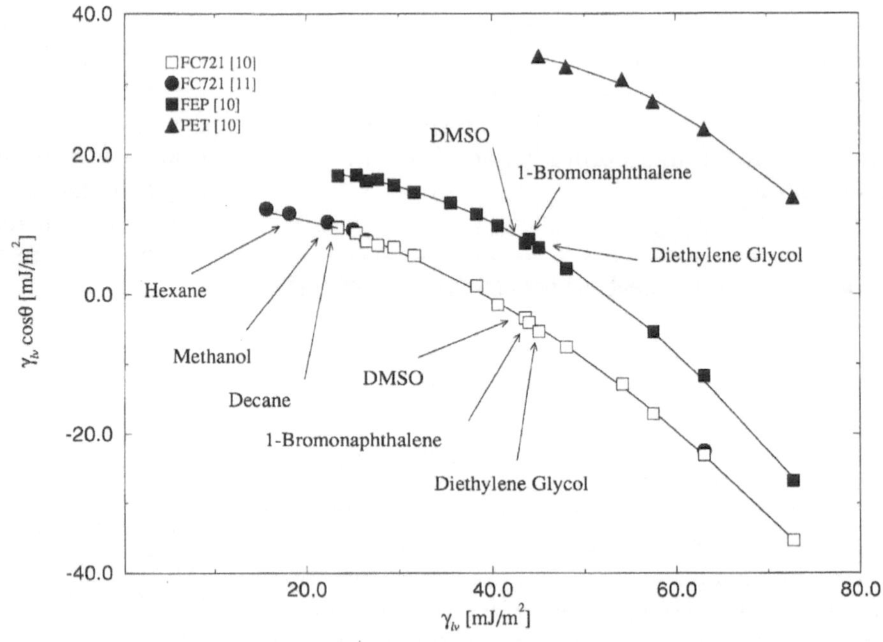

Fig. 3 A plot of $\gamma_{lv}\cos\Theta$ versus γ_{lv} for three carefully prepared solid surfaces: FC-721 coated on mica, heat pressed Teflon FEP and poly(ethylene terephthalate) PET. The smoothness of the curves indicates that $\gamma_{lv}\cos\Theta$ is a function of γ_{lv} and γ_{sv} only and not directly of molecular forces

Rearranging Eq. (4) gives an equation-of-state relationship as

$$\gamma_{sl} = f(\gamma_{lv}, \gamma_{sv}) \tag{5}$$

It should be noted that the above experimental results agree very well with the stipulation from the thermodynamic phase rule for capillary systems [12–14], which states that there are only two degrees of freedom for a two-component sessile drop system (see Fig. 1): The phase rule for capillary systems only allows one to search for a two-degree-of-freedom relation, e.g. Eq. (5), but not for others, such as the Lifshitz-van der Waals/acid-base approach [7] which would require six degrees of freedom.

The experimental results in Fig. 3 are also consistent with the equation of state approach for interfacial tensions [4, 5]: It was shown from thermodynamics that there exists an equation-of-state relationship between the three interfacial tensions, γ_{sl}, γ_{lv}, and γ_{sv}, as long as the solid is ideal, rigid and inert [15]. A recent formulation of the equation of state [5, 16] can be written as

$$\gamma_{sl} = \gamma_{lv} + \gamma_{sv} - 2\sqrt{\gamma_{lv}\gamma_{sv}}\, e^{-\beta(\gamma_{lv}-\gamma_{sv})^2} \quad (6)$$

where β is a constant, which can be obtained from the experimental liquid surface tensions and contact angles. Combining Eq. (6) with the Young equation (1) yields

$$\cos\Theta = -1 + 2\sqrt{\frac{\gamma_{sv}}{\gamma_{lv}}}\, e^{-\beta(\gamma_{lv}-\gamma_{sv})^2} \quad (7)$$

Based on the data in Fig. 3 the best experimental value of β is found to be $0.0001247\,(m^2/mJ)^2$ [10]. Knowing the value of β, Eq. (7) can be used to determine γ_{sv} when γ_{lv} and Θ are known from experiment. The solid–liquid surface tension γ_{sl} can then be obtained from the Young equation (1) or Eq. (6), using the value of γ_{sv} calculated from Eq. (7).

The experimental liquid surface tensions γ_{lv} and contact angles Θ for heat pressed Teflon (FEP) are shown in Table 1 [10]. The values of γ_{sv} in this table are calculated from the equation of state approach for interfacial tensions, Eq. (7). The concistency of the calculated solid-vapour surface tension γ_{sv} for differnt polar and non-polar liquids further highlights the validity of the equation of state approach for interfacial tensions.

It should be noted that the experimental patterns shown in Fig. 3 are not always observable in the literature: Curves far less smooth or no unique curves at all are sometimes reported. Such patterns can have a variety of causes. Accurate contact angle measurements require extreme experimental care. Even very minor vibrations can cause advancing contact angles to decrease, resulting in errors of several degrees. Surface roughness can affect contact angles and make Young's equation inapplicable. Swelling of solids by liquids can change the chemistry of the solid and hence the values of γ_{sv} and Θ. All that experience has shown is that, if ideal solid surfaces have been studied carefully with very exacting methodology, the patterns shown in Fig. 3 emerge.

Non-ideal surface systems

To illustrate the complexity of contact angle phenomena, Table 2 shows the contact angles of three different liquids on PTFE surfaces with increasing roughness [17]: a smooth surface, a surface roughened by 600 grit sandpaper and a surface roughened by 220 grit sandpaper. The values of γ_{sv}, for the smooth PTFE surfaces, calculated from the equation of state approach, are essentially constant for the three liquids. However, as surface roughness increases, the contact angle tends to increase, making the calculated values of γ_{sv} inconsistent. This, of course, does not mean that the equation of state approach is invalid, but simply that the Young equation is inapplicable for these roughened surfaces. Thus, the observed contact angles for these roughened surfaces are no longer Young contact angles.

Another example is shown in Fig. 4 for dynamic contact angle measurements of thiodiethanol on propene-hexylmaleimide copolymer coated on a silicon wafer surface (Fig. 4A) and thiodiethanol on FC-722 coated on a mica surface (Fig. 4B).

Table 1 Contact angles for Teflon (FEP) are reproduced from ref. [10]; the values of γ_{sv} are calculated from the equation of state approach for interfacial tension, i.e. Eq. (7)

Liquid	γ_{lv} (mJ/m^2)	Contact angle (deg.)	γ_{sv} (mJ/m^2)
Decane	23.43	43.70	17.54
Dodecane	25.44	47.96	17.98
Tetradecane	26.55	52.51	17.53
Hexadecane	27.76	53.75	17.99
trans-Decalin	29.50	58.14	17.81
cis-Decalin	31.65	62.60	17.71
Dimethylformamide	35.57	68.52	17.94
Ethylcinnamate	38.37	72.61	17.96
Dibenzylamine	40.63	75.99	17.84
DMSO	43.58	80.35	17.58
1-Bromonaphthalene	44.01	79.70	18.08
Diethylene glycol	45.04	81.48	17.85
Ethylene glycol	47.99	85.56	17.54
Formamide	57.49	95.38	17.57
Glycerol	63.11	100.63	17.60
Water	72.75	111.59	16.16

Table 2 Contact angles on Teflon PTFE surfaces of different roughnesses are reproduced from ref. [17]; the apparent surface tensions γ_{sv} of Teflon PTFE are calculated from the contact angles, using the equation of state approach for interfacial tensions

Surface finish	Contact angle Θ [deg.]			Solid surface tension γ_{sv} [mJ/m^2]		
	Ethylene Glycol	Glycerol	Water	Ethylene Glycol	Glycerol	Water
Roughened surface 220 grit	133.0	–	155.2	2.2	–	0.6
Roughened surface 600 grit	99.5	–	121.8	11.3	–	9.7
Smooth surface	81.0	96.0	104.0	20.0	19.6	20.0

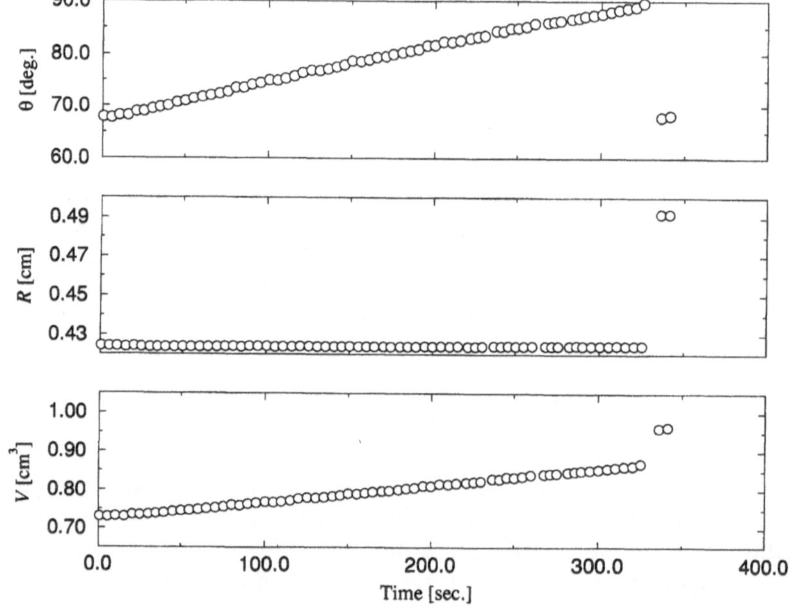

Fig. 4A Dynamic contact angle of thiodiethanol on propene-hexylmaleimide copolymer. The increase in the contact angles at constant radius of the three-phase contact line indicates a slip and stick condition

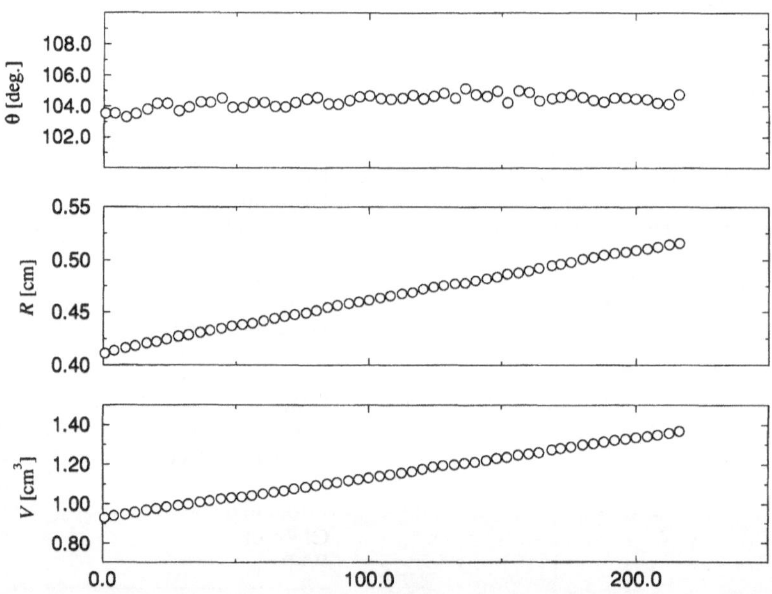

Fig. 4B Dynamic contact angle of thiodiethanol on FC-722 dip-coated on mica surface

These experiments were performed by ADSA-P, using a motorized syringe mechanism. By increasing the drop volume, the contact angle of thiodiethanol on propene-hexylmaleimide copolymer increases linearly from about 68° to 90° at essentially constant radius of the three-phase contact line (Fig. 4A). The increase in the contact angle at constant radius indicates that the drop is sticking on the surface. Suddenly, as the drop volume continues to increase, the drop jumps to a new location and the contact angle decreases from about 90° to 68°. Similar patterns can also be observed for many other systems. Obviously, these observed contact angles cannot all be the Young contact angles, since γ_{sv}, γ_{sl} and γ_{lv} are constants, so that, because of Young's equation, Θ ought to be a constant. In the case of propene-hexylmaleimide copolymer, it is assumed that this surfaces is non-inert in contact with thiodiethanol due to swelling of the solid.

Let us compare the results in Fig. 4A with those in Fig. 4B for a more inert surface (FC-722 dip-coated on mica surface): As the drop volume increases linearly with time, the contact angle remains essentially constant while the radius of the three-phase contact line is increasing, indicating that slip/stick does not occur.

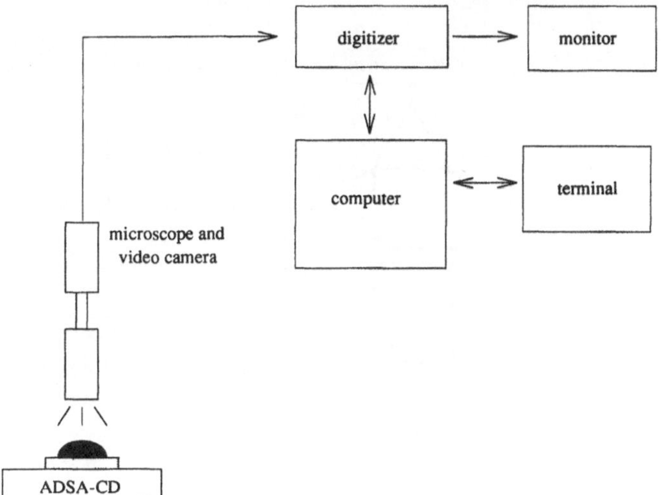

Fig. 5 A schematic of the experimental set-up for ADSA-CD contact angle measurements

Measurements of low contact angles on flat solid surfaces

For the case of low contact angles (e.g. below 20°), it becomes increasingly difficult to measure contact angles with most techniques. The precision of ADSA-P is also decreased since it becomes more difficult to acquire coordinate points along the edge of the drop profile. This deficiency can be overcome by using ADSA-CD (contact diameter) [18], where the drop is viewed from above. For the ADSA-CD program, the contact angle is determined from a numerical integration of the Laplace equation of capillarity, Eq. (2), when the contact diameter, surface tension and drop volume are known. A computer assisted digitization procedure of the drop perimeter was used to measure the effective contact diameter of the drop on an image taken from above. Figure 5 shows a schematic of the experimental set-up for ADSA-CD.

Table 3 shows the experimental contact angles of eleven liquids on a PET surface measured by ADSA-CD and γ_{lv} values measured by ADSA-P pendant drop experiments. In Fig. 6, these contact angle data, together with the data from Fig. 3, are plotted as $\gamma_{lv}\cos\Theta$ versus γ_{lv} for the PET surface. As γ_{lv} decreases, the curve changes smoothly, reaching a maximum and then decreases linearly with the values of γ_{lv}. This can be explained by the schematic shown in Fig. 7, which indicates that the maximum in Fig. 6 is indeed the γ_{sv} for the PET surface: According to the curve in Fig. 6, the γ_{sv} for the PET surface is about 35.5 mJ/m^2, which is in excellent agreement with that calculated from the equation of state approach, $\gamma_{sv} = 35.63 \pm 0.39$ mJ/m^2 [10].

Table 3 Contact angles of different liquids on PET surface measured by ADSA-CD

Liquids	Contact angle Θ (deg.)	$\gamma_{lv}\cos\Theta$	No. of drops	Surface tension γ_{lv} (mJ/m^2)
Hexane	11.5	18.1	7	18.50
Decane	2.1	23.9	8	23.90
Dodecane	6.1	25.3	8	25.44
Pentanol	1.7	25.6	8	25.61
trans-Decalin	4.0	27.1	8	27.20
Hexadecane	6.3	27.5	8	27.62
Decanol	3.3	28.9	8	28.99
cis-Decalin	5.4	32.2	8	32.20
Ethylcinamate	4.2	37.1	8	37.17
Dimethylformamide	14.4	35.5	8	36.65
DMSO	36.0	34.5	8	42.70

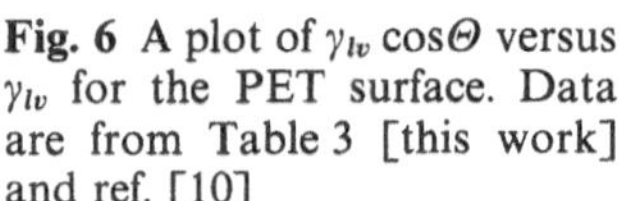

Fig. 6 A plot of $\gamma_{lv}\cos\Theta$ versus γ_{lv} for the PET surface. Data are from Table 3 [this work] and ref. [10]

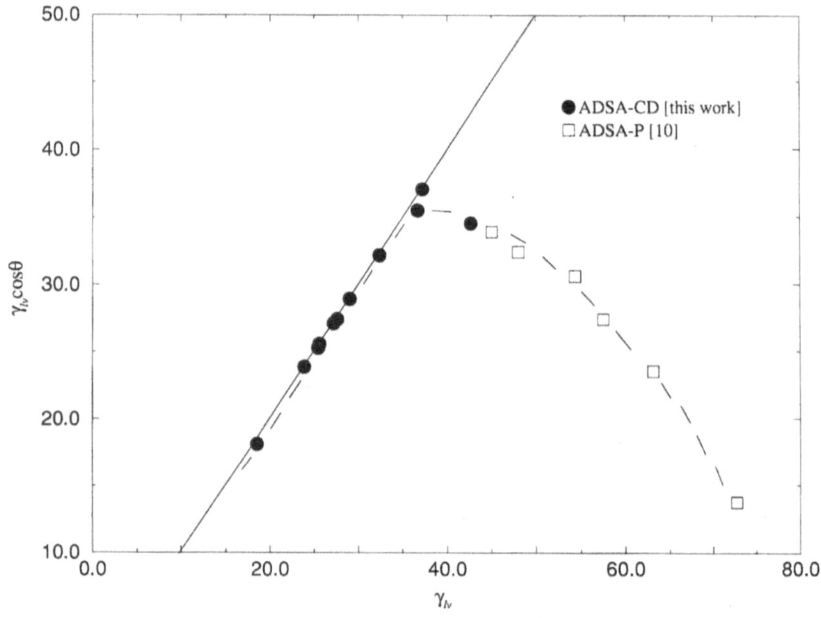

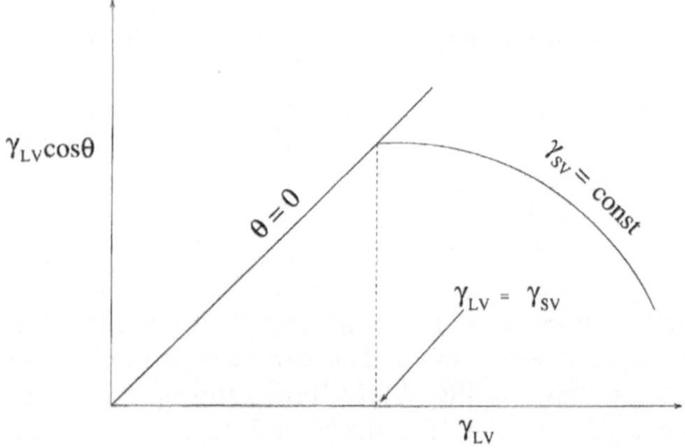

Fig. 7 A schematic contact angle plot for an ideal solid surface. Along the 45° straight line, $\gamma_{lv} < \gamma_{sv}$, $\Theta = 0$ and hence $\gamma_{lv} = \gamma_{lv}\cos\Theta$. As γ_{lv} increases beyond $\gamma_{lv} = \gamma_{sv}$, Θ becomes finite and $\gamma_{lv}\cos\Theta < \gamma_{lv}$

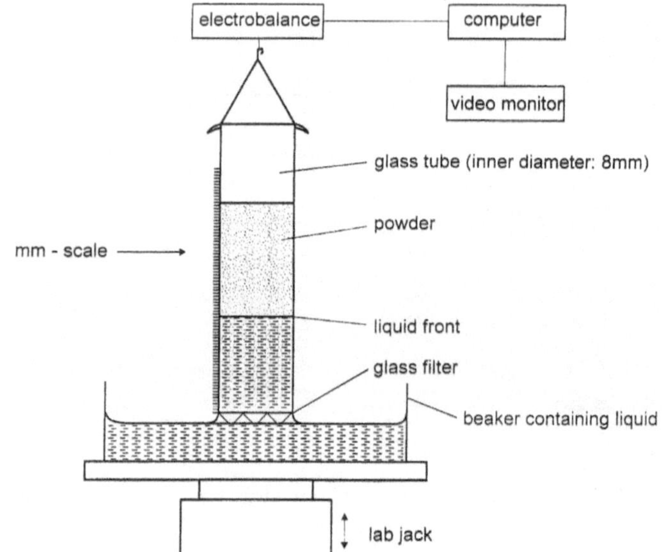

Fig. 8 A schematic of set-up for capillary penetration experiment

In the following it will be shown that similar curves can be obtained if the capillary penetration of pure test liquids in a loosely packed column of powder or in membranes is measured.

Capillary penetration of liquids in porous solids

Experimental set-up

Figure 8 shows the experimental set-up for the capillary penetration measurements. Similar arrangements for capillary penetration tests were used by Bruil [19], Cheever [20], and Kilau [21]. The powder is packed in a glass tube at which the lower end is closed with a glass filter. Considerable care is necessary to obtain a constant and homogeneous powder packing. A precisely weighed quantity of the powder has to fill up to the same height in the glass tube by manually tapping the powder. The filled columns are attached to an electrobalance and brought into contact with several test liquids. Their penetration velocities are determined by measuring the weight gain with the electrobalance as a function of time. A tensiometer can be used for these gravimetric experiments as described in an earlier paper of the authors [22]. The penetration velocities reported in this paper are average values of at least six individual measurements with a maximum deviation of 9% for hydrophobic PTFE powder and 14% for hydrophilic cellulose membranes within the 95% confidence limit. It was found that the main source of error is the geometry of the porous system which has to be reproduced for each measurement. The physical properties of the liquids used for the capillary penetration measurements are given in Table 4.

Theory

The theoretical basis to interpret these gravimetric wetting experiments is derived in the following section.

By combining the Laplace equation of capillarity for a vertical capillary of circular cross-section

$$\Delta P = 2\,\gamma_{lv}\cos\Theta\,\frac{1}{r} \tag{8}$$

with the Hagen-Poiseuille equation for steady flow

$$\frac{dV}{dt} = \frac{\pi\,\Delta P\,r^4}{8\eta h} \tag{9}$$

and

$$dV = \pi r^2 dh \tag{10}$$

Washburn [23] obtained an equation for the flow in a single capillary as

$$\frac{dh}{dt} = \frac{r\gamma_{lv}\cos\Theta}{4\eta h} \tag{11}$$

and after integration

$$h^2 = \frac{rt}{2\eta}\gamma_{lv}\cos\Theta, \tag{12}$$

where ΔP is the Laplace pressure, r is the radius of the capillary, V is the volume of the liquid, t is the time, η is the viscosity of the liquid, and h is the capillary height of the liquid front. If h is replaced by the weight M of the liquid which penetrates into the capillary

$$M = \rho V = \rho h A, \tag{13}$$

where A is the cross-sectional area of the capillary and ρ is the density of the liquid

$$h = \frac{M}{\rho A}$$

$$h^2 = \frac{M^2}{\rho^2 A^2},$$

it follows that

$$\frac{M^2}{\rho^2 A^2} = \frac{rt}{2\eta}\gamma_{lv}\cos\Theta, \tag{14}$$

and by rearranging

$$\gamma_{lv}\cos\Theta = \left[\frac{2}{A^2 r}\right]\left[\frac{\eta}{\rho^2}\right]\left[\frac{M^2}{t}\right], \tag{15}$$

where $[2/A^2r]$ is the geometry factor of the capillary, $[\eta/\rho^2]$ reflects properties of the test liquid and $[M^2/t]$ is determined in the experiment.

In the case of powder packings or other porous solids, such as membranes, the geometry of the capillary system is not known. The value of $[2/A^2r]$ in Eq. (15) is therefore

Table 4 Physical properties of the liquids used for the capillary penetration measurements

Liquid	γ_{lv} (mJ/m^2)[a]	η (mPa sec)[b]	ρ (g/cm^3)[b]
Perfluropolyether	14.31	1.427	1.720
Isopentan	17.13	0.223	0.624
Hexane	18.30	0.308	0.659
Heptane	20.50	0.413	0.683
Octane	21.42	0.546	0.702
Decane	23.22	0.907	0.730
Dodecane	24.69	1.383	0.751
Tetradecane	26.40	2.128	0.761
Hexadecane	27.90	3.032	0.775
Tetralin	34.54	2.020	0.976
Benzylalcohol	39.00	5.474	1.044
1-Bromonaphthalene	44.53	4.520	1.483
Ethylene glycol	48.00	19.900	1.113
Formamide	58.20	3.300	1.133
Water	72.50	1.001	0.998

[a]measured by the Ring Method at room temperature
[b]tabulated values from the literature (D.R. Lide, Handbook of Chemistry and Physics, CRC Press, 1992)

replaced by an unknown factor $1/K$:

$$\gamma_{lv}\cos\Theta = \frac{1}{K}\left[\frac{\eta}{\rho^2}\right]\left[\frac{M^2}{t}\right] \tag{16}$$

or

$$K\gamma_{lv}\cos\Theta = \left[\frac{\eta}{\rho^2}\right]\left[\frac{M^2}{t}\right]. \tag{17}$$

This modified Washburn equation is applied with the following assumptions: 1) laminar flow predominates in the pore spaces; 2) gravity can be neglected; and 3) the geometry of the porous solid is constant.

Experimental results

Figure 9 shows experimental data for a PTFE (Teflon 807-N) powder plotted as M versus $\sqrt{t}$. By determining the slope of the linear part of these plots, the experimental quantity $[M^2/t]$ can be obtained.

If the penetration velocity of a series of test liquids is measured $[\eta/\rho^2]$ $[M^2/t]$ can be determined for each test liquid. Because $[\eta/\rho^2]$ $[M^2/t]$ equals $K\gamma_{lv}\cos\Theta$ according to Eq. (17) we can plot $K\gamma_{lv}\cos\Theta$ versus the surface tension γ_{lv} of the test liquids. This was done for the PTFE powder with nine liquids. The experimental results are shown in Fig. 10. As can be seen, the curve obtained by capillary penetration measurements shows a maximum just as Figs. 6 and 7. The maximum is reached at 20.4 mJ/m^2 which equals the solid-vapour surface tension of the PTFE powder. For such a surface with $\gamma_{sv} = 20.4$ mJ/m^2 and water with $\gamma_{lv} = 72.5$ mJ/m^2, the equation of state (Eq. (7)) predicts a contact angle of 104°. Remarkably, this is exactly what is observed on a smooth Teflon surface, cf. Table 2. Two important conclusions can be drawn from this result: 1) The capillary penetration phenomena are entirely consistent with the equation of state for interfacial tensions. 2) Although the packed powder bed certainly does not represent a flat and smooth solid surface, the derived value for γ_{sv} is that obtained by contact angle measurements of a flat and smooth surface. On the other hand, it is well known that even a highly compacted hydrophobic powder, presenting a seemingly flat and smooth solid surface, does not yield the same contact angle as truly smooth and coherent solid surfaces. It appears that an indirect method, such as capillary penetration, may provide much more relevant information than direct contact angle measurements on imperfect solid surfaces.

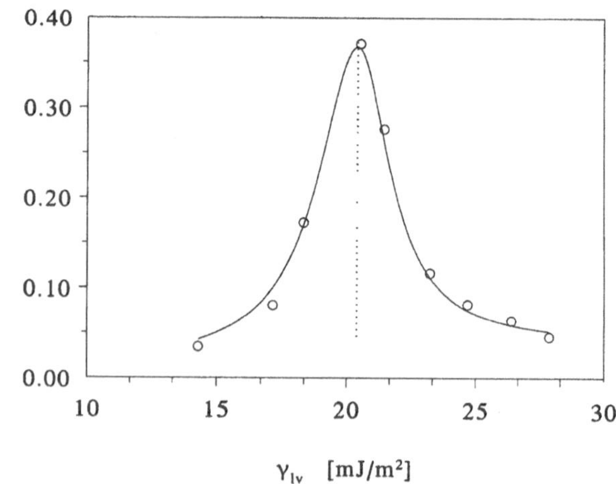

Fig. 10 A plot of $K\gamma_{lv}\cos\Theta$ versus γ_{lv} for nine liquids on the PTFE (Teflon 807-N) powder

Fig. 9 A plot of the weight M of a liquid (n-heptane) penetrating into the porous solid (PTFE powder) versus time $\sqrt{t}$

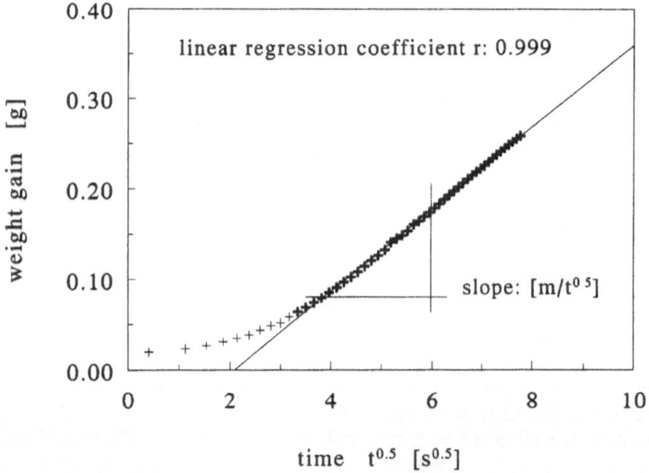

The different shape of the curves $\gamma_{lv}\cos\Theta$ versus γ_{lv} obtained by direct contact angle measurements on a flat surface (see Fig. 6) and by the capillary penetration experiments (see Fig. 10) may be attributable to the unknown effect of the constant K. It was shown above that this constant reflects the geometry of the porous solid which may change in a non-predictable way during the penetration of different liquids. From the experiments, it can be concluded that we do not need any information about the geometric constant K because the position of the maximum, which is expected to reflect the solid-vapour surface tension of the powder, is not affected by the powder geometry. Thus, these studies show that solid-vapour surface tensions of rough and porous materials can be determined in an indirect way by measuring the capillary penetration velocity of liquids in these solids.

At present, many authors working in this field follow another concept [24–27]. From the plot $[\eta/\rho^2]$ $[M^2/t]$

versus the surface tension γ_{lv} of the liquids, the geometric factor K is calculated for those liquids which should wet the solid completely, i.e. which lie to the left of the maximum in Fig. 10. By inserting this K value in Eq. (17) and $[\eta/\rho^2]$ $[M^2/t]$ for each test liquid to the right of the maximum in Fig. 10, their contact angles Θ are calculated. These contact angles are used to calculate the solid-vapour surface tension of the porous material [24, 25, 27]. This procedure is dubious because it can be expected that the contact angles, calculated from the Washburn equation, are affected by roughness and porosity. If we apply this procedure to the PTFE powder shown in Fig. 10, we get $\Theta = 88°$ for hexadecane ($K = 5.34\ 10^{-4}\ \text{cm}^5$, $[M^2/t] = 4.71\ 10^{-5}\ \text{g}^2/\text{sec}$); however, this contact angle is well known to be 46° on flat and smooth PTFE surfaces. Similar results were obtained for octane, decane and tetradecane contact angles calculated in the same way.

Figure 11 shows another example of a capillary penetration experiment. In this case, single cellulose hollow fibres were used. They are extensively applied in hemodialysis to remove metabolic waste from the blood [28]. To improve the blood compatibility the cellulose is chemically modified with the aim of rendering the cellulose material less hydrophilic. Wetting measurements can give quantitative information about the hydrophilic-hydrophobic surface properties of the unmodified and modified cellulose.

The cellulose hollow fibre is a capillary membrane which consists of a single capillary with a diameter of about 200 μm and a porous shell (thickness: 8 μm). The liquid penetrates into the single capillary and into the porous fibre shell. These different processes result in different penetration rates (see Fig. 11). Additionally, swelling of

Fig. 11 A plot of the weight M of a liquid (1-bromonaphthalene) penetrating into the porous solid (cellulose hollow fibre) versus time $\sqrt{t}$

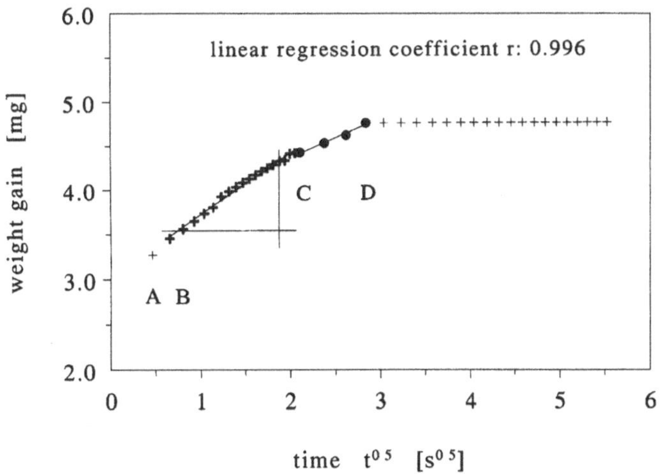

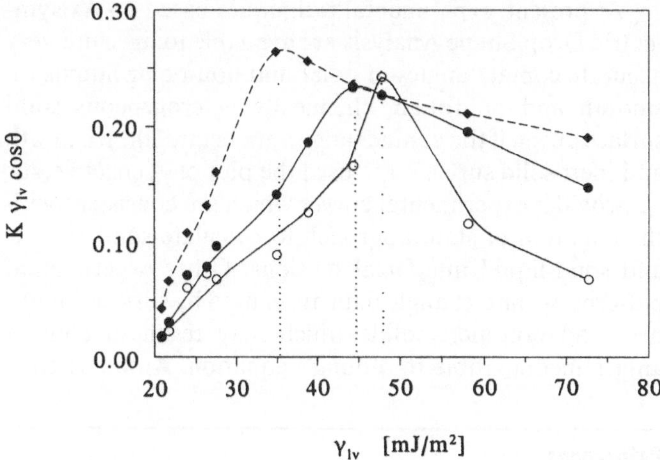

Fig. 12 A plot of $K\gamma_{lv}\cos\Theta$ versus γ_{lv} for eleven liquids on three differently modified cellulose hollow fibres: ○ unmodified Cellulose CUPROPHAN; ● chemically modified cellulose M1; ◆ chemically modified cellulose M2

the hydrophilic cellulose membranes may change the geometry of the fibre in an unknown way. Regardless of this complicated nature of the penetration process we can determine the penetration velocities for several liquids from the linear part B-C of the plot M versus $\sqrt{t}$ and obtain curves shown in Fig. 12 for differently modified cellulose fibres.

As in the case of the PTFE powder (Fig. 10) we obtain curves with a maximum if $K\gamma_{lv}\cos\Theta$ versus the surface tension γ_{lv} of the test liquids is plotted. It can be seen that the unmodified cellulose fibre indicates the highest γ_{sv} value whereas the modified types have a lower γ_{sv}, i.e. a less hydrophilic surface. We can predict a water contact angle of 54° for an ideally smooth unmodified cellulose hollow fibre CUPROPHAN, if we input $\gamma_{sv} = 48\ \text{mJ/m}^2$ and $\gamma_{lv} = 72.5\ \text{mJ/m}^2$ for water in Eq. (7). It has to be considered that processing agents are applied during the manufacture of the CUPROPHAN fibres which can be expected to influence even the surface properties of the unmodified cellulose material [29]. By chemical modification of the cellulose the γ_{sv} could be decreased to 35 mJ/m², as can be seen from Fig. 12; a water contact angle of 71° can be predicted for this modified cellulose material by Eq. (7).

Conclusions

Theoretical concepts and experimental methods for contact angle measurements were presented to determine the solid-vapour and solid-liquid interfacial tensions which are important thermodynamic parameters in many areas of applied science and technology.

At present, experimental techniques based on Axisymmetric Drop Shape Analysis are available to measure very accurate contact angles of polar and non-polar liquids on smooth and on rough, chemically heterogeneous solid surfaces, even if the contact angles are below 20°. If smooth and inert solid surfaces are used the plot of $\gamma_{lv}\cos\Theta$ versus γ_{lv} provides experimental curves which are consistent with the equation of state approach to calculate solid-vapour and solid-liquid interfacial tensions. Other experimental patterns of contact angle data are caused by surface roughness and non-inert solids which may result in contact angles incompatible to Young's equation. An alternative way to obtain the solid-vapour surface tension of rough and porous solids are capillary penetration experiments. Determination of the penetration velocity of liquids into rough and porous solids allows $K\gamma_{lv}\cos\Theta$ versus γ_{lv} plots, which provide γ_{sv} values for these systems. The application of this concept was demonstrated for a hydrophobic PTFE powder and for hydrophilic cellulose membranes.

Acknowledgments This research was supported by the Natural Science and Engineering Research Council of Canada (Grants: No. A8278 and No. EQP173469) and a University of Toronto Open Fellowship (D.Y.K.). The research was also supported by BMBF of Germany (Grants: No. 03N 4001/7).

References

1. De Gennes PG (1985) Reviews Mod Phys 57:827
2. Misev TA (ed) (1991) Powder Coatings Chemistry and Technology. John Wiley, New York, pp 206
3. Garbassi F, Morra M, Occhiello E (eds) (1994) Polymer Surfaces From Physics to Technology, Chapter 9. John Wiley, New York
4. Neumann AW, Good RJ, Hope CJ, Sejpal M (1974) J Colloid Interface Sci 49:291
5. Spelt JK, Li D, Neumann AW (1992) Chapter 5: The Equation of State Approach to Interfacial Tensions In: Schrader ME, Loeb GI (eds) Modern Approaches to Wettability. Plenum Press, New York, pp 101
6. Fowkes FM (1964) Ind Eng Chem 12:40
7. van Oss CJ, Chaudhury MK, Good RJ (1988) Chem Rev 88:927
8. Rotenberg Y, Boruvka L, Neumann AW (1983) J Colloid Interface Sci 93:169
9. Cheng P, Li D, Boruvka L, Rotenberg Y, Neumann AW (1990) Colloids and Surfaces 43:151
10. Li D, Neumann AW (1992) J Colloid Interface Sci 148:190
11. Li D, Xie M, Neumann AW (1993) Colloids Polym Sci 271:573
12. Li D, Gaydos J, Neumann AW (1989) Langmuir 5:1133
13. Defay R (ed) (1934) Etude Thermodynamique de la Tension Superficielle, Gauthier-Villars, Paris
14. Defay R, Priogine (1966) In: Bellemans A, Everett DH (eds) Surface Tension and Adsorption. Longmans-Green, London
15. Ward CA, Neumann AW (1974) J Colloid Interface Sci 49:286
16. Li D, Neumann AW (1990) J Colloid Interface Sci 137:304
17. Vargha-Butler EI, Absolom DR, Neumann AW, Hamza HA (1989) In: Botsaris GD, Glazman YM (eds) Interfacial Phenomena in Coal Technology, Chapter 2: Characterization of Coal by Contact Angle and Surface Tension Measurements. Marcel Dekker, Inc New York
18. Skinner FK, Rotenberg Y, Neumann AW (1989) J Colloid Interface Sci 130:25
19. Bruil HG (1974) Colloid Polym Sci 252:32
20. Cheever GD (1983) J Coat Technol 55:53
21. Kilau HW (1987) Colloids and Surfaces 26:217
22. Grundke K, Boerner M, Jacobasch HJ (1991) Colloids and Surfaces 58:47
23. Washburn EW (1921) Phys Rev 17:273
24. Costanzo PM, Giese RF, van Oss CJ (1990) J Adhesion Sci Technol 4:267
25. van Oss CJ, Giese RF, Li Z, Murphy K, Norris J, Chaudhury MK, Good RJ (1992) J Adhesion Sci Technol 6:413
26. Buckton G (1993) J Adhesion Sci Technol 7:205
27. Chibowski E, Holysz L (1992) Langmuir 8:710
28. Bambauer R, Weber U, Lissmann J (1990) Hämostasiologie 10:84
29. Jacobasch HJ (1984) Oberflächenchemie faserbildender Polymerer, Akademieverlag, Berlin

Progr Colloid Polym Sci (1996) 101:69–74

K. Holmberg

Unsaturated monoethanolamide ethoxylates as paint surfactants

Prof. K. Holmberg (✉)
Institute for Surface Chemistry,
P.O. Box 56 07
114 86 Stockholm, Sweden

Abstract The paper discusses use of ethoxylated monoethanolamide of highly unsaturated fatty acids for replacement for nonylphenol ethoxylates in coatings. The double bonds in the hydrophobic tail of the ethanolamide ethoxylates impart both bulkiness and polarizability, properties of value with respect to packing at interfaces and interaction with many pigment surfaces. These ethanolamide ethoxylates may also undergo autoxidation. When used as emulsifiers in alkyd emulsions, drying and film properties are improved as compared to formulations based on ethoxylated nonylphenol of the same HLB number. ESCA analysis shows that the polymerizable surfactant migrates to a lesser extent to the film surfaces than does the reference surfactant.

Key words Polymerizable surfactant – monoethanolamide ethoxylate – alkyd emulsion – nonylphenol ethoxylate

Introduction

Improved film properties have traditionally been the main driving force for the development of new surfactants for coatings. In recent years, environmental concern has emerged as a governing factor of at least equal importance. The concern relates to the whole formulation with emphasis on reduction of organic volatiles, but also the individual surfactants used are being critically evaluated. As in all surfactant applications, biodegradability and ecotoxicity are issues of increasing importance and one of the most commonly used surfactant types, nonylphenol ethoxylates (NPEs), have come in focus. NPEs have a long tradition as emulsifier, dispersant and wetting agent in paint formulations. Slow biodegradation and considerable biotoxicity of this class of substances have prompted a need to replace them with more environment-friendly surfactants. This paper discusses one option for replacement of NPEs.

Most water-borne coating systems contain several different types of surface active agents. Low molecular weight surfactants are used as emulsifier for the binder, as dispersant for the pigment, as wetting and anti-foaming agents, etc. Surface active polymers are frequently used for rheology control. The variety of surface active agents present in the formulation makes the situation with regard to surface interactions very complex. Any surfactant introduced into the system will, in principle, be in dynamic equilibrium between the bulk liquid phase and all liquid-solid, liquid-liquid and liquid-air interfaces that are present in the formulation. A seemingly harmless change in a paint recipe may, therefore, give unexpected effects. For instance, a nonionic surfactant of low HLB number, added to improve spreading of a latex paint on a hard-to-wet surface, may totally replace the anionic surfactant stabilizing the binder droplets, leading to deterioration of the emulsion stability [1]. In other instances, addition of a new surfactant may create unexpected rheological changes due to interaction between the surfactant and the surface active associative thickener [2]. Poor film properties will be the net result in both cases.

One way to circumvent the problem of uncontrolled, competitive adsorption of surfactants in a coatings system is to use polymerizable surface active agents, i.e. surfactants which are capable of homo- or copolymerization. Polymerization may take place either in dispersion – usually on latex particles – or during the curing process. Polymerization in the curing step is particularly relevant to alkyd emulsions and an example from that field is presented in this paper.

Materials and methods

Two nonionic surfactants were used:

1) Ethoxylated (14 EO) monoethanolamide of a fatty acid fraction obtained from linseed oil. Relative proportions of the main fatty acid components in the triglyceride were linolenic acid, 50%; linoleic acid, 20%; and oleic acid, 20%. The product was prepared by aminolysis of the triglyceride followed by ethoxylation (3).

2) Ethoxylated (12 EO) nonylphenol (NP 12). Starting nonylphenol had the isomer composition characteristic of the propane trimer route with the methyl-branched nonyl group preferentially attached at a tertiary carbon *para* to the phenolic OH.

The number of oxyethylene units stated are average figures. Both compounds had the broad homologue distribution typical of ethoxylates prepared by alkali-initiated alkoxylation.

Drying of lacquer films was monitored as described previously (4). ESCA was conducted using a Kratos AXIS-HS instrument with deconvolution of spectra performed by means of a least-square Gaussian curve-fitting program. The carbon 1 s peak of the polyoxyethylene chain (O–C–C–O carbon) at 286.5 eV was used for quantifying surfactant at the surface.

Results and discussion

Comparison between nonylphenol ethoxylates and alcohol ethoxylates

Replacement of an NPE with a conventional alcohol ethoxylate of the same HLB number is often not satisfactory. Neither latex particle stabilization, nor pigment dispersion efficiency is equivalent. Since the polar groups of the two types of surfactants are the same, the difference in performance must be related to the hydrophobic tails of the surfactant molecules. These differ in two respects:

Hydrophobe geometry

Fatty alcohols have linear alkyl groups with a high degree of flexibility in the hydrocarbon chain. Nonylphenyl, on the other hand, is a compact, partly rigid and bulky molecule. Using the packing parameter terminology (5), alcohol ethoxylates have small hydrophobe volume, v, and large hydrophobe critical chain length, l_c, whereas NPEs have larger v and smaller l_c. The optimal head group area, a_0, is the same if the number of oxyethylene groups is the same.

The packing parameter concept can be used to predict what type of aggregate will spontaneously form in solution:

$$\frac{v}{a_0 \times l_c} < \frac{1}{3} \text{ leads to spherical micelles}$$

$$\frac{1}{3} < \frac{v}{a_0 \times l_c} < \frac{1}{2} \text{ leads to non-spherical micelles}$$

$$\frac{1}{2} < \frac{v}{a_0 \times l_c} < 1 \text{ leads to bilayers}$$

$$\frac{v}{a_0 \times l_c} > 1 \text{ leads to inverted micelles}$$

For surfactants based on saturated hydrocarbons, v and l_c can readily be calculated using the expressions:

$$v = (27.4 + 26.9n) \times 10^{-3}\ (\text{nm}^3)$$

and

$$l_c < l_{max} = (0.15 + 0.1265n)\ (\text{nm})$$

per saturated hydrocarbon chain with n carbon atoms.

The optimal head group area, a_o, can be calculated from the expression:

$$M = \frac{4\pi R^2}{a_0} = \frac{4\pi R^3}{3v},$$

where M is the aggregation number and R the radius of the surfactant micelle.

Simple calculations show that dodecyl alcohol ethoxylates of the polyoxyethylene chain length of interest for latex stabilization and pigment dispersion, i.e. around 15 ethylene oxide units or more, are in the $v/a_0 \times l_c < 1/3$ category. The 15 EO ethoxylate has an experimentally determined aggregation number, M, of 58. Calculations of v and l_c using the above formulas with n equals 12 give the values 0.350 nm^3 and 1.67 nm, respectively. This leads to R = 1.7 nm which, in turn, gives $a_0 = 0.62\ \text{nm}^2$. Thus, $v/a_0 \times l_c$ is 0.33 for the 15 EO ethoxylate. Higher ethoxylates have lower M, meaning smaller R, meaning higher a_0. Since v and l_c are the same, $v/a_0 \times l_c$ will be smaller than 0.33.

There is no straight-forward way to obtain numerical values of v and l_c for irregularly shaped hydrophobes, such as nonylphenol. However, a rough estimate of v and l_c values can be obtained by inspection of molecular models. The approximate values of v/l_c obtained are accurate enough for most practical purposes and are useful for comparative studies.

Calculations of this type show that v/l_c for nonylphenol is about 1.5 times that of linear dodecanol. Using the same a_0 values as for the alcohol ethoxylates above – which should be correct since the polar group is the same – NP 15 has a $v/a_0 \times l_c$ value of approximately 0.5. This indicates that NPEs do not form spherical micelles which has, in fact, been demonstrated experimentally by means of a fluorescence probing technique (6).

These calculations – although qualitative – demonstrate that alcohol ethoxylates pack less efficiently than NP ethoxylates at surfaces unless these are strongly convex curved against water. Another way of viewing this difference is that the driving force for alcohol ethoxylates to align at surfaces is smaller than for NP ethoxylates because the interchain attraction between the hydrophobic tails is smaller.

Electronic effects

The 6π electrons of the aromatic ring of nonylphenol render the hydrophobe polarizable (7) which affects the interaction with many surfaces. Phenols are known to be able to act as donors in electron donor-acceptor (EDA) complexes, donating π electrons to suitable acceptor molecules. This type of interaction can be of considerable magnitude but the exact nature of the bonding, called charge transfer bonding, is not well understood (8). However, it is reasonable to assume that EDA complex formation will play a role in the interaction between NPEs and many surfaces. No such contribution is, of course, present in the case of alcohol ethoxylates.

The electronic effect is not as general as the effect of the geometrical packing. Charge transfer interactions will only take place with surfaces that can function as acceptors in EDA complexes. Olefins and aromatics that contain electron-withdrawing substituents act as acceptor molecules. Tetracyanoethylene, quinones, and picric acid are typical examples of acceptors in EDA complexes. Most likely, many organic pigment surfaces also belong to this category. Actually, one alkylphenol ethoxylate, Triton X-100, has been found to form strong 1:1 complexes with a range of organic dye molecules, including Methylene Blue, rhodamin B, fuchsin and crystal violet. Interactions in micelles and reverse micelles resulted in the pronounced change in absorption maximum that is indicative of charge-transfer interactions (9, 10). Even if the magnitude of such interactions with organic pigments is uncertain, it is likely that electronic effects contribute to the strong attraction seen between NPEs and many surfaces (Fig. 1).

Unsaturated fatty acid amide ethoxylates as NPE substitutes

The monoethanolamide of unsaturated fatty acids is an interesting alcohol alternative to alkylphenol. Each *cis* double bond causes the acyl chain to bend. Since there is free rotation about the carbon-carbon bonds, a high degree of bulkiness of the chain is obtained. Fatty acids with varying degree of unsaturation are available from natural triglycerides, although all technical products are blends of acids with different numbers of olefinic bonds. Amide ethoxylates based on highly unsaturated vegetable oils, such as soyabeen oil or linseed oil, have been found to be good substitutes for NPEs as dispersants in tinting colors. The structure of ethoxylated ethanolamide of linoleic acid is schematically shown in Fig. 2. Since the technical products are always mixtures of fatty acids with varying numbers of double bonds and with varying chain lengths, estimations of $v/a_0 \times l_c$ are difficult to make. However, it is obvious that the considerable bulkiness imparted by the unsaturation should be favorable with respect to geometrical packing properties which, in turn, should lead to good performance in applications where the arrangement of

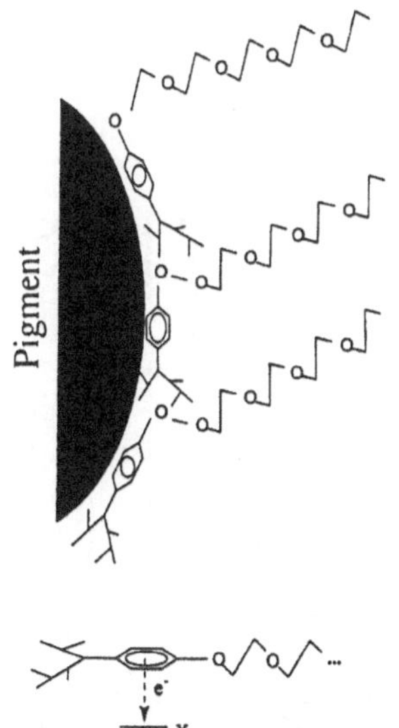

Fig. 1 Schematic illustration of an electron donor-acceptor complex between an alkylphenol ethoxylate and unsaturated moieties on a pigment surface

surfactants at interfaces are crucial. This type of surfactant has been found to be superior to alcohol ethoxylates as emulsifier for alkyd resins (11).

Due to the presence of double bonds in the acyl chains this class of surfactant may resemble NPEs also with regard to ability to participate in EDA complexes. Although aromatic rings are likely to be better donors than the homoconjugated double bonds present in most fatty acids, such as linoleic acid and linolenic acid, it may well be that also the latter structures can form charge-transfer complexes with powerful acceptor molecules. This is an area that warrants further attention.

Unsaturated fatty acid amide ethyoxylates as polymerizable surfactants

Alkyd emulsions are gaining in importance as a consequence of environmental demands. Stable emulsions can

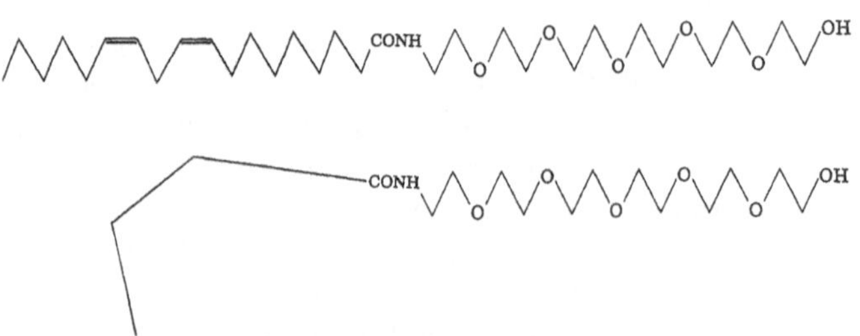

Fig. 2 Structure of ethoxylated monoethanolamide of linoleic acid (top); schematic illustration of the bulkiness of the hydrophobic tail imparted by the two olefinic bonds (bottom)

be made from most alkyds, provided the resin viscosity is not too high and sufficient shear forces are applied in the emulsification (12, 13). Prior to emulsification, alkyd carboxylic groups are, at least partly, neutralized with an amine, thus making the resin itself slightly surface active. The surfactant used should have a HLB number that matches that of the neutralized resin. Best results are normally obtained with nonionic surfactants, although they tend to give less temperature stable emulsions than anionic emulsifiers.

Compared with white spirit-based alkyd formulations, emulsions in general give slower drying and exhibit somewhat impaired film properties. The dried film is usually softer and more water sensitive than a film obtained from the same resin formulated in an organic solvent. The difference in film properties is at least partly caused by the surfactant which migrates to the surface and acts as a water soluble external plasticizer.

It has been found that by using emulsifiers capable of participating in the autoxidative drying of the binder, the film properties can be considerably improved (4). Figure 3 depicts schematic pictures of the curing process with a conventional emulsifier and with a monoethanolamide ethoxylate based on unsaturated fatty acids, capable of copolymerization with the binder.

Table 1 shows the influence of a polymerizable monoethanolamide ethoxylate and a NP ethoxylate, both surfactants having HLB numbers around 14, on the drying rate and hardness of an alkyd film. As can be seen from the table, film hardness is almost unaffected by the reactive surfactant and much inferior when the conventional surfactant has been added.

Fig. 3 Curing of an alkyd emulsion based on a conventional surfactant (left) and on a reactive surfactant (right)

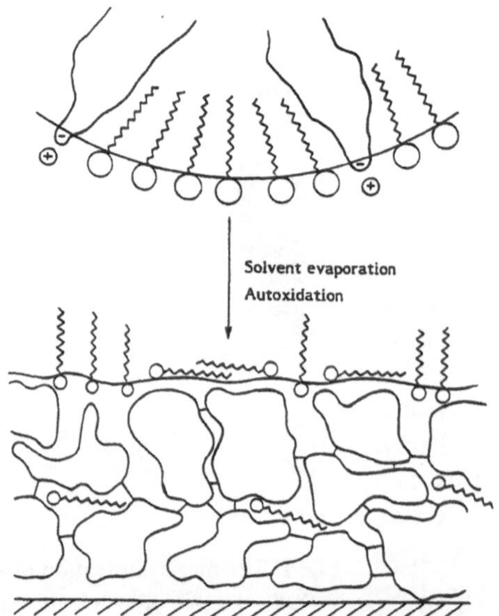

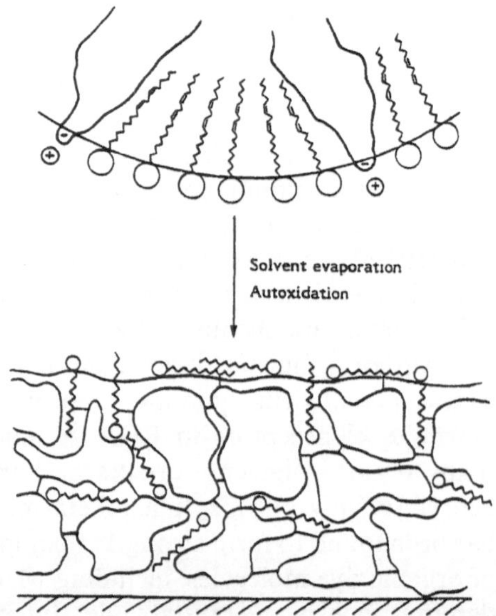

Table 1 Properties of films obtained from a Soya alkyd dissolved in white spirit with varying amounts of surfactant added*

Surfactant added	Drying (Beck-Koller) of 20 μm films (h)					Hardness (pendulum) of 40 μm films (s) after			
	1	2	3	4	5	2d	7d	15d	30d
No surfactant	0.3	0.6	0.8	2.1	6.1	6	14	22	30
4% reactive	0.3	0.6	0.9	2.3	6.8	5	12	21	28
8% reactive	0.3	0.8	0.9	2.1	7.2	5	12	22	27
4% non-reactive	0.3	0.6	1.0	2.4	8.1	5	8	14	20
8% non-reactive	0.4	0.7	1.1	2.6	8.9	4	7	12	17

* Reactive, ethoxylated (14 EO) linseed oil acid monoethanolamide, and non-reactive, ethoxylated (12 EO) nonyl phenol, surfactant were used. The alkyd had an oil length of 62% and an acid value of 6 mg KOH/g. 0.05% Co and 0.1% Zr were used as driers.

Surfactant accumulation at the film surface

Surfactant concentrations at the film-air interface was measured by ESCA and the results are shown in Fig. 4. As can be seen, both the NP ethyoxylate and the amide ethoxylate accumulate at the surface and the concentration increases with time. Whereas the concentration vs. time curve is almost linear for the NP ethoxylate, it levels off for the amide ethoxylate. For the latter species, the distribution of surfactant in the film seems to be established within a 3-day drying period.

Studies of the distribution of non-reactive surfactants in coating films during drying have been performed before using FTIR, ESCA and SIMS as analysis techniques (14–16). For instance, NP 10 was found to accumulate at both the film-air and the film-substrate interface and the process of accumulation continued for at least 60 days (16). Polymerizable surfactants seem not to have been investigated in this respect before.

The difference in behavior between the NP ethoxylate and the amide ethoxylate is probably due to the fact that the latter surfactant becomes immobilized through coupling to binder molecules during the drying process. Once covalently incorporated into the network, the migration process will cease. Another contributing factor for the poor migration of the amide ethoxylate could be that this surfactant is likely to be very compatible with the binder, a long oil alkyd resin. Surfactant-polymer compatibility has previously been found to be decisive for surfactant distribution in films (14, 17). Surfactants are carried towards the surface by the flux of water during film drying and this process is particularly effective when there is poor compatibility between surfactant and polymer.

The effect on surface composition of soaking the dried film in water is also shown in Fig. 4. Whereas more than half of the NP ethoxylate disappears from the outermost surface layer (approximately 50 Å), the effect on the amide ethoxylate is small, in spite of the fact that both surfactants have about equal water solubility. This is a further indication of the amide ethoxylate being immobilized during the drying process.

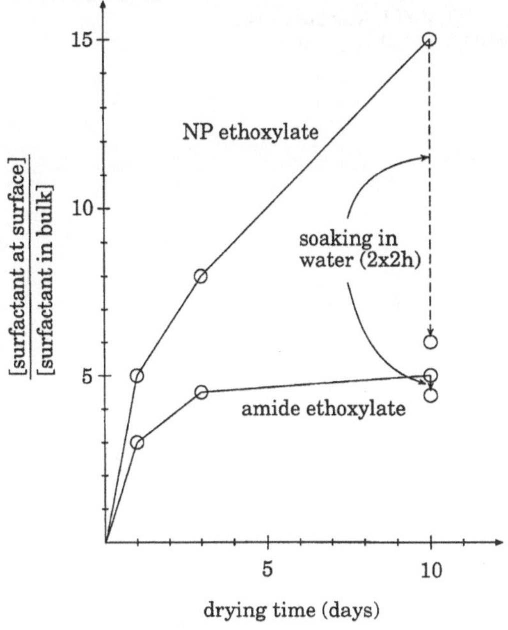

Fig. 4 Relative surface concentration of surfactant as a function of drying time as revealed by ESCA. The experimental conditions were those given in Table 1 but alkyd emulsions containing 4 wt% surfactant calculated on the binder were used instead of white spirit solutions

The evidence presented in this paper of the surfactant participating in the autoxidation is only indirect. The sensitivity of ESCA is not sufficient for monitoring disappearance of carbon-carbon double bonds, which would have been the most direct way of studying surfactant polymerization. However, studies on cobalt initiated autoxidation of ethyl esters of unsaturated fatty acids under conditions similar to the ones used here have shown that whereas not much happens with oleate ester during a period of 110 days, linoleate ester polymerizes almost completely in 3 days (18). These findings lend support to our statement that amide ethoxylates based on fatty acids with a high degree of unsaturation become covalently incorporated in the dried film.

References

1. Kronberg B, Kuortti J, Stenius P (1986) Colloids Surfaces 18:411
2. Huldén M (1994) Dissertation, Åbo Akademi, Finland
3. Sandberg E (1992) PCT Int Patent Appl WO 9208690
4. Holmberg K (1992) Prog Org Coatings 20:325
5. Israelachvili JN (1985) Intermolecular and Surface Forces. Academic Press, New York
6. Zana R, Weill C (1985) J Physique Lett 46:L 963
7. Urbina-Villalba G, Reif I, Márques ML, Rogel E (1995) Colloids Surfaces A 99:207
8. Foster F (1969) Organic Charge-Transfer Complexes. Academic Press, New York
9. Rohatgi-Mukherjee KK, Chaudhuri R, Bhowmik BB (1985) J Colloid Interface Sci 106:45
10. Pramanick D, Mukherjee D (1993) J Colloid Interface Sci 157:131
11. Östberg G, Huldén M, Bergenståhl B, Holmberg K (1994) Prog Org Coatings 24:281
12. Bufkin BG, Grawe JR (1978) J Coat Technol 50:66
13. Hofland R, Schaap F (1988) Polym Paint Colour J 178:620
14. Zhao CL, Dobler F, Pith T, Holl Y, Lambla M (1989) J Colloid Interface Sci 128: 437
15. Torstensson M, Rånby B, Hult A (1990) Macromolecules 23:126
16. Kientz E, Holl Y (1993) Colloids Surfaces A 78:255
17. Evanson KW, Urban MW (1991) J Appl Polym Sci 42:2287
18. Muizebelt WJ, Hubert JC, Venderbosch RAM (1994) Prog Org Coat 24:263

Progr Colloid Polym Sci (1996) 101:75–85

D. Nees
M. Blenkle
A. Koschade
T. Wolff
P. Baglioni
L. Dei

Switching flow and phase behavior in surfactant systems via photochemical reactions

D. Nees
Universität-GH-Siegen
Physikalische Chemie
57068 Siegen, FRG

M. Blenkle · A. Koschade
Prof. Dr. T. Wolff (✉)
Technische Universität Dresden,
Institut für Physikalische Chemie
01062 Dresden, Germany

P. Baglioni · L. Dei
Università degli Studi di Firenze
Dipartimento di Chimica,
50121 Firenze, Italy

Abstract Photochemically induced viscosity changes in dilute aqueous micellar solutions of cetyltrimethylammonium bromide (CTAB) and of Triton X-100 (TRI) are brought about *via* the photoconversion of solubilized *N*-methyldiphenylamine (MDPA) to *N*-methylcarbazole (MC). Previously, these photorheological effects were ascribed to transformations of the micellar hydration shell. New support for this rationalization was gained from ESR-spectra of various nitroxide radical spin probes measured in the respective systems. Coupling constant variations of probes associated with the micelles indicated that MDPA is solubilized near the micelle/water interface while MC is located deeper in the micellar core. Probes residing in the bulk aqueous phase were not affected by the presence of solubilizates. Only spin probes located in the aqueous phase just outside the micelle, i.e., in the hydration layer, responded to observed macroscopic viscosity changes by accordingly varying motional correlation times. – In lyotropic liquid crystalline systems phase transitions can be induced by photochemically transforming certain solubilizates, when photoeducts induce transition temperatures distinct from those induced by the photoproducts. Series of bis-(thienyl)cyclopentenes (BTCPs) (capable of photocyclization) and of hydroxy-stilbazolium bromides (capable of *cis-trans* isomerizations and dimerizations) were tested as photochemically reactive solubilizates in CTAB-water (transition nematic-isotropic), AOT-glycerol, and AOT-formamide (transition lamellar-isotropic). Shifts of transition temperatures up to 10 °C upon irradiating systems containing these solubilizates were found.

Key words Rheology – spin probes – lyotropic liquid crystals – phase transitions

Introduction

Rheological effects

Considerable macroscopic viscosity changes in dilute micellar solutions are known to occur upon solubilizing certain aromatic compounds at low concentrations, such as anthracene derivatives [1, 2], stilbene derivatives [3], or *N*-methyldiphenylamine (MDPA) [4]. Upon irradiating such systems further variations of the flow behavior take place, e.g., additional viscosity changes or transfer from Newtonian to non-Newtonian flow and *vice versa*. Therefore, the flow properties can be altered *in situ* by exposing samples to suitable light, provided the photoeduct induces a rheological behavior differing from that induced by the photoproduct (photorheological effect).

The irradiations do not affect the surfactant materials but cause well known photochemical reactions of the solubilizates, i.e. photodimerizations, *cis-trans*-isomerizations or cyclizations. An example is the photocyclization of *N*-methyldiphenylamine (MDPA) to *N*-methylcarbazole (MC) [5], see Scheme 1.

The general rheological behavior of aqueous micellar solutions of cetyltrimethylammonium bromide (CTAB) and of the nonionic surfactant Triton X-100 in the presence of MDPA and/or MC is Newtonian. Viscosity changes upon solubilizing and photochemically converting *N*-methyldiphenylamine are comprised in Table 1. Inspection of the table reveals that in both systems the solubilization of MDPA leads to a viscosity increase which in the CTAB system is partially reverted upon photocyclization of MDPA to MC (performed *in situ* or in separate experiments). In the Triton X-100 system, however, photocyclization induces a further increase of viscosity. Rheological and light-scattering studies [4] as well as electron micrographs [6] have revealed that MDPA in the CTAB system induces the formation of globular giant micelles with aggregations numbers z of several thousands exceeding those of pure CTAB solutions ($z \simeq 80$) by up to two orders of magnitude. A moderate further increase of z occurs after photocyclization which is accompanied by a viscosity decrease. In the Triton X-100 system no significant change of z is observed after solubilization of MDPA and after its photocyclization to MC (in contrast to most of the systems showing photorheological effects; for a review see [7]). Therefore, the photorheological changes in both the systems cannot be consequences of changes in the pure dilute CTAB-H_2O system [8], the effects were tentatively ascribed [4] to the extent, the degree of order and the rigidity of the micellar hydration shell (2 through 4 water layers). This was corroborated by viscosity measurements [4] in the presence of "structure making" and "structure breaking" additives [9], which are known to increase or decrease, respectively, the number of hydrogen bridge bonds in water.

In order to test the interpretation further, we carried out spin probe experiments using the solutions whose compositions, macroscopic viscosities and aggregation numbers are given in Table 1. Spin probes were chosen which are known to probe the microviscosity at different sites of the microheterogeneous micellar solutions, such as bulk water, micellar interior, and the interfacial region [10–12].

$$\text{MDPA} \xrightarrow[-H_2O_2]{h\nu,\ O_2} \text{MC}$$

Scheme 1: Photocyclization of *N*-methyldiphenylamine (MDPA) to *N*-methylcarbazole (MC)

Phase behavior

In more concentrated surfactant systems lyotropic liquid crystalline (LLC) phases form [13], e.g., a nematic phase between 20 and 25% w/w CTAB-H_2O, adjacent to a hexagonal phase at higher CTAB-concentrations. Transitions from the hexagonal to cubic and lamellar phases take place at about 80 and 85%, respectively (a CTAB-H_2O phase diagram is reproduced in [14]). In other surfactant systems inverse hexagonal structures may form in the water-poor region, found favorably in surfactants bearing branched or doubled alkane chains, such as Aerosol-OT (AOT). A similar series order of lyotropic liquid crystalline phases exists in non-aqueous surfactant systems when water is substituted by other polar solvents, e.g. hydrazin, formamide, glycol or glycerol.

Table 1 Compositions and viscosities of the solutions used in the spin probe experiments

Sample	Concentration/mmol/dm^3				η/mPa s	Aggregation no. z
	CTAB	Triton X-100	MDPA	MC	at 25 °C[a]	
Water	–	–	–	–	0.98	–
CTAB	250	–	–	–	1.67	80[b]
CTAB/MDPA	250	–	20	–	11.5	3100[b,c]
CTAB/MC	250	–	–	20	5.86	3600[b,c]
TRI	–	250	–	–	4.26	111[d]
TRI/MDPA	–	250	35	–	12.3	70[d,e]
TRI/MC	–	250	–	35	15.5	81[d,e]

[a] from ref. [4]; [b] at [CTAB] = 1 – 4 mmol/dm^3; [c] at [MDPA or MC]/[CTAB] = 0.074; [d] at [Triton X-100] = 3 – 36 mmol/l; [e] at [MDPA or MC]/[Triton X-100] = 16.7

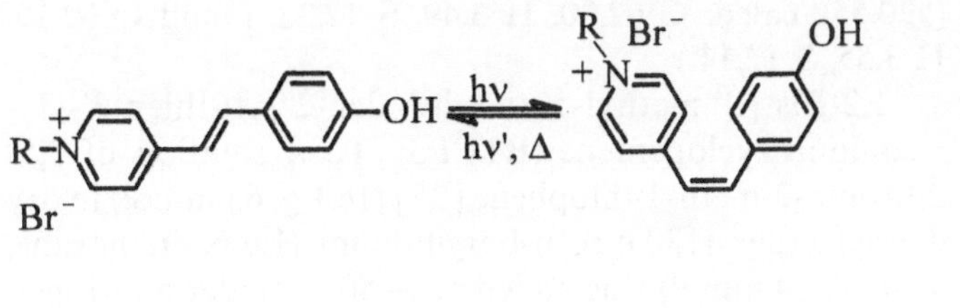

R = CH_3 (C1-HSB), R = $C_{16}H_{33}$ (16-HSB)

Scheme 2: Photoisomerization of *N*-alkyl hydroxystilbazolium bromides.

Most of the compounds and photoreactions exhibiting photorheological effects [7] also affect phase equilibria in LC systems, i.e., photoeducts induce transition temperatures differing from those in the presence of the the respective photoproducts. Again, a variety of photoreactions can be used to produce these effects: photodimerizations [15], *cis-trans*-isomerizations [14, 16], and photocyclizations [17]. Thus, phase transitions can be brought about by performing the photoreactions *in situ* under conditions near to the phase equilibria (for a review see [18]). In principle, the effect can be exploited for storing and erasing information. Incipient experiments along this line [19] revealed quite long periods required to switch phases in CTAB-H_2O, but photochemical phase transitions within milliseconds in non-aqueous AOT systems, on which, therefore, recent efforts were concentrated.

Here we present experiments in AOT-glycerol and AOT-formamide using the following compounds as photoisomerizable solubilizates: hydroxystilbazolium salts (Scheme 2) and bis-(thienyl)cyclopentenes (Scheme 3).

R= −H BTCP1
= (phenyl) BTCP2
= (naphthyl) BTCP3
= $-Si(CH_3)_3$ BTCP4
R = $-C_{12}H_{25}$ BTCP5
= (oxazoline) BTCP6

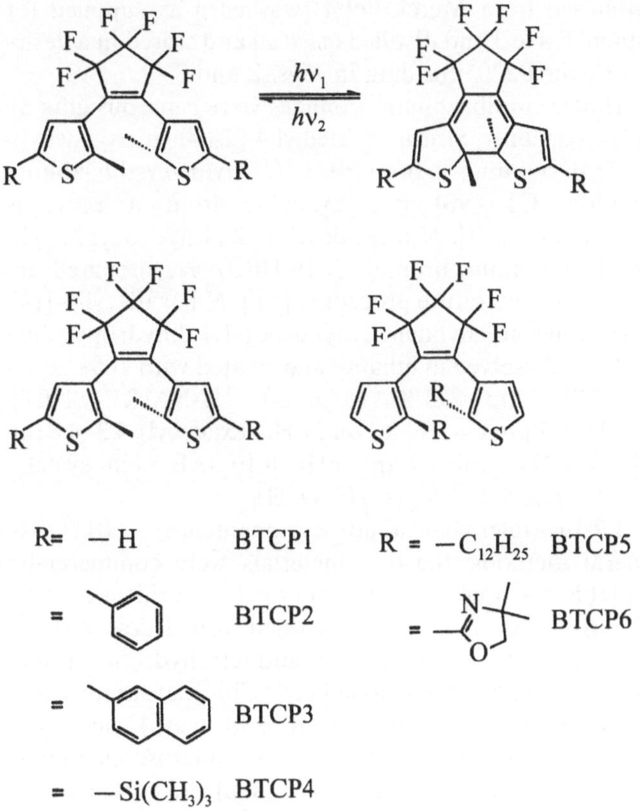

Scheme 3: Photocyclization of bis-(thienyl)cyclopentenes (BTCPs)

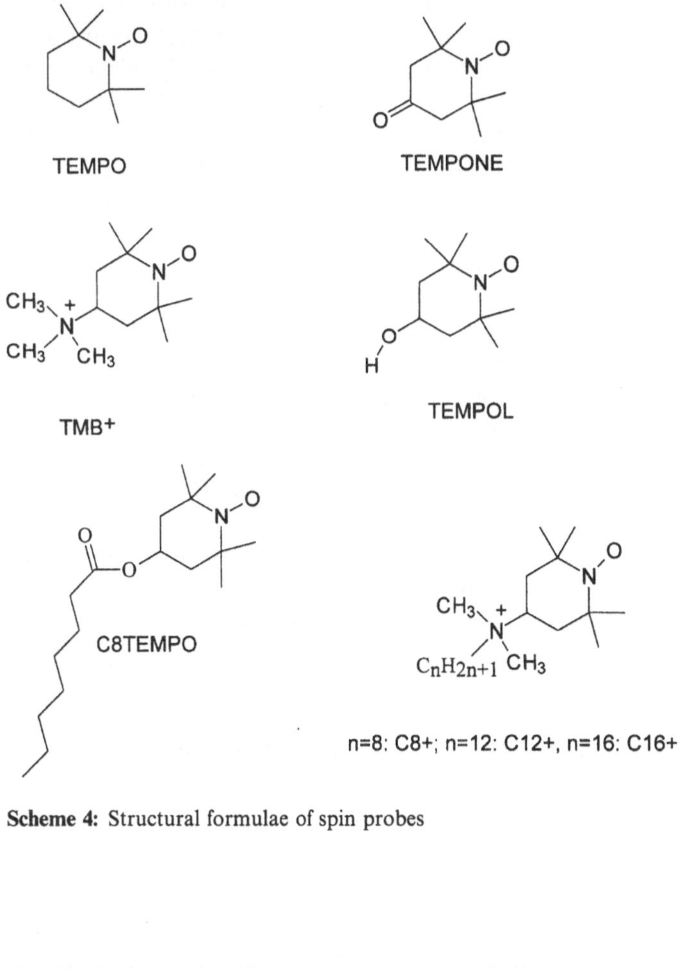

TEMPO
TEMPONE
TMB+
TEMPOL
C8TEMPO
n=8: C8+; n=12: C12+, n=16: C16+

Scheme 4: Structural formulae of spin probes

Experimental

Materials

Water was triply distilled in a quartz apparatus (or Millipore filtered for the ESR Experiments). Cetyltrimethylammonium bromide (CTAB) was purchased from Merck (p.a. grade) and recrystallized three times from acetone-methanol (9:1 mixture), Triton X-100 (Merck, p.a.) was used as supplied. *N*-Methyldiphenylamine (Kodak) was distilled *in vacuo*. N-Methylcarbazole was available from a previous investigation [4]. Spin probes were purchased from Molecular Probes, Inc. Structural formulae of the spin probes used are given in Scheme 4. Sodium di-2-ethylhexylsulfosuccinate (Aerosol OT, AOT) was obtained from Fluka (99%, for data in Figs. 2 and 3) or from Aldrich (98%, for data in Table 3, purified according to [20]). Formamide from Merck (99.5%), glycerol purchased from Merck (99%) was used as supplied for data in Table 3 and distilled *in vacuo* and stored in a desiccator over P2205 for data in Figs. 2 and 3.

Hydroxystilbazolium bromides were generous gifts by Dr. B. Klaußner, Siegen. *N*-Methyl-4-[2-(4-hydroxyphenyl)-vinyl]-pyridinium bromide (C1-hydroxystilbazolium bromide, C1-HSB) was available from a previous investigation [19]. *N*-n-hexdecyl-4-[2-(4-hydroxyphenyl)-vinyl]-pyridinium bromide (C16-HSB) was prepared according to a literature procedure [21]: *N*-hexadecyl-4-[(4′-oxocyclohexadienylidene)ethylidene]-1,4-dihydropyridine [22] was dissolved in ethanol and treated with 10% aqueous HBr. mp 225 °C (decomp). ^{1}H-NMR ($CDCl_3$, 80 MHz): δ/ppm = 0.8–2,1 (m 31 H, hexadecyl), 4.3–4.8 (m, 2H, N–CH_2), 6.6–7.4 (m, 6H), 8.10 (AB spin system, $^3J = 6.8$ Hz, 4H), 8.93 (s, 1H, O–H).

1,2-Bis-(thienyl)hexafluorocyclopentenes (BTCPs), general methods: Starting materials were commercially available and were used without further purification. Octafluorocylcopentene was purchased from Heraeus (Karlsruhe, Germany). Diethyl ether and tetrahydrofuran were distilled from benzophenone ketyl. Chromatography was performed on Kieselgel 60 (70–230 mesh) silica gel. Melting points were determined on a microscope apparatus > BOETIUS < (Kofler, Dresden) and are uncorrected. NMR spectra were recorded on a Bruker WH-90 (^{1}H) and Bruker MSL-300 (^{13}C). ^{1}H-NMR spectra were obtained at 90.023 MHz with the residual proton signals of the solvent as reference. ^{13}C-NMR spectra were obtained at 75.475 MHz with the solvent signals as reference.

1,2-Bis-(2′-methyl-thien-3′-yl)-hexafluorocyclopentene (BTCP1) and 1,2-Bis-(5′-trimethyl-silanyl-2′-methyl-thien-3′-yl)-hexafluorocyclopentene (BTCP4) were prepared by a described method [23].

1,2-Bis-(2′-methyl-5′-phenyl-thien-3′-yl)-hexafluorocyclopentene (BTCP2): s-Butyllithium (1.3 M cyclohexane/hexane, 10.5 ml, 13.7 mmol) was slowly added to a stirred solution of 2-methyl-3-bromo-5-phenyl-thiophene [24] (3.45 g, 13.6 mmol) in dry diethyl ether (50 ml) at −70 °C under a nitrogen atmosphere and stirred for 60 min. Precooled octafluorocyclopentene (1.4 g, 6.6 mmol) was then added and the solution was kept at −70 °C for 2 h. The mixture was allowed to warm to room temperature and hydrochloric acid (0.1 N, 10 ml) was added. Usual workup with diethyl ether, chromatography (cyclohexane/dichloromethane 4:1) and recrystallization from acetone yielded BTCP2 (2.2 g, 4.1 mmol, 64%) as colorless crystals (m.p. 137 °C-^{1}H-NMR ($CDCl_3$): $\delta = 1.97$ (s, 6H, CH_3), 7.40 (m, 12H, aryl-H).-^{13}C-NMR ($CDCl_3$): $\delta = 14.5$ (CH_3), 111.1 (C-4), 116.2 (C-3,-5), 122.8 (C-4′), 125.6 (C-2″,-6″), 125.9 (C-3′), 127.9 (C-4″), 129.0 (C-3″,-5″), 133.4 (C-1″), 136.2 (C-1,-2), 141.3 (C-5′), 142.3 (C-2′). -$C_{27}H_{18}F_6S_2$ (520.55): calcd. C 62.30, H 3.49, S 12.32; found C 62.29, H 3.55, S 12.11.)

1,2-Bis-[2′-methyl-5′-(naphthalen-2″-yl)-thien-3′-yl]-hexafluorocyclopentene (BTCP3): To a solution of 3,5-dibromo-2-methyl-thiophene [25] (16.1 g, 63 mmol) in dry diethyl ether (130 ml), n-butyllithium (1.6 N in hexane, 40.0 ml, 64 mmol) was added at −50 °C under a nitrogen atmosphere and stirred for 30 min. A solution of 2-tetralone (9.5 g, 95 mmol) in dry diethyl ether (30 ml) was then added and the mixture was allowed to reach room temperature, whereupon hydrochloric acid (5 N, 100 ml) was added. The mixture was refluxed for 3 h and cooled. Usual workup with diethyl ether, chromatography (*n*-hexane) and recrystallization from n-hexane yielded 3-bromo-2-methyl-5-(3′,4′-dihydro-naphthalen-2′-yl)-thiophene (7.8 g, 25.6 mmol, 41%), (m.p. 108 °C-^{1}H-NMR ($CDCl_3$): $\delta = 2.39$ (s, 3H, CH_3), 2.78 (m, 4-H, CH_2), 6.75 (m, 1H, 1′-H), 6.93 (s, 1H, 4-H), 7.13 (m, 4H, 5′-,6′-,7′-,8′-H). -^{13}C-NMR ($CDCl_3$): $\delta = 14.9$ (CH_3), 25.8, 27.8 (CH_2), 109.4 (C-3), 122.4 (C-1′), 125.6, 126.5, 126.7, 127.1, 127.3 (C-4, C-5′,-6′,-7′,-8′), 131.6, 133.0, 134.0, 134.7 (C-5, -2′,-4a′,-8a′), 142.0 (C-2). -$C_{15}H_{13}BrS$ (305.23): calcd. C 59.03, H 4.29, S 10.50: found C 59.30, H 4.32, S 10.66).

3-Bromo-2-methyl-5-(3′,4′-dihydro-naphthalen-2′-yl)-thiophene (6.7 g, 22 mmol) was aromatized with 2,3-dichloro-5,6-dicyanobenzoquinone (DDQ) (6.0 g, 26.4 mmol) in toluene (120 ml) at reflux temperature for 1 h. The mixture was cooled and filtered. The filtrate was washed with aqueous sodium hydroxide (5N, 3 × 50 ml), water and dried ($MgSO_4$). Evaporation of the solvent, chromatography (n-hexane) and recrystallization and cyclohexane yielded 3-bromo-2-methyl-5-(naphthalen-2′-yl)-thiophene (5.2 g, 17.1 mmol, 78%), (m.p. 117 °C-^{1}H-NMR ($CDCl_3$): $\delta = 2.44$ (s, 3H, CH_3), 7.23 (s, 1H, 4-H), 7.46 (m, 2H, 6′-, 7′-H), 7.62 (m, 1H, 3′-H), 7.82 (m, 3H, 4′-, 5′-, 8′-H), 7.94 (m, 1H, 1′-H).-^{13}C-NMR ($CDCl_3$): $\delta = 14.9$ (CH_3), 110.0 (C-3), 123.7, 123.8, 126.0, 126.1, 126.7, 127.7, 128.0, 128.7 (C-4, -1′,-3′,-4′,-5′,-6′,-7′,-8′), 130.9, 132.9, 133.6, 133.9 (C-5,-2′,-4a′,-8a′), 141.2 (C-2).-$C_{15}H_{11}BrS$ (303.21): calcd. C 59.42, H 3.66, S 10.57; found C 59.34, H 3.69, S 10.50).

s-Butyllithium (1.3 M in hexane/cyclohexane, 5.3 ml, 6.9 mmol), 3-Bromo-2-methyl-5-(naphthalen-2′-yl)-thiophene (2.0 g, 6.6 mmol) in dry diethyl ether (50 ml) and octafluorocyclopentene (0.70 g, 3.3 mmol) were reacted as described for the preparation of BTCP2. Workup, purification by chromatography (cyclohexane/dichloromethane 4:1) and recrystallization from cyclohexane yielded BTCP3 (1.25 g, 2.0 mmol, 61%) as colorless crystals (m.p. 157 °C-^{1}H-NMR ($CDCl_3$): $\delta = 2.04$ (s, 6H, CH_3, 7.43 (s, 2H, 4′-H), 7.53 (m, 4H, aryl-H), 7.62 (m, 2H, aryl-H), 7.82 (m, 6H, aryl-H), 7.98 (m, 2H, aryl-H).-^{13}C-NMR ($CDCl_3$): $\delta = 14.6$ (CH_3), 111.2 (C-4), 116.3 (C-3,-5), 122.9 (C-4′), 126.1 (C-3′), 123.8, 124.1, 126.3, 126.7, 127.8, 128.0,

128.8, 130.8, 132.9, 133.6 (C-naphthalenyl), 136.4 (C-1,-2) 141.5 (C-5′), 142.4 (C-2′).-$C_{35}H_{22}F_6S_2$ (620.67): calcd. C 67.73, H 3.57, S 10.33; found C 67.83, H 3.63, S 10.32).

1,2-Bis-(2′-dodecyl-thien-3′-yl)-hexafluorocyclopentene (BTCP5): Diisopropylamine (2.5 g, 25 mol) was added to a stirred solution of n-butyllithium (1.6 N in hexane, 13.8 ml, 22 mmol) in dry tetrahydrofuran (30 ml) at −30 °C under a nitrogen atmosphere. After 10 min 3-bromo-thiophene (3.0 g, 18.14 mmol) was added and the mixture was allowed to reach 0 °C, whereupon n-dodecyl iodide (5.5 g, 18.5 mmol) was added and stirring was continued for 12 h at room temperature. Usual workup with diethyl ether and distillation yielded 3-bromo-2-dodecyl-thiophene (4.4 g, 13.3 mmol, 72%) as viscous oil. (b.p. 145 °C/0.1 torr-^{1}H-NMR ($CDCl_3$): $\delta = 0.88$ (m, 3H, 12′-H), 1.26 (m, 18H, 3′-11′-H), 1.64 (m, 2H, 2′-H), 2.77 (m, 2H, 1′-H), 6.87 (d, $J = 5$ Hz, 1H, 4-H), 7.09 (d, $J = 5$ Hz, 1H, 5-H).-^{13}C-NMR ($CDCl_3$): $\delta = 14.1$ (C-12′), 22.7 (C-11′), 29.0–29.7 (C-2′-9′), 30.7 (C-10′), 31.9 (C-1′), 108.4 (C-3), 122.4 (C-5), 129.8 (C-4), 139.9 (C-2).-$C_{16}H_{27}BrS$ (331.35): calcd. C 58.00, H 8.21, S 9.68; found C 58.31, H 8.52, S 9.49).

A solution of 3-bromo-2-dodecyl-thiophene (4.0 g, 12.0 mmol) in dry diethyl ether (20 ml) was slowly added to a stirred solution of s-butyllithium (1.3 M in hexane/cyclohexane, 9.5 ml, 12.4 mmol) in dry diethyl ether (20 ml) at −70 °C under a nitrogen atmosphere and stirred for 30 min. Precooled octafluorocyclopentene (1.4 g, 6.6 mmol) was then added and the solution was kept at −70 °C for 2 h. The mixture was allowed to warm to room temperature and hydrochloric acid (0.1 N, 10 ml) was added. Usual workup with diethyl ether, chromatography (n-hexane) yielded BTCP5 (2.1 g, 3.1 mmol, 52%) as viscous oil. (^{1}H-NMR: ($CDCl_3$): $\delta = 0.88$ (m, 6H, 12″-H), 1.26 (m, 36H, 3″-11″-H), 2.17 (m, 4H, 2″-H), 2.75 (m, 4H, 1″-H), 7.03 (d, $J = 5$ Hz, 2H, 4′-H), 7.20 (d, $J = 5$ Hz, 2H, 5′-H).-^{13}C-NMR ($CDCl_3$): $\delta = 14.1$ (C-12″), 22.7 (C-11″), 28.8 (C-2″), 29.2–29.6 (C-3″-9″), 31.3 (C-10′), 31.9 (C-1″), 111.1 (C-4), 116.2 (C-3,-5), 123.5 (C-5′), 123.9 (C-3′), 126.9 (C-4′), 136.7 (C-1,-2,), 148.2 (C-2′). -$C_{37}H_{54}F_6S_2$ (676.94): calcd. C 65.65, H 8.04, S 9.47; found C 65.45, H 8.16, S 9.40).

1,2-Bis-[2′-(4″,4″-dimethyl-4″,5″-dihydro-oxazol-2″-yl)-thien-3′-yl)]-hexafluoro-cyclopentene (BTCP6): n-Butyllithium (1.6 M in hexane, 15.0 ml, 24 mmol) was slowly added to a stirred solution of 4,4-dimethyl-2-(thiophen-2′-yl)-4,5-dihydro-oxazole [26] (4.1 g, 22.6 mmol) in dry diethyl ether (80 ml) at −70 °C under a nitrogen atmosphere, stirred for 15 min at −70 °C and for 30 min at 0 °C. Precooled octafluorocyclopentene (2.4 g, 11 mmol) was then added at −70 °C and the solution was kept at −70 °C for 2 h. The mixture was allowed to warm to room temperature and water (80 ml) was added. Usual workup with diethyl ether, chromatography (dichloromethane/ethyl acetate 3:1) and recrystallization from cyclohexane yielded BTCP6 (2.5 g, 4.7 mmol, 43%) as colorless crystals (m.p. 180 °C-^{1}H-NMR ($CDCl_3$): $\delta = 1.28$ (s, 6H, CH_3), 4.00 (s, 4H, CH_2), 6.97 (d, $J = 5$ Hz, 2H, C-4′), 7.34 (d, $J = 5$ Hz, 2H, C-5′).-^{13}C-NMR ($CDCl_3$): $\delta = 28.1$ (CH_3), 67.8 (C-4″), 79.5 (C-5″), 111.2 (C-4), 115.6 (C-3,-5), 128.6, 129.1 (C-4′,-5′), 128.7, 130.4 (C-2′,-3′), 138.8 (C-1,-2), 156.4 (C-2″). -$C_{23}H_{20}N_2O_2F_6S_2$ (534.53): calcd. C 51.68, H 3.77, N 5.24, S 12.00; found C 51.93, H 3.83, N 5.16, S 11.81).

Solution and samples

The aqueous micellar solutions used in the spin probe experiments (cf. Table 1) were prepared in 5 ml volumetric flasks by weighing in the necessary ingredients, adding the desired amount of water, and sonicating until a clear solution appeared. 300 µl of these solutions were placed in small vials already containing the spin probes according to the following procedure: Appropriate amounts of ca. 10^{-4} molar chloroform solutions of the probes were filled into the vials and the solvent evaporated in a gentle stream of nitrogen. Volatile spin probes (Tempo) were added directly to the surfactant solutions. Final spin probe concentrations were about 10^{-5} molar.

Lyotropic liquid crystalline samples were prepared in sealed vials after weighing in the necessary ingredients. The sealed vials were kept at ca. 70 °C until a homogeneous solution was formed.

ESR-equipment

ESR-spectra were recorded on a Bruker 200 D (X-Band) spectrometer at a modulation frequency of 100 kHz. Parameters for measuring spin probe spectra the following parameters were chosen: modulation amplitude 1.25 Gpp, signal gain $1.25 \cdot 10^5$–$3.2 \cdot 10^5$, time constant 2 s, sweep time 1000 s, microwave frequency of 9.44 GHz. The desired temperature was controlled by a Bruker variable temperature unit (ST 100/700).

Irradiations

A 100 W high pressure mercury lamp or a 150 W xenon lamp (Osram) served as irradiation source. Irradiations were performed either directly in an apparatus for determining transition temperatures (clearing points) [14] or separately in cuvettes or in cells used for the polarizing microscope. Irradiation times were chosen so that photochemical equilibria of the photochromic systems were reached (about 30 min, depending on the irradiation wavelengths, on optical path lengths, on absorption coefficients of the respective photoisomers and on quantum yields of forward and back reactions). Appropriate filters were used to exclude side and back reactions as far as possible. For the ESR experiments photochemical conversion was simulated by solubilizing photoeduct and externally prepared photoproduct (leading to the same viscosities as *in situ* irradiations).

Determination of phase transition temperatures (clearing points)

Transition temperatures were measured in a computer-controlled apparatus described previously [14] or in a similar non-automatic apparatus [17] or with a polarizing microscope (Carl Zeiss, Jena) watching LLC textures [27] as a function of temperature.

Results

Spin probe studies on photorheological effects induced by the photocyclization of *N*-methyldiphenylamine

A typical ESR-spectrum of a nitroxide radical in aqueous micellar solution consists of three lines due to coupling of the spin of the free electron with the nuclear spin of the nitrogen atom. The coupling constant $\langle A_N \rangle$ in nitroxide radicals depends on the polarity of the environment, i.e., polar solvents increase $\langle A_N \rangle$ as compared to nonpolar one [28, 29]. Thus, changes of $\langle A_N \rangle$ as comprised in Table 2 reflect the polarity of the microenvironment of the probes.

Since ESR-spectra (absorption vs. magnetic field strength) are recorded as first derivatives, the heights of the observed three lines give the widths ΔH of the respective hyperfine transitions, which can be expressed by

$$\Delta H(m_N) = A + B(m_N) + C(m_N), \tag{1}$$

where superhyperfine transitions (due to coupling with nuclear spins of hydrogen atoms) are neglected [30, 31]. A is equal to the central line width $\Delta H(0)$ and contains non motional terms only, while B and C are parameters describing differential broadening and motional correlation times τ_B and τ_C which in the limit of isotropic Brownian motion agree with the Debye-Stokes-Einstein reorientational correlation time

$$\tau = 4\pi a^3/(3kT) = \tau_B = \tau_C. \tag{2}$$

When the motion is anisotropic τ_B and τ_C may differ from τ and from each other. B and C are related to experimental line height h_+, h_0, and h_- (upfield, center, and downfield line of the triplet signal) according to

$$B = \frac{1}{2}\Delta H(0)[\sqrt{h_0/h_+} - \sqrt{h_0/h_-}] \tag{3}$$

and

$$C = \frac{1}{2}\Delta H(0)[\sqrt{h_0/h_+} + \sqrt{h_0/h_-} - 2] \tag{4}$$

In the limits $2 \cdot 19^{-9}\,\text{s} > \tau > 5 \cdot 10^{-11}\,\text{s}$ approximate relations for τ_B and τ_C [30–33] for nitroxide radicals are

$$\tau_B = -5.85 \times 10^{-10} (\text{s}/T)\,\Delta H(0)\,[\sqrt{h_0/h_+} - \sqrt{h_0/h_-}] = (-5.85 \cdot 10^{-10} \times 2B)\,\text{s} \tag{5}$$

and

$$\tau_C = 6.6 \times 10^{-10} (\text{s}/T)\,\Delta H\,[\sqrt{h_0/h_+} + \sqrt{h_0/h_-} - 2] = (6.6 \times 10^{-10} \times 2C)\text{s}. \tag{6}$$

Values for τ_B and τ_C are collected in Table 2 for the spin probes investigated in pure water and in the aqueous micellar systems. Inspection of the table reveals that (with the exception of the probe TMB +) all the correlation times increase in micellar media by at least an order of magnitude indicating an accordingly increased microviscosity in the molecular vicinity of the probes.

Photochemically induced phase transitions in LLC systems

Cuts of the phase diagrams of the system AOT-formamide and AOT-glycerol were measured and are reproduced in Figs. 1 and 2, respectively. The AOT-formamide-diagram coincides fairly well with a literature diagram [34], in which Bergenståhl et al. assign the anisotropic phase lamellar at around 70% w/w AOT. The anisotropic phase in the AOT-glycerol system was assigned lamellar according to textures obtained using a polarizing microscope.

The solubilization of 0.5% w/w of various BTCPs and their photoisomerization in aqueous CTAB, AOT-formamide, and AOT-glycerol changes transition temperatures as listed in Table 3. In the case of CTAB 0.5% of BTCP is already near to the solubility limit. Irradiation of the systems, i.e. converting the bis-(thienyl)cyclopentenes (BTCPs) to the colored isomers, causes only small changes of transition temperatures (in both directions). Although BTCPs are famous for their clean, photochemically reversible cyclization, we observed a side reaction in BTCP1, BTCP2, and BTCP3 when irradiated polychromatically in aqueous CTAB. The side product is characterized by an intensive absorption band around 260 nm and its formation prevents photocyclization or causes decoloration of systems containing cyclized BTCP. The formation of the

Table 2 Coupling constants $\langle A_N \rangle$, central linewidths $\Delta H(0)$, and motional correlation times τ_B and τ_C in pure water and in the micellar systems characterized in Table 1

Sample	$\langle A_N \rangle/10^4$ T	$\Delta H/10^4$ T	$\tau_B/10^{-9}$ s	$\tau_C/10^{-9}$ s
Water/Tempo	17.23	1.3	0.0129	0.0635
Water/Tempol	17.00	1.6	0.0270	0.0422
Water/Tempone	17.07	0.5	0.00737	0.0122
Water/C8Tempo	16.92	1.4	0.0367	0.0854
Water/TMA +	16.75	1.6	0.0719	0.105
Water/C8 +	16.76	1.7	0.0490	0.242
Water/C16 +	16.74	1.8	0.0777	0.015
CTAB/Tempo	16.46	1.33	1.54	1.37
CTAB/MDPA/Tempo	16.40	1.37	1.79	1.42
CTAB/MC/Tempo	16.41	1.34	1.78	1.52
TRI/Tempo	16.46	1.36	0.300	0.225
TRI/MDPA/Tempo	16.37	1.39	0.287	0.239
TRI/MC/Tempo	16.40	1.39	0.299	0.237
CTAB/Tempol	16.69	1.63	0.280	0.248
CTAB/MDPA/Tempol	16.72	1.70	0.276	0.233
CTAB/MC/Tempol	16.70	1.72	0.291	0.263
TRI/Tempol	16.74	1.58	0.125	0.0887
TRI/MDPA/Tempol	16.70	1.61	0.121	0.0892
TRI/MC/Tempol	16.70	1.65	0.121	0.0958
CTAB/Tempone	15.68	0.64	0.294	0.272
CTAB/MDPA/Tempone	15.65	0.68	0.306	0.279
CTAB/MC/Tempone	15.67	0.67	0.295	0.272
TRI/Tempone	15.75	0.42	0.104	0.0817
TRI/MDPA/Tempone	15.73	0.50	0.134	0.108
TRI/MC/Tempone	15.74	0.55	0.138	0.111
CTAB/C8 Tempo	16.21	1.66	0.600	0.591
CTAB/MDPA/C8Tempo	16.12	1.68	0.652	0.618
CTAB/MC/C8Tempo	16.20	1.68	0.653	0.634
TRI/C8 Tempo	16.34	1.90	0.806	0.996
TRI/MDPA/C8Tempo	15.96	1.93	0.771	0.954
TRI/MC/C8Tempo	16.01	1.96	0.897	1.11
CTAB/TMA +	16.75	1.61	0.0719	0.105
CTAB/MDPA/TMA +	16.75	1.61	0.0719	0.105
TRI/TMA +	16.75	1.60	0.0719	0.015
TRI/MDPA/TMA +	16.76	1.60	0.0719	0.015
TRI/MC/TMA +	16.74	1.60	0.0719	0.015
CTAB/C8 +	16.48	1.82	0.505	0.434
CTAB/MDPA/C8 +	16.65	2.09	0.481	0.411
CTAB/MC/C8 +	16.54	1.81	0.419	0.365
TRI/C8 +	16.53	1.82	0.431	0.343
TRI/MDPA/C8 +	16.50	1.90	0.470	0.377
TRI/MC/C8 +	16.47	1.88	0.476	0.372
CTAB/C12 +	16.19	1.93	1.11	0.95
CTAB/MDPA/C12 +	16.05	2.25	1.44	1.24
CTAB/MC/C12 +	16.13	2.28	1.48	1.27
TRI/C12 +	16.23	2.25	1.52	1.16
TRI/MDPA/C12 +	16.08	2.40	1.83	1.43
TRI/MC/C12 +	16.23	2.25	1.54	1.19
CTAB/C16 +	16.16	2.25	1.46	1.22
CTAB/MDPA/C16 +	16.07	2.30	1.51	1.28
CTAB/MC/C16 +	16.12	2.30	1.53	1.27
TRI/C16 +	16.20	2.32	1.60	1.23
TRI/MDPA/C16 +	16.18	2.33	1.58	1.19
TRI/MC/C16 +	16.15	2.36	1.58	1.15

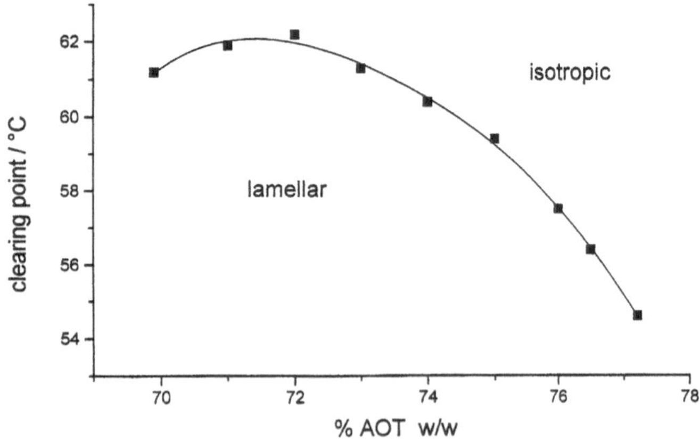

Fig. 1 Cut of the binary diagram AOT-formamide

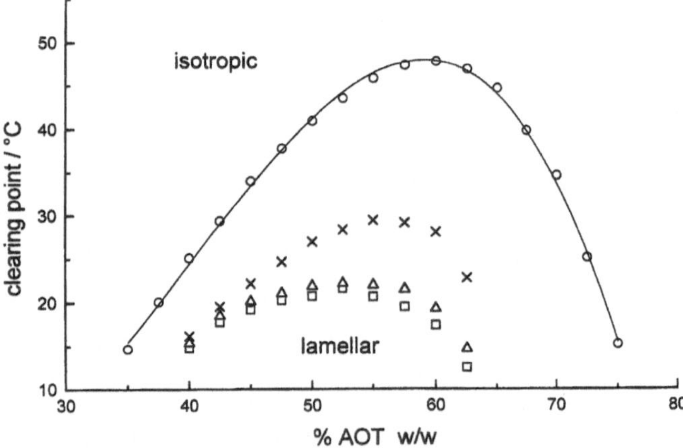

Fig. 2 Cut of the binary diagram AOT-glycerol. ○: Pure AOT-glycerol; □: AOT-glycerol + 3% w/w C1-HSB; ×: after irradiation, i.e photochemical *trans* → *cis*-isomerization of C1-HSB; △: after thermal re-isomerization at T: 80 °C

side product is strongly suppressed by the use of a Schott-UG 5 band pass filter, which precludes visible irradiation.

In the non-aqueous systems AOT-glycerol and AOT-formamide effects of solubilizing BTCPs and photo-isomerizing them are also quite limited (Table 3). In some of the systems demixing of samples and formation of colored zones occur after irradiation indicating that the colored ring closed from of BTCP is less soluble than the ring opened isomer.

As illustrated by Fig. 2, much more sensitivity to the presence of solubilizates is found when hydroxystilbazolium bromides are solubilized in AOT-glycerol: the phase transition temperature is depressed by more than 30 °C in the presence of 3% methylhydroxystilbazolium bromide (C1-HSB); the maximum of the phase equilibrium line is shifted to lower AOT content. Photochemical *trans* → *cis* isomerization reverts the effect, i.e. clearing points rise upon irradiation by up to 10 °C. This effect is more pronounced in the AOT-rich part of the lamellar region. Upon keeping samples at 80 °C over night the photoreaction thermally reverts to a great extent, indicating that *cis-trans* isomerization is the main photoreaction in the lamellar systems – rather than $2\pi + 2\pi$-dimerization. This can be taken as an indirect support for the presence of a lamellar phase, since in inverse hexagonal AOT systems dimerization competes more favorably due to a closer packing of the chromophore (as revealed by UV/VIS-absorption spectra [35]).

An interesting difference becomes apparent, when in the AOT-glycerol system the solubilizate C1-HSB is substituted by *N*-cetyl-hydroxystilbazolium bromide (C16-HSB, see Fig. 3): Irradiation, i.e., *trans* → *cis* isomerization,

Table 3 Transition temperature lamellar-isotropic (clearing points) in CTAB-water (23:77 w/w), AOT-glycerol (60:40 w/w) and AOT-formamide (72:28 w/w) in the presence of 0.5% w/w of various bis-(thienyl)-hexafluorocyclopentenes (BTCPs)

Solubilizate	Clearing point/°C non irradiated	Transition temperature/°C irradiated
	CTAB	
–	33.0	–
BTCP1	34.8	34.3
BTCP2	44.0	43.6
BTCP3[1]	39.4	38.9
	AOT-glycerol	
–	53.4	–
BTCP1	51.4	51.6
BTCP2	51.2	50.6[2]
BTCP3	51.2	50.6[2]
BTCP5	49.0	–[3]
BTCP6	51.4	51.0
BTCP4	49.9	50.2
	AOT-formamide	
–	61.9	–
BTCP1	60.8	–[3]
BTCP2	61.4	–[3]

[1] < 0.5% soluble; [2] partial demixing [3] demixing

causes an increase of transition temperature at high AOT/glycerol ratios and a decrease at low ratios.

Discussion

Spin probe investigations

The spin probe Tempone is clearly dissolved in the aqueous phase, since coupling constants $\langle A_N \rangle$ coincide in all micellar systems investigated. However, in contrast to other probes residing in the bulk phase (e.g., TMA +) the motional correlation times τ_B and τ_C are larger by an order of magnitude as compared to the values in pure water. We, therefore, can conclude that Tempone is solubilized in an aqueous surrounding which differs from bulk water such as the hydration water adjacent to the micelle surface. A comparison of viscosities in Table 1 and correlation times in Table 2 reveals the latter follow tendentiously the former only in the case of Tempone (and the correlation time variation is well out of error limits for TRI/Tempone). This result is in keeping with the previous interpretation of photorheological effects that extent and/or rigidity of the hydration water is responsible for the observed viscosity effects. A quantitative agreement between macroscopic viscosities and microviscosities (to be calculated from correlation times using Eq. (5)) cannot be expected as bulk viscosity is not affected by the solubilizates (cf. τ-values for TMA + in water and micellar solutions).

Tempol is another probe whose coupling constant is not significantly affected by the presence of MDPA or MC indicating its location in the aqueous phase. Due to the polar OH-group the molecule can be expected to reside in the vicinity of the micelle surface. This is in agreement with the strong correlation time increase in micellar solutions as compared to the pure aqueous solution. However, the correlation times vary only within errors limits ($\pm$ 10%). The reason for that can be the capability of Tempol to form hydrogen bridge bonds, thereby interfering with the hydration layer (ordered by hydrogen bridges) via specific interactions.

All the probes not jet discussed seem to be associated with the micelles as revealed by their differing coupling constants in the various micellar media and in some cases also by distorted triplet signals indicating slow and anisotropic motion. The nitroxide radical moieties of the probe molecules can be expected near the surface either due to their inherent polarity or due to the positive charge in the vicinity of the NO-group. With two exceptions (CTAB/C8 + and TRI/C16 +), in these cases a significant decrease of the coupling constant is observed in the presence of MDPA but not when MC is added. We may conclude that MDPA is solubilized in the surface region of the micelles and decreases the polarity there while MC is buried deeper in the micelles. This conclusion is strongly supported by the fact that MC is a much less polar molecule than MDPA [5].

Phase transitions

In a previous study we found an influence of the solubilizate shape on transition temperatures in CTAB-H_2O and potassium myristate systems [36] and suggested this effect to be one the origins of photochemically switching phases: photoisomers of globular shape induce transition temperatures exceeding those induced by according longish isomers. The results obtained using BTCPs in AOT-glycerol seem to corroborate this interpretation (Table 3) since the photoisomers differ only moderately in shape as revealed by evaluation of force field calculations (MM2 algorithm). The comparatively small amplitudes of photochemically induced clearing point shifts probably prevent the practical use of these systems in spite of the known fortunate features of the BTCP photochromes: thermal irreversibility accompanied by excellent multiple photochemical reversibility [37]. The observed side-reaction in CTAB systems upon polychromatic irradiation may be a $2\pi + 2\pi$-photodimerization of the ring opened form at the cyclopentene double bond, caused by specific packing of BTCP in the CTAB aggregates. This reaction would be favored in polychromatic irradiations since ring opened molecules consumed by dimerization are replenished by the visible part of the irradiation light, i.e., the dynamic photochemical equilibrium of ring closed and ring opened material is readjusted. In keeping is the reduction of this latter effect when an UG-5 filter (not transparent to visible light) is used.

C1-HSB is well soluble in glycerol and can be expected to be solubilized in the glycerol layers of the lamellar phases. Therefore, its generally destabilizing influence on the lamellar phase is less in the glycerol-rich region (left side) of the diagrams in Fig. 2 as compared to the glycerol-poor region (right side), where the solubilizate has to interact more strongly with the AOT-aggregates. Accordingly, the most marked effects of the photochemical *trans* → *cis* isomerization of C1-HSB on clearing points are found in the AOT-rich region. In contrast, C16-HSB, which is much less soluble in glycerol, should be solubilized associated to the AOT double layers with at least the cetyl chain incorporated. Thus, in the glycerol rich region *trans*-C16-HSB can act like a cotenside stabilizing the anisotropic phase as shown by a slight clearing point increase at 40% AOT-glycerol (Fig. 3). Consequently, *trans* → *cis* isomerization causes a destabilization of the lamellar phase. By that C16-HSB in this

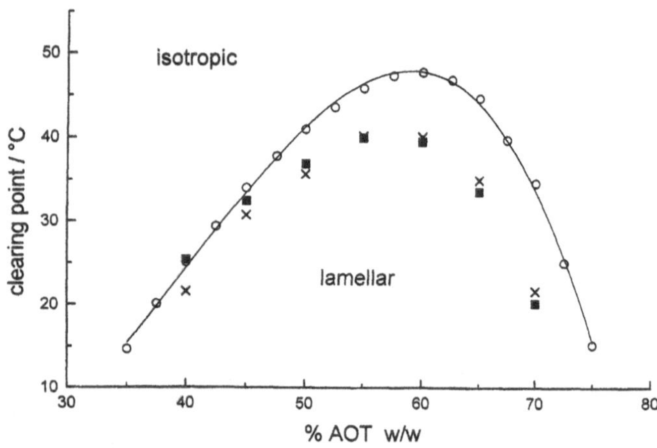

Fig. 3 Cut of the phase diagram AOT-glyerol. O: Pure AOT-glycerol; ■: AOT-glycerol + 1% w/w C16-HSB; ×: after irradiation, i.e photochemical *trans* → *cis*-isomerization of C16-HSB

specific system contrasts to all *cis-trans*-isomerizing photochromes (stilbene derivatives, etc.) investigated in lyotropic liquid crystalline systems so far [14, 16, 19, 36]. In the AOT-rich region, however, C16-HSB behaves like other stilbenes, i.e., conversion to the *cis*-form increases transition temperatures.

Acknowledgements Financial support by the Deutsche Forschungsgemeinschaft (DFG) and by the Fonds er Chemischen Industrie is gratefully acknowledged. A part of this work stems from a project of the Graduiertenkolleg "Struktur-Eigenschaftsbeziehungen bei Heterocyclen" (financed by the DFG and the Sächsisches Ministerium für Wissenschaft und Kunst). One of us (TW) is grateful to CNRI for the grant of a stay in Florence, Italy. Thanks are due to Professors H. Hoffmann, Bayreuth, and H.-D. Dörfler, Dresden, for helpful discussions concerning the assignment of LLC phase structures.

Appendix

TEMPO, TEMPOL, TEMPONE spin probes

These probes have coupling constant $\langle A_N \rangle$ independent of the micellar system investigated. However, micellization leads to a decrease of $\langle A_N \rangle$ with respect to pure water, indicating a decrease of the probe environment polarity. This result is also confirmed by the motional correlation times, τ_b and τ_c, which differ for TEMPO, TEMPOL, and TEMPONE by one order of magnitude or more when micellar and pure aqueous systems are compared. For CTAB and TRITON micelles, the polarity changes "seen" by the probes after micellization decrease in the order TEMPONE > TEMPO > TEMPOL and do not depend on the presence of MDPA and MC. Therefore, the analysis of the coupling constants, $\langle A_N \rangle$, and the evaluation of the correlation times, τ_b and τ_c, of these probes suggest the following consideration: the three spin probes "see" the micellar surface, but their average distance from this interface is in the order TEMPONE < TEMPO < TEMPOL, i.e. TEMPONE is the probe nearest to the micellar surface. In order to explain this we calculated the dipolar moments (by using the software Hyperchem, supplied by Autodesk, Sausalito, CA, U.S.A.) of the three spin probes, TEMPO, TEMPOL, and TEMPONE. The dipolar moments are in the order TEMPONE (1.135 D) < TEMPOL (2.556 D) < TEMPO (2.682 D). TEMPONE being significantly less polar than the other molecules can interact and better detect the decrease of polarity due to the micellization process, in agreement with the experimental results. Thus TEMPONE is the best probe to detect microviscosity changes induced by MDPA or MC.

References

1. Wolff T, Emming C-S, Suck TA, von Bünau G (1989) J Phys Chem 93:4984
2. Wolff T, Kerperin KJ (1993) J Colloid Interface Sci 157:185
3. Kerperin KJ, Wolff T (1993) Ber Bunsenges Phys Chem 97:36
4. Wolff T, Emming C-S, von Bünau G (1991) J Phys Chem 95:3731
5. Grellmann K-H, Kühnle W, Weller H, Wolff T (1981) J Am Chem Soc 103:6889
6. Wolff T, Emming C-S, von Bünau G, Zierold K (1992) Colloid Polym Sci 270:1222
7. Wolff T (1994) in "Structure and Flow in Surfactant Solutions", Herb CA and Prud'homme RK (eds), ACS Symposium Series No 578, chapter 12, pp 181–191
8. Ekwall P, Mandell L, Solyom P (1971) J Colloid Interface Sci 35:519
9. Mandal AB, Ray S, Biswas AM, Moulik SP (1980) J Phys Chem 84:856
10. Ottaviani MF, Baglioni P, Martini G (1983) J Phys Chem 87:3146
11. Baglioni P, Ferroni E, Martini G, Ottaviani MF (1984) J Phys Chem 88:5107
12. Baglioni P, Ottaviani MF, Martini G (1986) J Phys Chem 90:5878
13. Tiddy GJT (1980) Phys Rep 57:1
14. Wolff T, Klaußner B, von Bünau G (1990) Progr Colloid Polym Sci 83:176
15. Wolff T, von Bünau G (1984) Ber Bunsenges Phys Chem 88:1098
16. Wolff T, Seim S, Klaußner B (1991) Liq Cryst 9:839
17. Wolff T (1989) Colloid Polym Sci 267:345
18. Wolff T, Klaußner B (1995) Adv Colloid Interface Sci 59:31
19. Wolff T, Klaußner B, Nees D (1993) Ber Bunsenges Phys Chem 97:1407
20. Park D, Rogers J, Toft RW, Winsor PA (1970) J Colloid Interface Sci 32:81
21. Minch MJ, Sadiq Shah S (1977) J Chem Educ 54:709
22. Donchi KF, Robert GP, Ternai B, Derrick PJ (1980) Austr J Chem 33:2199
23. Saika T, Irie M, Shimidzu T (1994) J Chem Soc Chem Commun 2123
24. Gronowitz S, Frejd T (1976) Acta Chem Scand B 30:485
25. Lantz B, Hörnfeldt A-B (1972) Chem Scr 2:9

26. Della Vecchia L, Vlattas I (1977) J Org Chem 42:2649
27. Rosevear FB (1954) J Am Oil Chem Soc 31:628
28. Stout G, Engberts JBFN (1974) J Org Chem 39:3800
29. Knauer RR, Naples JJ (1976) J Am Chem Soc 98:4395
30. Kivelson D (1960) J Chem Phys 33:1094
31. Jolicoeur C, Friedman HL (1971) Ber Bunsenges Phys Chem 75:248
32. Ernandes JR, Schreier S, Chaimovich Y (1976) Chem Phys Lipids 16:14
33. Martinie J, Michou J, Rassat A (1975) J Am Chem Soc 97:1818
34. Bergenståhl B, Jønsson A, Sjøblom J, Stenius P, Wärnheim T (1987) Progr Colloid Polym Sci 74:108
35. Nees D, Wolff T to be published
36. Klaußner B, Nees D, Schmidt F, Wolff T (1994) J Colloid Interface Sci 162:481
37. Nakamura S, Irie M (1988) J Org Chem 53:6136

Progr Colloid Polym Sci (1996) 101:86–92

A. Zosel
G. Ley

Film formation from polymer latices

Dr. A. Zosel (✉) · G. Ley
Kunststofflaboratorium
BASF AG
67056 Ludwigshafen, FRG

Abstract In most of their applications, emulsion polymers are in the form of a coherent film. Consequently, the formation of films from polymer latices is a subject of great importance. The present report reviews the current state of the art and recent work on the subject.

Film formation from aqueous latices can be regarded as a process consisting of three stages: concentration, deformation, and coalescence as a result of interdiffusion. It is evident from recent literature that the second stage, i.e. the deformation of particles into polyhedrons and their dense packing, is still the subject of particularly controversial discussion, whereas substantial progress has been achieved in studies on the interdiffusion phase. Other questions that remain open are the formation of films from latices with structured particles, the location of auxiliaries in the film, and the formation of films from latices containing pigments or fillers.

Key words Polymer latices – film formation – coalescence – interdiffusion

Introduction

Although they are produced by emulsion polymerization in the aqueous phase, polymer latices are encountered as coherent films in almost all their applications. The films are transparent in many cases, and the high level of their mechanical properties is frequently on a par with that of films applied in the form of a polymer solution or melt. The great significance that is thus attached to film formation from aqueous dispersions can be explained by the fact that the final performance characteristics of the emulsion polymers are developed while the film is still being formed.

It can be seen from the model presented in Fig. 1 that film formation can be regarded as a three-stage process consisting of

1) evaporation of the water leading to concentration and dense packing of the latex particles

2) deformation of the particles into polyhedrons under the action of surface and capillary forces

3) coalescence of the particles as a result of interdiffusion of chain segments and ends across the particle boundaries.

All three stages are discussed below, and particular attention has been devoted to particle coalescence. Other subjects that are briefly dealt with are the whereabouts of the low-molecular-weight auxiliaries and water-soluble components, formation of films from dispersions consisting of structured particles and from pigmented latices.

Concentration and packing of latex particles

The first stage in film formation is the increase in the concentration of the latex as a result of the evaporation of water. Ever since Luck, Klier and Wesslau [1] reported on

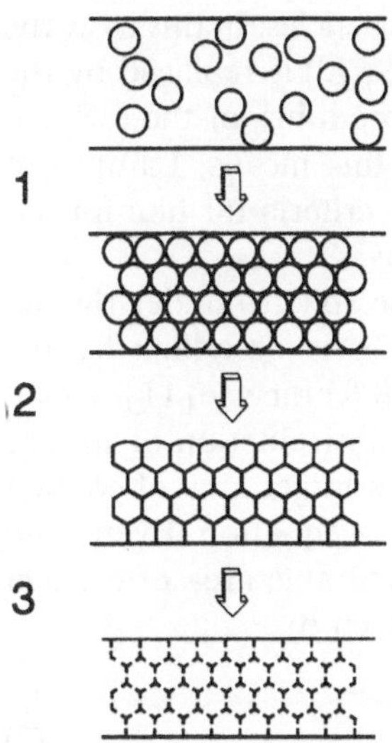

Fig. 1 Schematic diagram illustrating the formation of a film from an aqueous polymer latex

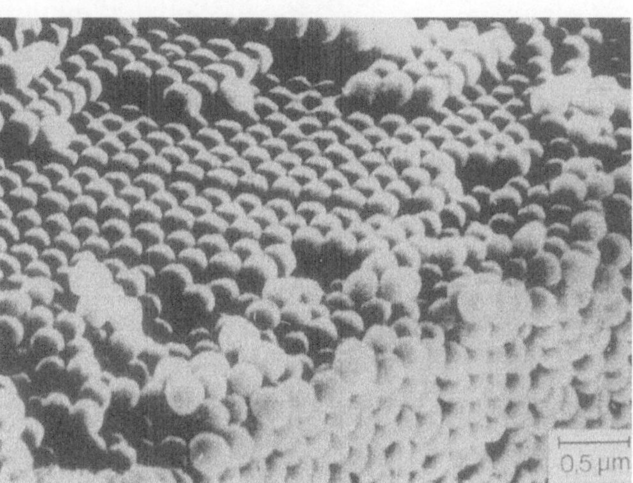

Fig. 2 Scanning electron micrograph of packed polystyrene latex particles

Bragg scattering in opalescent dispersions, it has been known that regularly spaced, crystalline zones may be formed during this stage. The example given in Fig. 2 is a scanning electron microscope photograph in which the dense packing of the particles of a polystyrene latex with a relatively narrow particle size distribution can be seen. To a certain extent, this structure is fixed, because the polystyrene particles cannot coalesce owing to their high glass transition temperature. Regular crystal planes separated by dislocations can be clearly seen alongside zones of greater lattice disorder. Recently, crystalline zones were also detected by small-angle neutron scattering (SANS) in D_2O containing latices during drying [2] and in films into which heavy water diffused after drying [3, 4].

Obviously, the development of crystalline phases is disturbed as the particle size distribution becomes more polydisperse and would even be impossible once the polydispersity exceeded a "critical" limit. At the instant of its immobilization, a film obtained from a polymer latex consists of

- crystalline zones of various sizes consisting predominantly of face-centered cubic structures
- crystal boundaries
- disordered zones solidified to a glassy state
- defects and holes.

As film formation proceeds, this structure can shrink further only as a whole. The spatial arrangement of the latex particles and the relevant position of the interstitial spaces, defects and holes are retained [5].

In other words, the structure of the film that is subsequently formed and that exerts an influence on the final properties is already fixed in rough outlines in this early stage of film formation.

Deformation and sintering of the latex particles

The second stage in the film-forming process described in the simplified model shown in Fig. 1 comprises the period of time that elapses between the instant in which the particles first contact one another until the instant in which they are completely deformed into polyhedrons and the film is compacted into the "anhydrous" state. The starting point in studies on this phase is the empirically determined fact that polymer latices have what is known as a minimum film-forming temperature (MFT), below which a cohesive film cannot be formed. Numerous investigations have demonstrated that the MFT is connected with the glass transition temperature of the polymer or, to be more exact, that of the polymer that has been swollen by water in the latex up to the saturation point.

The search for models to represent the deformation phase commenced in 1951 with the work of Dillon, Matheson and Bradford [6], who described particle coalescence as a deformation process in which the driving force is the surface tension of the particles. This force counteracts the viscose resistance of the polymer. According to the theory propounded by Frenkel [7], the increase in the radius a of the circle of contact for the coalescence of two liquid droplets (Fig. 3) is given by

$$a^2 = \frac{3R\gamma}{2\pi\eta} \cdot t, \quad (1)$$

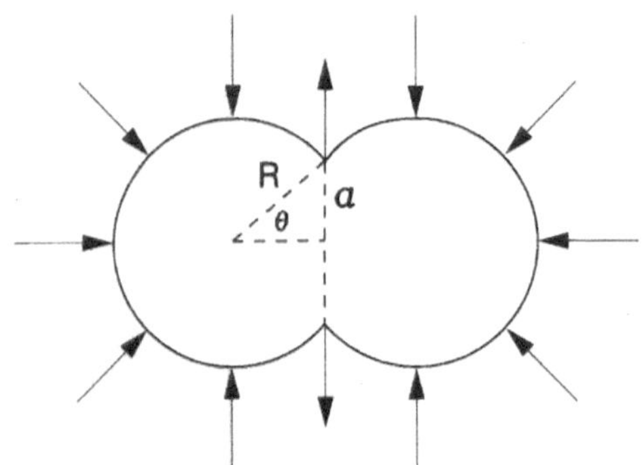

Fig. 3 Coalescence of two latex particles according to the Frenkel model [7]

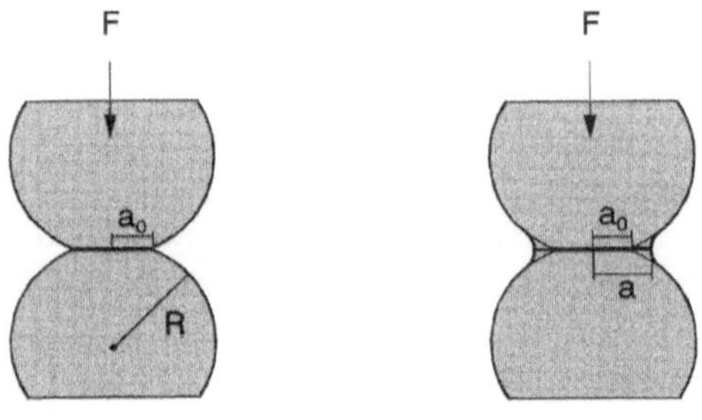

Fig. 4 Contact between two elastic spheres according to the Hertz [9] and the Johnson, Kendall and Roberts (JKR) theories [11]

where R is the particle radius, η is the viscosity, and γ is the surface tension of the polymer. In this model, the water is of no significance, and coalescence can thus be regarded as "dry sintering" of the particles. In the light of experimental evidence, Brown [8] raised two substantial objections against the idea of representing the deformation stage as dry sintering caused by viscous flow.

The first objection is that the resistance to particle coalescence originates not so much from viscous flow as from elastic deformation of the particles. Thus even (slightly) crosslinked particles also form films. The premise in this case is the Hertz theory [9] for the contact of elastic spheres under the action of an external force F. This is illustrated on the left-hand side of Fig. 4. For the case of identical spheres, the radius of the circle of contact a_0 is given by the following simplified equation

$$a_0^3 = \frac{3R}{4E} \cdot F, \tag{2}$$

where E is Young's modulus of the polymer.

The Hertz theory was subsequently extended to embrace viscoelastic spheres, a case that corresponds more to the conditions that relate to latex particles. In this case, the constant Young's modulus E in Eq. (2) is replaced by the time-dependent stress relaxation modulus, or the reciprocal of the creep compliance. By this means, Lamprecht [10] succeeded in formulating the criteria for film formation in more sharply defined terms.

The deformation of two elastic spheres under the action of external and surface forces is described by the Johnson, Kendall and Roberts (JKR) theory [11], which generalized the Hertz model. The introduction of surface and interfacial forces gives rise to a larger area of contact than that calculated from the Hertz equation. Even if no external force F is applied, a measurable area of contact can be found whose radius a is given by

$$a^3 = \frac{9\pi R^2 \gamma}{2E}, \tag{3}$$

where γ is the surface tension of the polymer.

The JKR model was successfully applied by Kendall [12] to the dry sintering of polystyrene latex particles above their glass transition temperature.

The second objection raised by Brown concerns the viewpoint that film formation can be regarded as dry sintering. It is based on the experimental finding that film formation progresses in step with the evaporation of water and is concluded on complete evaporation. The driving force postulated by Brown is the pressure that exists if the particles in the upper layers are tightly packed and the curvature of the serum meniscus between the particles is concave.

This "capillary pressure model" was afterwards corrected and advanced by various authors, e.g. Mason [13]. More recently, Eckersley and Rudin [14] formulated a "comprehensive" model for film formation, in which the radius of the circle of contact is the sum of those affected by the capillary forces and the surface tensions. Despite all efforts, it has not yet been clarified whether the deformation stage in film formation should be regarded as dry or wet sintering and what significance should be attached to the capillary forces. The fact that considerable discrepancies occur in the experimental evidence makes it all the more difficult to find a solution. A root of the trouble is undoubtedly the fact that studies of this stage in film formation are difficult: the temperature and the relative humidity must be accurately set, and the rates of evaporation must be exactly defined. Swelling and plasticizing of the polymers by water are major factors.

An example is given by two recent reports on the significance of water that have led to diametrically opposite conclusions. Sperry et al. [15] determined minimum film-forming temperatures (MFT) under various conditions, e.g. dried dispersion powder in a dry atmosphere and a latex in a humid atmosphere. They found that the conditions for film formation from latex particles containing hydrophobic polymers did not have any significant effect on the MFT, and regarded this finding as "compelling evidence for the lack of a special role of water in film formation".

As opposed to this, Lin and Meier [16] discovered in AFM studies on dispersion particle monolayers that film formation under moist conditions is about ten times more rapid than that under dry conditions. A question that remains unanswered in this case is whether the results thus obtained can be applied directly to normal films, which consist of many layers of particles. Further studies are required to clarify this discrepancy.

Coalescence of particles by interdiffusion

Interdiffusion in latex films is a subject that falls under the heading of diffusion in polymers and the attendant healing of fracture surfaces. This field of activity is mainly associated with the name of Voyutskii [17] in its early stages. In the last few years, it has great progress as a result of the work of de Gennes, Kausch, Tirrell and Wool [18–21], and the de Gennes [22] reptation model has provided a sound theoretical basis.

In investigations on latex films, it is difficult to regard interdiffusion separately from the preceding establishment of contacts between adjacent particles. This is because diffusion processes set in immediately after contact has been made if the polymers have adequate segmental mobility. In other words, interdiffusion may commence at different times at different parts on the surface of particles and different zones of the entire film.

One means of sharply delineating the commencement of the diffusion stage is to form, as far as possible, a film from the latex below the glass transition temperature of its polymer and, subsequently, to initiate interdiffusion at a defined instant of time by heating to a temperature above the glass transition.

In the last few years, two methods have become established for studying diffusion in latex films: small-angle neutron scattering (SANS) and fluorescence decay measurements. Originally, two groups resorted to SANS: Hahn, Ley et al. in BASF [23, 24] and Klein, Sperling et al. in Lehigh University, USA [25]. They were later joined by a third group in Rhône-Poulenc [26, 27]. Films with a small fraction, e.g. 1%, of deuterated particles are investigated in this technique. Owing to their low concentration, the radius of gyration of these deuterated particles can be measured as a function of the annealing time. The thickness d of the interdiffusion layer can be obtained from the increase in the radius of gyration.

In the second method, two latices with closely identical particles are mixed together in a ratio of 1: 1, and a film is formed from the mixture [28, 29]. The one kind of particle contains an energy donor with a fluorescent group, e.g. a phenanthrene; and the other, the corresponding acceptor, e.g. an anthracene compound. The nonradiative transfer of energy from the phenanthrene excited by a light source to the anthracene depends greatly on the spacing of both groups. If the groups interdiffuse, the rate of energy transfer increases and leads to a more rapid fluorescence decay, from which the length of interdiffusion path can be calculated.

The increase in mechanical strength that results from interdiffusion can be determined by mechanical measurements. A direct correlation between the length of the interdiffusion path and the mechanical strength has been obtained by measuring the mechanical strength of poly-*n*-butylmethacrylate (PBMA) latices which had also been subjected to SANS experiments [30, 31]. PBMA has proved to be a suitable model polymer for this purpose, because it forms a coherent, although very brittle film despite the fact that its glass transition temperature is 29 °C. It has been determined by neutron scattering experiments that measurable diffusion still does not occur at 23 °C [23].

Reports published by other working groups concern measurements on films obtained from polystyrene (PS) latices [25] and n-butylacrylate/methyl methacrylate copolymer (BA-MMA) latices [32] and also demonstrate the effect of interdiffusion on mechanical strength.

The procedure adopted for the experiments was to keep PBMA films that were produced at 23 °C at temperatures above the glass transition, e.g. 90 °C, for defined periods of time. The films were then subjected to tensile tests with uniaxial strain. The stress-strain ($\sigma - \varepsilon$) curves thus obtained allowed the tensile stress and elongation at break to be determined. The curves can be integrated to yield the failure energy w_B per unit volume, which describes the toughness of the material, i.e.

$$w_B = \int \sigma d\varepsilon \tag{4}$$

Various failure mechanisms can be deduced from the shape of the stress-strain curve. Figure 5 shows curves for PBMA films that were annealed at 90 °C for different periods of time. The unannealed film displays typical brittle fracture with low elongation at break and failure energy. This behavior is completely altered after annealing for merely 5 min. In this case, the film undergoes plastic deformation, and the elongation at break is greater by about two orders of magnitude. After brief annealing at a temperature above the glass transition, there is an abrupt transition from brittle to tough behavior. If the annealing time is lengthened, only a slight change takes place in the shape of the stress-strain curve.

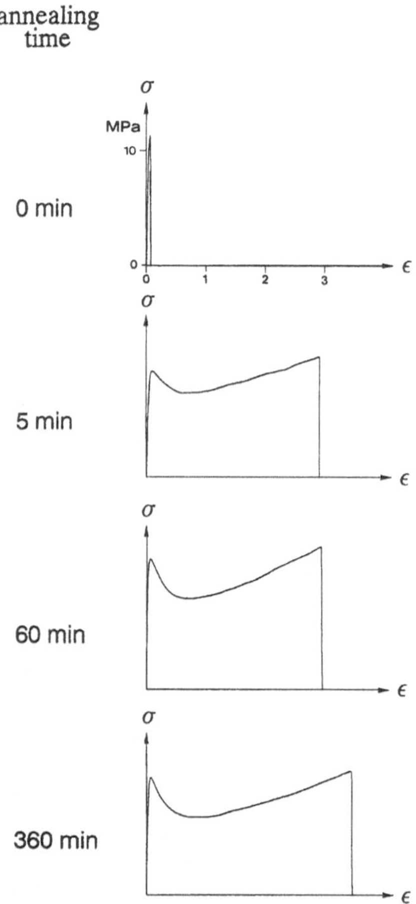

Fig. 5 Stress-strain diagrams for films obtained from a polybutylmethacrylate (PBMA) latex after annealing for various periods of time at 90 °C

The change in mechanical behavior brought about by annealing can be quantified in terms of the failure energy and equated to the results of SANS experiments performed on films obtained from the same latex. Thus the failure energy w_B and the length of the interdiffusion path d have been plotted against the square root of the annealing time at 90 °C in Fig. 6. It can be seen that the length of the diffusion path is proportional to $t^{1/2}$ and thus agrees with the de Gennes reptation theory. Within the first 5 min, w_B increases suddenly by more than two orders of magnitude, corresponding to the brittle-tough transition. Afterwards, it also rises at a rate proportional to the square root of the annealing time and becomes constant after a period of about 10 h.

The assumption that no significant interdiffusion takes place below the glass transition temperature has been confirmed by SANS experiments. Hence the unconditioned film consisting of hard latex particles packed together has a certain ultimate strength that results from interparticulate Van der Waals interactions. However, it does not have any measurable toughness, i.e., the capacity to store and dissipate mechanical energy during deformation.

The toughness is connected with the slipping and entanglement of macromolecules and necessitates interdiffusion of chain segments and the formation of entanglements beyond the particle boundaries. Hence it is not the further dry sintering of dispersion particles that is responsible for the initial rapid increase in w_B in the brittle-tough transition but the incipient interdiffusion of chain ends, which leads to a diffusion path of about 2–3 nm length after five minutes. This diffusion path is comparable to the length of a stretched chain with an entanglement molar mass M_e, for which a value of about $2 \cdot 10^4$ g/mol has been found for PBMA. It is therefore reasonable to assume that the first interparticulate entanglements are formed during this period of time. The subsequent slow rise in the film toughness is related to the progressive interdiffusion of long chain molecules beyond the particle boundaries and their entanglement. The failure energy becomes constant at an interdiffusion length of about 40 nm. This value lies within the order of magnitude for the radius of gyration of the PBMA molecules, for which an average molar mass of $5 \cdot 10^5$ g/mol has been determined.

It has been assumed that not only a diffuse boundary layer between the particles is the main factor responsible for high mechanical toughness but also interparticulate entanglements. This assumption explains the well-known empirical finding that latices whose particles are too densely crosslinked form films of poor strength. In the diffusion model for film formation presented above, it is presumed that crosslinkages impede or even prevent interdiffusion and the formation of entanglements if the average molar mass M_c between two crosslinks is less than the entanglement length. This hypothesis was checked by mechanical measurements on films obtained from PBMA dispersions containing a bifunctional, crosslinkable monomer, viz. methylallyl methacrylate, in various concentrations. It follows from these measurements, some results of which are shown in Fig. 7, that films of densely crosslinked particles remain brittle even after long annealing periods and have only very poor strength, i.e, orders of magnitude less than that of films consisting of uncrosslinked particles [30, 31].

It is evident from the above remarks that at least the fundamentals of the third phase in film formation are well understood. Nevertheless, some questions still remain open. On the one hand, SANS and mechanical measurements permit the conclusion that the entangled structure of films obtained from polymer latices is similar to that obtained from polymer solutions or melts. On the other hand, it has been revealed by many electron-microscope photographs that latex films may still display the original

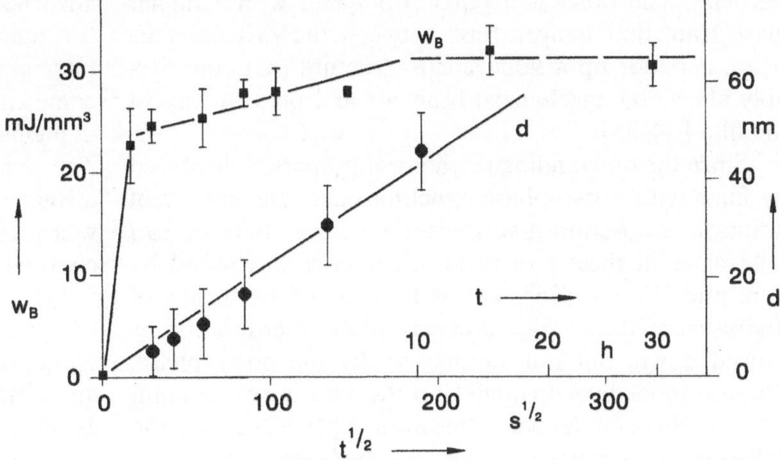

Fig. 6 Failure energy w_B and length *d* of interdiffusion path for PBMA as a function of the square root of the annealing time at 90 °C

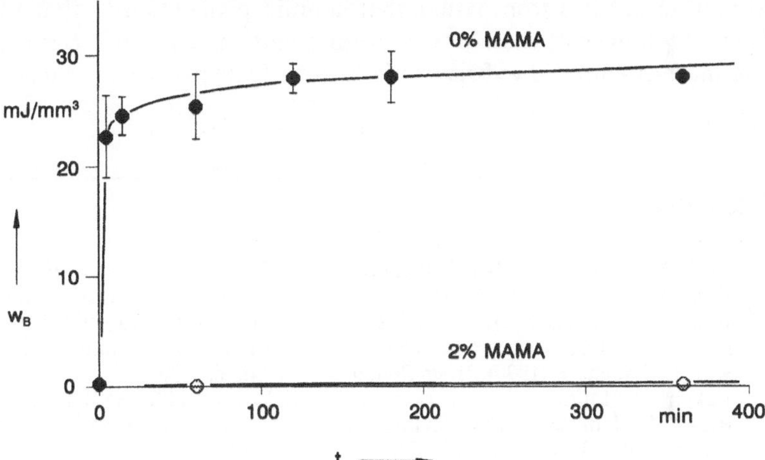

Fig. 7 Failure energy w_B for films from PBMA latices with and without 2% of methallylmethacrylate (MAMA) as a function of the annealing time at 90 °C

structure of the individual particles, yet have high mechanical strength, even after adequately long periods of diffusion. In this case, the original particle structure is apparently retained in the film; and the greater contrast in the boundary layers between the particles indicates that the chemical composition there differs from that in the interior of the particles. Evidently, these differences do not impede the interdiffusion of polymers.

Auxiliaries in dispersion films and films consisting of structured particles

As has already been mentioned, polymer latices contain low-molecular-weight auxiliaries and components such as emulsifying agents, salts, etc., which may be dissolved in the serum or adsorbed on the surface of the particles. Up to now, very little research has been devoted to the whereabouts of these additives in the film. It can be indirectly concluded from SANS studies on films that were swollen with D_2O after they had been formed [4] that many of these hydrophilic components fo the dispersion are concentrated in the interstices between the original particles and account for an increase in the concentration of heavy water at these points. Recent atomic force microscopy (AFM) studies have been concerned with the exudation of emulsifying agents to the surface of a film [33], a subject that is also of practical relevance.

It has been known for some time that if comonomers of different polarity are emulsion polymerized they will be inhomogeneously distributed in the particles so that the more hydrophilic monomers are more strongly concentrated in the outer particle shell. The term "core shell model" is often associated with this finding. Mechanical measurements combined with electron microscopy have proved that films obtained from these particles have a pronounced two-phase structure. The one phase consists of the main copolymer, which is located in the interior of the particles. The other is a polar copolymer with a higher glass transition temperature; between the original particles, it builds up a honeycomb structure that considerably affects the mechanical behavior and performance of the film [34, 35].

Since the outstanding mechanical properties displayed by films with a two-phase structure imply the agency of diffusion, a question that arises is how interdiffusion can take place in these structures. An answer was found by Kim and Winnik [36], who demonstrated by means of fluorescence decay measurements that interdiffusion is slowed down, but not suppressed, by the polar phase. These authors drew up models for the diffusion of the main polymer through the polar "membrane phase" and for the diffusion of the polar polymer into the honeycomb structure itself [26, 27].

Another question of great practical significance is the formation of films from latices that contain pigments or fillers. Examples of these are emulsion paints and paper coatings. Attempts have been made to investigate particles adsorbed on pigment surfaces by means of cryogenic fracture mechanics and scanning electron microscopy [37]. An interesting methodological approach that has recently become known is to determine the structure of latex particles in pigmented films by small-angle neutron scattering [38]. The experiments concerned were performed on acrylic dispersions with calcium carbonate as the filler. The serum was a mixture of water and heavy water that had the same contrast ratio to the neutrons as that of the calcium carbonate particles, which thus became 'invisible," in analogy to the case of an adaptation of the refractive indices. The scatter diagrams are produced solely by the latex particles and yield information on whether the particles are present as individuals or aggregates in the film, whether they are deformed by adsorption on the pigments, etc.

The review on the formation of films from aqueous dispersions, as presented in this publication, demonstrates that well-founded models exist on the mechanisms concerned, that open questions still arise, and that scope remains for further studies.

References

1. Luck W, Klier M, Wesslau H (1963) Ber Bunsenges Phys Chemie 67:75
2. Joanicot M, Wong K, Maquet J, Chevalier Y, Pichot C, Graillat C, Lindner P, Rios L, Cabane B (1990) Progr Colloid Polym Sci 81:175
3. Rieger J, Hädicke E, Ley G, Lindner P (1992) Phys Rew Letters 68:2782
4. Rieger J, Dippel O, Hädicke E, Ley G in Goodwin JW, Buscall R (ed) (1995) Colloidal Polymer Particles. Academic Press, London, p 29
5. Balik CM, Said MA, Hare TM (1989) J Appl Polym Sci 38:557
6. Dillon RE, Matheson LA, Bradford EB (1951) J Colloid Sci 6:108
7. Frankel J (1945) J Phys (USSR) 9:385
8. Brown GL (1956) J Polym Sci 22:423
9. Hertz H (1881) Jf reine u. angew. Mathematik (Crelle) 92:156
10. Lamprecht J (1980) Colloid Polymer Sci 258:960
11. Johnson KL, Kendall K, Roberts AD (1971) Proc R Soc Lond A 324:301
12. Kendall K, Padget JC (1982) Int J Adhesion a Adhesives 1:149
13. Mason G (1973) Br Polym J 5:101
14. Eckersley ST, Rudin A (1994) J Appl Polym Sci 53:1139
15. Sperry PR, Snyder BS, O'Dowd ML, Lesko PM (1994) Langmuir 10:2619
16. Lin F, Meier DJ (1995) Proc Intern Conference on Coatings Sci a Technol. Athen, p 297
17. Voyutskii SS (1963) Autohesion and Adhesion of High Polymers. Wiley, New York
18. de Gennes P-G (1980) C R Acad Sc (Paris) B291:219
19. Kausch HH (1981) Colloid Polymer Sci 259:917
20. Prager S, Tirrell M (1981) J Chem Phys 75:5194
21. Wool RP (1991) In: Lee H-H (ed) Fundamentals of Adhesion. Plenum Press, New York, p 207
22. de Gennes P-G (1971) J Chem Phys 55:572
23. Hahn K, Ley G, Schuller H, Oberthür R (1986) Colloid Polym Sci 264:1092
24. Hahn K, Ley G, Oberthür R (1988) Colloid Polym Sci 266:631
25. Yoo JN, Sperling LH, Glinka CJ, Klein A (1990) and (1991) Macromolecules 23:3962 and 24:2868
26. Joanicot M, Wong K, Richard J, Maquet J, Cabane B (1993) Macromolecules 26:3168
27. Richard J, Wong K (1995) J Polym Sci B: Polym Phys 33:1395
28. Winnik MA, Wang Y, Haley F (1992) J Coatings Technol 64(811) :51
29. Boczar EM, Dionne BC, Fu Z, Kirk AB, Lesko PM, Koller AD (1993) Macromolecules 26:5772
30. Zosel A, Ley G (1992) Polymer Bulletin 27:459
31. Zosel A, Ley G (1993) Macromolecules 26:2222
32. Eckersley ST, Plumtree A, Rudin A (1993) J Appl Polym Sci 48:1689
33. Juhué D, Wang Y, Lang J, Leung O-M, Groh MC, Winnik MA (1995) Proc ACS, Div Polym Mat: Sci Engng 73:86
34. Zosel A, Heckmann W, Ley G, Mächtle W (1986) Colloid Polym Sci 265:113
35. Zosel A, Heckmann W, Ley G, Mächtle W (1990) Makromol Chem, Macromol Symp 35/36:423
36. Kim H-B, Winnik MA (1995) Macromolecules 28:2033
37. Sheehan JG, Takamura K, Davis HT, Scriven LE (1993) Advanced Coating Fundamentals, p 109
38. Joanicot M, Cabane B, Wong K (1995) Proc ACS, Div Polymer Materials: Sci a Engng 73:143

Progr Colloid Polym Sci (1996) 101:93–96
© Steinkopff Verlag 1996

H. Rehage
A. Burger
H. Leonhard
G. Pieper

Flow-induced deformation of microcapsules

Abstract The radical polymerization of surface-active aminomethacrylates can be used to synthesize ultrathin crosslinked membranes at the interface between oil and water. We have systematically measured the kinetics of the surface gelation and the rheological properties of the stabilized films. The membranes, thus prepared, exhibit striking rubber-elastic properties. These are ideal conditions for the formation of a new type of microcapsule. In a series of experiments we have investigated the influence of the membrane elasticity on the shear-induced deformation of these artificial cells.

Key words Ultrathin films – surface polymerization – surface rheology – microcapsules – shear induced deformation

Prof. Dr. Heinz Rehage (✉) · A. Burger
H. Leonhard · G. Pieper
Institut für Umweltanalytik
Universitätsstraße 5-7
45141 Essen, FRG

Introduction

Ultrathin membranes are interesting model systems for numerous applications in industry and science. In former investigations, we have used the photopolymerization of surface active methacrylate diesters, in order to synthesize these structures at the interface between oil and water [1]. It turns out that these crosslinked films are very well characterized in respect to their rheological properties, but due to the rather long polymerization times of about 20 h, it is rather difficult to synthesize microcapsules with these techniques. A more successful way of forming artificial cells is based on a combined redox- and radical reaction which is induced by the addition of Ce^{4+}-salts to aminomethacrylates. Starting point of this polymerization are surface active methacrylate diesters, which are strongly adsorbed at the interface between oil and water. A typical molecule, showing this type of behavior, is N,N-bis-(2-methacryloyloxyethyl)-hexadecylamine. This compound is soluble in dodecane, but quasi insoluble in polar liquids like water. Due to the discrepancy between the hydrophilic oxygen atoms and the hydrophobic paraffin chains these molecules are strongly adsorbed at the interface between oil and water.

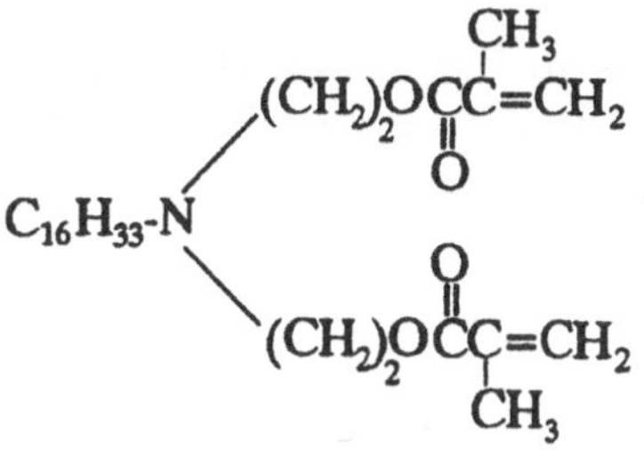

$C_{16}H_{33}$-N[–$(CH_2)_2$OC(=O)C(CH_3)=CH_2]$_2$

Fig. 1 Structural formula of N,N-bis-(2-methacryloyloxyethyl)-hexadecylamine

At the present state, the exact mechanism of the Ce^{4+}-polymerization seems still to be unknown. As alcohols are often used to form radicals with cerium salts, we first tested this type of reaction, but we did not get polymerized films with this technique [2]. With aminomethacrylates, however, a spontaneous gelation process sets in. In analogy to three-dimensional polymerizations, there might be an initiation process of the following type:

$$R\text{-}N[(CH_2)_2OCOC(CH_3)=CH_2]_2 + Ce^{4+} \longrightarrow R\text{-}N^{\dot{+}}[(CH_2)_2OCOC(CH_3)=CH_2]_2 + Ce^{3+}$$

As Ce^{4+}-ions do not react with pure methacrylates, we suppose that the radical is formed at the nitrogen atom. In this way, the monomers might be transformed to charged radicals, which can polymerize by further reaction steps with double bonds of the metacrylic groups. As the crosslinking process occurs within a few minutes, it is easy to form microcapsules using this polymerization technique.

Experimental part

The monomers were synthesized by reacting N,N-bis(2-hydroxyethyl)-hexadecylamine with methacrylic acid in tetrahydrofuran after addition of small amounts of dicyclohexylcarbodiimide. The molecules were purified by medium pressure liquid chromatography (MPLC). As solvents we used dodecane and water, which were carefully purified and stored under argon atmosphere. The water was twice distilled over potassium permanganate and kept under argon.

Microcapsules were prepared by the interfacial polymerization of oil droplets which are suspended in water. The chemical reactions are identical compared to that of the planar interface.

The rheological measurements were carried out in different types of surface rheometers, which are explained in detail in ref. [1]. The microcapsules were observed in a rheoscope, which is schematically shown in Fig. 2.

Experimental results

The kinetics of surface gelation were detected by measuring the two-dimensional storage modulus μ' as a function of the reaction time (time sweep). In these experiments, a sinusoidal deformation with small amplitude is applied to the sample at a constant angular velocity ω. As the two-dimensional shear modulus does not depend on the frequency, the data describe the formation of cross-linking points as a function of time. A typical curve, showing such data, is given in Fig. 3.

It is evident that the polymerization proceeds rapidly with time constants of the order of a few minutes. In contrast to crosslinking reactions initiated by UV-irradiation the kinetics are very fast. It is interesting to note that the crosslinked network structure exhibits rubber-elastic properties. Figure 4 gives some insight into the dynamic features of these membranes.

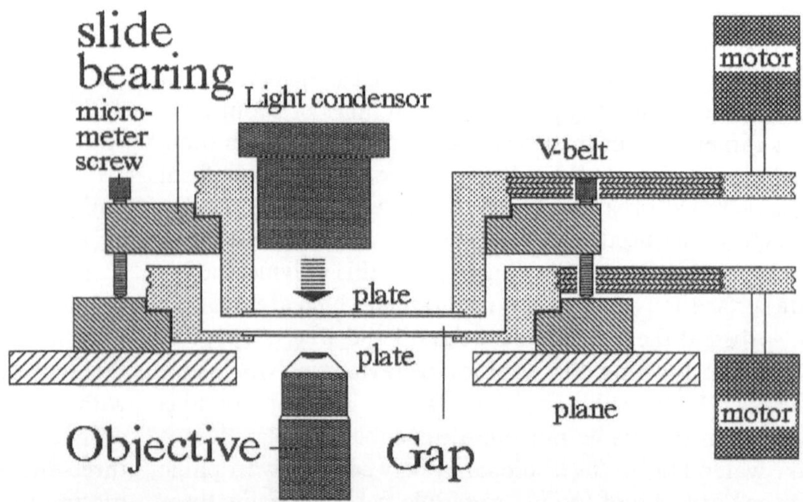

Fig. 2 Experimental set-up for the observation of microcapsules during flow. The two plates are rotating in an opposite direction. In this way a non-moving shear plane is generated in the central gap between the plates. The deformation and breaking process of the microcapsules can be observed with a video camera. The video tapes were digitized and interpreted by computer analysis

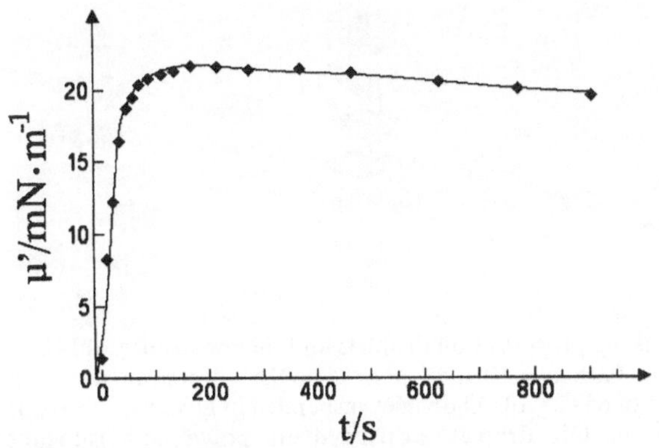

Fig. 3 The two-dimensional storage modulus of the ultrathin membrane as a function of the reaction time. Monomer concentration: 5 mmol/L, T = 20°C, $\omega = 0.1\ s^{-1}$. The polymerization was induced by the addition of 1 mmol/L $Ce(SO_4)_2$ in 10^{-3} mol H_2SO_4

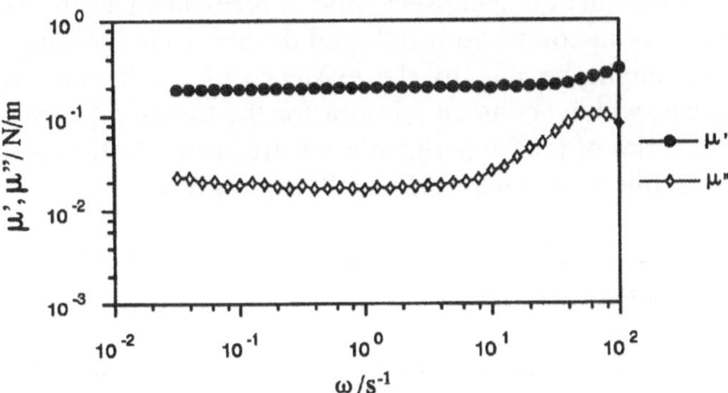

Fig. 4 The two-dimensional storage modulus and the two dimensional loss modulus as a function of the angular frequency for a thin membrane at the interface between dodecane and water (Aminomethacrylate concentration; 5 mmol/L, T = 20°C, $\gamma = 1\%$)

The two-dimensional storage modulus μ' describes the elastic properties of the sample and the two-dimensional loss modulus is proportional to the energy dissipated as heat. The constant plateau value points to the absence of any relaxation process. This phenomenon can be traced back to the existence of permanent crosslinking points, which remain stable as a function of time. Such curves are typical for the existence of rubber-elastic materials. By analogy with three-dimensional systems, the elastic modulus remains constant up to quite large deformations. This is clearly shown in Fig. 5 which gives some insight into results of strain sweep experiments.

It is interesting to note that the modulus remains nearly constant up to deformations of about 20%. This is typical for rubber-elastic materials, where breakdown of the structure generally starts to appear at deformations between 10% and 500%. In terms of molecular models, this phenomenon can be explained by the high flexibility of

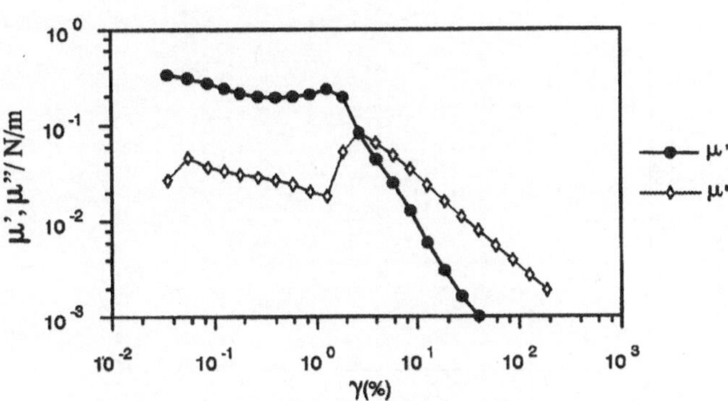

Fig. 5 The two dimensional storage modulus and the two dimensional loss modulus as a function of the deformation (Aminomethacrylate concentration c = 5 mmol/L, T = 20°C, $\omega = 5s^{-1}$)

the macromolecules. The elastic properties of the cross-linked membranes are, hence, a direct consequence of the molecular motion of these cross-linked molecules.

The interfacial polymerization is not restricted to flat surfaces, and oil droplets which are surrounded by a network of this kind are thereby stabilized against coalescence. During flow, these artificial cells are deformed, and at high values of the velocity gradient breaking processes are often observed. In the regime of small external forces it is possible to describe the shear induced deformation by a theory which was recently proposed by D. Barthès-Biesel [3]. According to this model, the deformation of the microcapsule can be described by:

$$D = \frac{L - B}{L + B} \tag{1}$$

In this equation, D denotes the diameter of the capsules and L the length. The ratio of viscous to elastic forces is usually described as the capillary number C which is defined by:

$$C = \frac{\eta_s \mathring{\gamma} a}{E^{2d}} \tag{2}$$

a is the radius of the capsules in the quiescent state, η_s denotes the solvent viscosity and $\mathring{\gamma}$ is the shear rate. The two-dimensional shear elastic modulus E^{2d} is in rubber-elastic materials identical with μ'. It is, therefore, possible to calculate the capillary number if the two-dimensional storage modulus is known from independent rheological measurements on the planar surfaces. Relevant results, which were obtained using the rheoscope, are summarized in Fig. 6.

According to the theory of Barthès-Biesel the deformation of oil droplets and microcapsules depends on the capillary number [3]. For liquid droplets, one obtains in

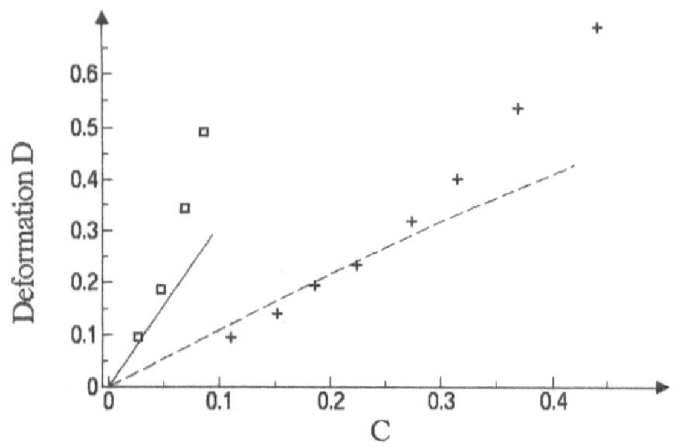

Fig. 6 The deformation of microcapsules and oil droplets as a function of the capillary number. The rectangular symbols represent measurements of microcapsules (aminomethacrylate c = 5 mmol/L, the density of the solvent and oil phase was adjusted by using mixtures of decahydronaphthaline (60%) and carbon tetrachloride (40%) for the unpolar phase and water-glycerin mixtures [(13%:87%] for the polar liquid; $\lambda = 0.01$, $\mu' = 15$ mN/m, a = 99.3 μm. For comparison purposes, the behavior of liquid oil droplets is plotted in the same diagram (cross symbols). As solvent, we used polydimethylsiloxane and the suspended liquid was n-butyl-diethanolamine, $\lambda = 0.147$, $\sigma = 3.4$ mN/m, a = 95.2 μm

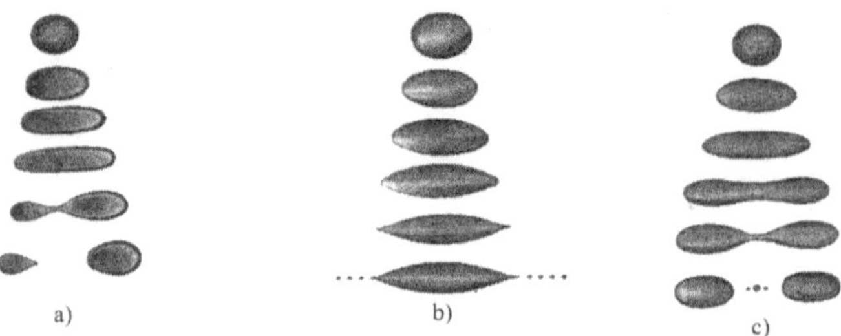

Fig. 7 Breaking process of oil droplets and microcapsules. a) Microcapsule, $\lambda = 0.01$, a = 105 μm, $\mu' = 15$ mN/m. Bursting sets in at a shear rate of 65 s^{-1}. b) Oil droplet suspended in glycerin, $\lambda = 0.001$, a = 67 μm. c) Oil droplet suspended in polydimethylsiloxane, $\lambda = 0.147$, a = 83 μm

the limit of small deformations:

$$D = C\frac{19\lambda + 16}{16\lambda + 16}. \tag{3}$$

Here, λ denotes the viscosity ratio between the two phases. For microcapsules, the corresponding relationship is simply given by [3]:

$$D = \frac{25C}{8}. \tag{4}$$

The drawn lines in Fig. 6 correspond to Eqs. (3) and (4). It is evident that the theoretical predictions hold in the regime of small capillary numbers. At elevated deformations, one obtains deviations from the linear relationships, which can be calculated at the present state to order C^2 [3]. It is interesting to note that the breaking process of capsules and liquid droplets is very different. This can easily be learned from Fig. 7, where some typical phenomena are summarized.

It is evident that capsules tend to break in an asymmetrical way in comparison to liquid droplets. This phenomenon might be due to the existence of small network defects, which act as an initiator for the bursting process. In a series of new experiments, we are currently trying to check this idea. This work is still in progress.

Conclusions

The polymerization of aminomethacrylates with Ce^{4+}-salts can be used to synthesize ultrathin networks at the interface between oil and water. The rheological properties of these cross-linked membranes are well defined and they exhibit rubber-elastic features. These are ideal conditions for the preparation of microcapsules, which may be used as simple model systems for investigations of biological systems.

Acknowledgment Financial support of this work by a grant of the Deutsche Forschungsgemeinschaft and the Fonds der Chemischen Industrie are gratefully acknowledged.

References

1. Burger A, Leonhard H, Rehage H, Wagner R, Schwoerer M (1995) Macromol Chem Phys 196:1–46
2. Kern W (1947) Makromol Chem 1:249
3. Barthès-Biesel D (1991) Physica A 172:103–124

Progr Colloid Polym Sci (1996) 101:97–100

G. Peschel
K. Tüssner
M. Droschinski

The problem of stability of a silica hydrosol in presence of various aliphatic alcohols

Prof. Dr. G. Peschel (✉) · K. Tüssner
M. Droschinski
Institut für Physikalische
und Theoretische Chemie
Universität Essen
Universitätsstraße
45117 Essen, FRG

Abstract The stability of a silica hydrosol in the presence of simple aliphatic alcohols was investigated via measuring the coagulation kinetics of the silica particles. Formation of aggregates was measured by photon correlation spectroscopy. The hydrosol stability decreased in the order of methanol, iso-propanol, ethanol, propanol and *n*-butanol added to the hydrosol. The effect is ascribed to hydrophobic particles–particle interaction. The exposed roll of isopropyl alcohol could be manifested by testing the water structure modifying properties of the alcohols using the Cu^{++}-ion as a spectroscopic probe for the state of water structure in the alcoholic solution.

Key words Hydrosol stability – coagulation kinetics – aliphatic alcohols – photon correlation spectroscopy – UV/VIS-spectroscopy

Coagulation tests with a hydrosol in the presence of aliphatic alcohols

Introduction

The stability of hydrosols in the presence of organic solutes has meanwhile gained considerable technical interest. But this does not exclude the fact that systematic and comprehensive studies in this field are still rather sparse. Adsorption of organic solutes onto the silica surfaces might render them hydrophobic to a certain extent which gives rise to hydrophobic particle–particle interaction and, hence, to coagulation effects provided the attractive interactions are strong enough [1]. More quantitative considerations concerning this particular phenomenon are due to Lu and coworkers [2–5]. Recently, we have treated a related problem when investigating the coagulation kinetics of a silica hydrosol containing particles which were hydrophobized by adsorbed tetraalkylammonium cations [6]. For performing these experiments we have applied photon correlation spectroscopy (PCS) as a light scattering method as was likewise the case in our previous coagulation tests [7–9].

The purpose of the present work is to study the influence of various aliphatic alcohols on a silica hydrosol stability with regard to hydrophobic interaction effects of further interest is use of a spectroscopic Cu^{++}-probe focused on the state of water in these solutions possibly contributing to sol stability.

Theory

Coagulation effects can according to the height of an existing coagulation barrier occur as rapid (characterized by the rate constant k_r) and slow (characterized by the rate constant k_s) process, respectively [10]. In the latter case repulsion forces are effective created by overlap of electric double layers near the surfaces of encountering particles.

The stability ratio of a colloidal system is defined by

$$W = \frac{k_r}{k_s} = 2a \int_{2a}^{\infty} \exp(V_{tot}(H_0)/k_B T) \frac{dH_0}{H_0^2} \quad (1)$$

a is the radius of the spherical particles and H_0 the shortest distance between them; V_{tot} is the total particle interaction potential [1].

In the stability diagram log W is plotted against log$c$, when c is the electrolyte concentration. The less concentrated region for slow coagulation is separated from that for rapid coagulation by a distinct break indicating the critical coagulation concentration (c.c.c.).

In the present paper, however, the concentration of NH_4Cl as a basic electrolyte in the aqueous solution was kept constant at 0.10 mol dm^{-3}. Nevertheless stability diagrams could be constructed by varying on the abscissa the concentration of selected alcohols.

Experimental method

The coagulation tests with a silica hydrosol (particle diameter circa 283 nm) were performed in the presence of NH_4Cl ($c = 0.10$ M) and various differently concentrated aliphatic alcohols (methanol, ethanol, n-propanol, iso-propanol, and n-butanol) by using, as in our recent work [6–9], PCS by which the particle diffusion constants and, hence, on the basis of the theory by Versmold and Härtl [12], the rate constants k_r and k_s could be obtained.

The PCS experiments were carried out by aid of a Nicomp Laser Particle Sizer Model 200 [7]. The silica particles were prepared by applying the well-known method of Stöber et al. [13]. The alcohols and NH_4Cl used were of high analytical quality.

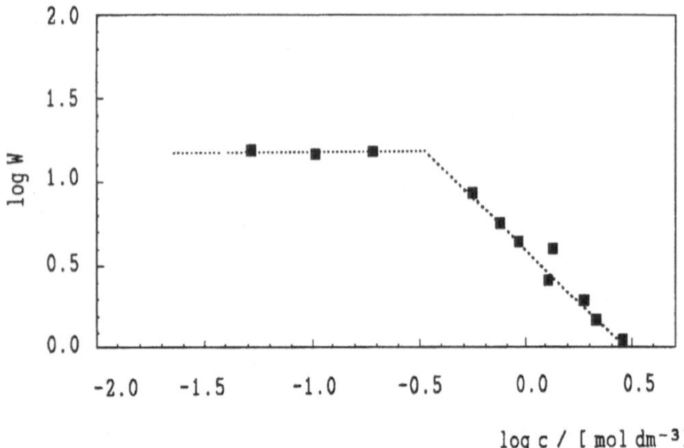

Fig. 1 Stability diagram for ethanol dissolved in an aqueous solution of NH_4Cl ($c = 0.10$ mol dm^{-3})

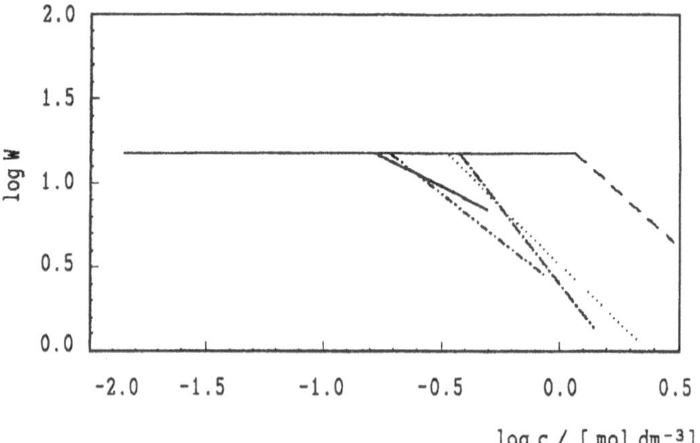

Fig. 2 Stability diagram for methanol (---), ethanol (······), n-propanol (-··-), iso-propanol (·-·-·) and n-butanol (——) dissolved in aqueous solution of NH_4Cl ($c = 0.10$ mol dm^{-3})

Results and discussion

In Fig. 1a stability diagram for ethanol at various concentrations and constantly concentrated NH_4Cl is shown, giving confirmatory evidence for the existence of a definable c.c.c. at about 0.34 mol dm^{-3}. That of the pure NH_4Cl-solution lay at about 0.34 mol dm^{-1}. For the other alcohols investigated diagrams of similar quality were derived from experiment. Figure 2 displays the respective stability diagrams for comparison.

In order to explain this evidence let us assume that the different alcohols will adsorb onto the silica particle surfaces and make them hydrophobic to some degree. Particle encounters will then effect hydrophobic interaction which should be strongest for n-butanol [14] as is, in fact, the case, evidenced by the rather low c.c.c. The coagulation processes probably occurred in the secondary potential minimum as in our former tests with different electrolyte solutions [6–9].

A spectroscopic method for monitoring the state of water structure in aqueous solutions

Introduction

It is well-known that a number of different substances dissolved in water can according to their electrolytic or polar character, respectively, more or less significantly modify the structure of surrounding water [14–16]. Of course, a selection of accurate instrumental methods, particularly of spectroscopic nature, is available to monitor water structure in aqueous solutions. The present paper aims at details about water structure in aqueous solutions of some aliphatic alcohols.

To that end, we have, as a special application, developed an additional spectroscopic test method using the Cu^{++}-ion as a spectroscopic probe embedded in the solution to be investigated.

Theory

The electronic configuration of the sixfold copper complex in the fundamental state is given by the term $(t_{2g})^6(e_g)^3$ [17]. Different electronic occupation of both the e_g-orbitals (d_z^2) and $(d_x^2 - y^2)$ can effect canceling of orbital symmetry generating a distortion along the z-axis (Jahn–Tellner-effect) by which the orbitals are further split into two degenerated and three nondegenerated terms.

The energy of electronic transition $d_z^2 \rightarrow d_{x^2-y^2}^2$ at about 800 nm can most easily be affected by changes of the bonding lengths of the water molecule ligands. According to Clack and Farrimond [18] the equatorial bonding length lies at 0.186 nm and the axial one at 0.195 nm.

Considering the ligands as components of the surrounding solvent water lattice, one should realize that these, when involved in the formation of ordered water aggregates extending from organic solutes nearby, might be more weakly bonded to the central Cu^{++}-ion.

Hence, water strengthening in the $CuCl_2$-solution should be reflected in an increase of the absorption wavelength and destruction of water structure by the respective decrease.

Results and discussion

Figure 3 displays the temperature dependence of the absorption wavelength of the hexaaquocomplex dissolved in pure water and in an aqueous solution of ethanol, respectively. Typically the absorption wavelength becomes larger with increasing temperature indicating weakening of bonding strength. Surprisingly ethanol (4.00 mol dm^{-3}) evidently builds up water structure below 303 K, above this temperature the opposite seems to be the case. (303 K is according to Drost–Hansen [18] a characteristic temperature at which water structure near interfaces or solutes may undergo a structural transition of higher order.)

Figure 4 shows how water structure is enhanced in the aqueous solutions of various differently concentrated aliphatic alcohols. Principally the effect becomes larger with higher concentrations as is expected in view of the existence of hydrophobic hydration. For sufficiently low concentrations the water structure enhancing effect clearly increases in the usual order from methanol to n-butanol since hydrophobic hydration becomes larger. Isopropanol, however, does not fit this pattern; it even exhibits a water-structure-destroying ability for low concentrations. We cannot exclude the possibility that this evidence has some reference to the coagulation results with isopropanol as depicted in Fig. 2 and might provide an explanation for that irregular evidence.

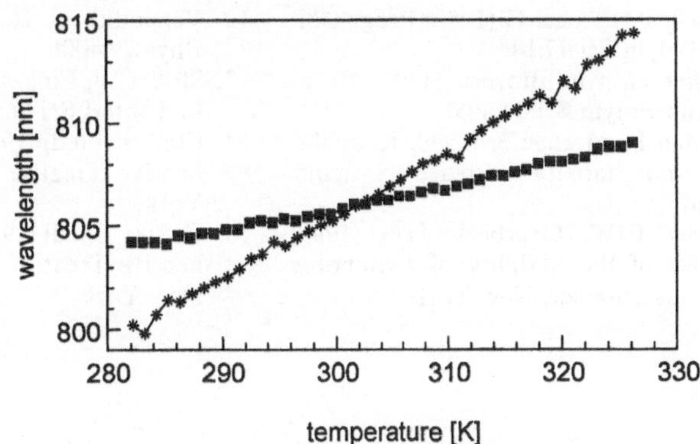

Fig. 3 Wavelength of maximum light absorption of an aqueous $CuCl_2$-solution ($c = 5 \cdot 10^{-3}$ mol dm^{-3}) with (■) and without (∗) addition of ethanol over a broad temperature range. Concentration of ethanol $c = 4.00$ mol dm^{-3}

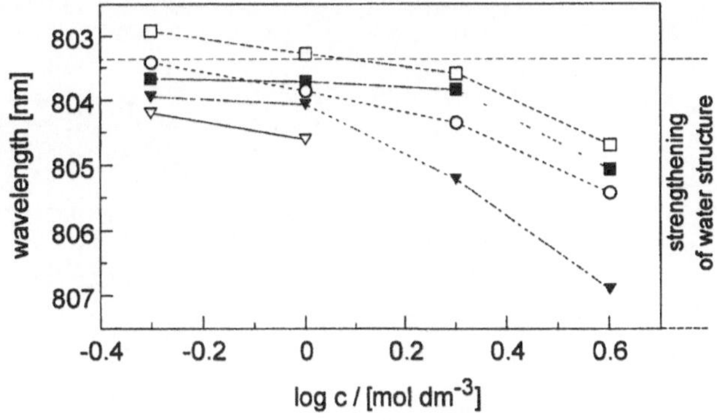

Fig. 4 Wavelength of maximum light absorption of an aqueous $CuCl_2$-solution with addition of differently concentrated methanol (○), ethanol (■), n-propanol (▼), iso-propanol (□) and n-butanol (▽). $T = 293$ K. The horizontal dotted line indicates the absorption wavelength of a pure aqueous solution of $CuCl_2$. Below this line water structure might be enhanced, above this line breakdown of water structure is assumed

References

1. Sonntag H (1977) Lehrbuch der Kolloidwissenschaft, VEB Deutscher Verlag der Wissenschaften, Berlin, p 128
2. Lu SC (1990) Colloid J USSR 52:743
3. Lu SC (1991) Colloids Surfaces 57:49
4. Dai Z, Lu SC (1991) Colloids Surfaces 57:61
5. Lu SC (1991) Colloids Surfaces 57:73
6. Peschel G, von Teubern, being prepared
7. Ludwig P, Peschel G (1988) Progr Colloid Polym Sci 76:42

8. Ludwig P, Peschel G (1988) Progr Colloid Polym Sci 77:146
9. Peschel G, von Brevern (1991) Progr Colloid Polym Sci 84:405
10. Sonntag H, Strenge K (1970) Koagulation und Stabilität disperser Systeme, Berlin
11. Verwey EJW, Overbeek JTG (1948) Theory of the Stability of Lyophobic Colloids. Elsevier, New York
12. Versmold H, Härtl W (1983) J Chem Phys 79:4006
13. Stöber W, Fink A, Bohn E (1968) J Colloid Interf Sci 26:62
14. Franks F (ed) (1975) Water. A Comprehensive Treatise. Vol 4, Plenum, New York
15. Franks F (ed) (1973) Water. A Comprehensive Treatise. Vol 2 and 3, Plenum, New York
16. Luck W (1964) Fortchr Chem Forsch 64:653
17. Kober F (1988) Grundlagen der Komplexchemie, de Gruyter, Berlin
18. Drost-Hansen W (1971) In: Brown HD (ed) Chemistry of the Cell Interface, Academic Press, New York, p 3

Progr Colloid Polym Sci (1996) 101:101–104

H.-J. Neumann
B. Paczynska-Lahme

Stability and demulsification of petroleum emulsions

Prof. Dr. H.-J. Neumann (✉)
B. Paczynska-Lahme
CONTEXT
Am Bergwäldchen 22
37520 Osterode am Harz, FRG

Abstract Petroleum is predominantly recovered in form of W/O-emulsions which are stabilized by petroleum resins and asphaltenes, colloidal disperse components of petroleum. Both of these substances are polydisperse, resoluble, oleophilic micellar colloids, ocurring in dilutions in spherical form. In concentrated solutions, however, they form larger, inter-micellar structures. These structures generally form with the enrichment of resins and asphaltenes at oil/water interfaces.

Petroleum emulsions are stabilized by "thick films" according to Gibbs which are composed from different petroleum-internal tensides and have a Gibb's elasticity. The more the interfacial activities of the components differ from each other, the higher is the elasticity module of the multi-component films.

We found optically anisotropic interface films in petroleum emulsions. The observed structure-formation in interface films of emulsions which are mostly stabilized by petroleum resins, increases continuously with raising temperature. Even isolated petroleum resins in volume-phases show such a structure-formation. Lyotropic, liquid–crystalline mesophases develop. This proves that the "enhanced emulsion stability" according to Friberg is valid for petroleum resins as multi-component systems as well.

The demulsification of emulsions after addition of demulsifiers as a result of a shift of the balance lyophilic–lyophobic is being discussed.

Key words Petroleum emulsions – emulsion stability – demulsification – liquid crystals – interfacial films

Introduction

In 1958, van der Waarden [1] stated: "*mineral oil–water-emulsions are stable if there are asphaltenes in the oil-phase which are near the state of beginning flocculation.*" This statement can still be used as the basis for investigation of petroleum emulsions. In the meantime, however, we have distinguished that additional to the asphaltenes, petroleum resins play a particularly important role in the stability of petroleum emulsions.

Petroleum is predominantly recovered in the form of W/O-emulsions. In the course of the necessary demulsification, very often a water-rich O/W-emulsion is obtained – especially in a process with high local demulsifier dosage.

Colloids in petroleum

Petroleums are regular systems containing asphaltics [2] which are dissolved in a continuous oilphase in a colloid-disperse manner. The asphaltics contain the two groups: the asphaltenes and the petroleum resins [3, 4]. Both groups can be separated by multi-stage ultra-filtration from the emulsion and from each other [3]; the asphaltenes are retained on filters with pores of 35 to 10 nm, the petroleum resins on filters with pores of 5 nm.

In the past, attention was mainly directed to the asphaltenes whereas the petroleum resins were hardly being dealt with. However, the petroleum resins play the significant role in stability of petroleum emulsions.

Both groups of asphaltics are polydisperse, resoluble, oleophilic micellar colloids [3]. The micellas are spherical in dilutions (i.e., in most of the petroleums): they are able to form larger inter-micellar structures without any spherical particles.

The petroleum resins as well as the asphaltenes enrich at oil/water interfaces while changing their colloidal structures. They form mechanically stable interface films which are non-soluble neither real nor colloidal nor in the oil or water phase. They are oil-wetted and thus stabilize W/O emulsions, according to the Rule of Bancroft.

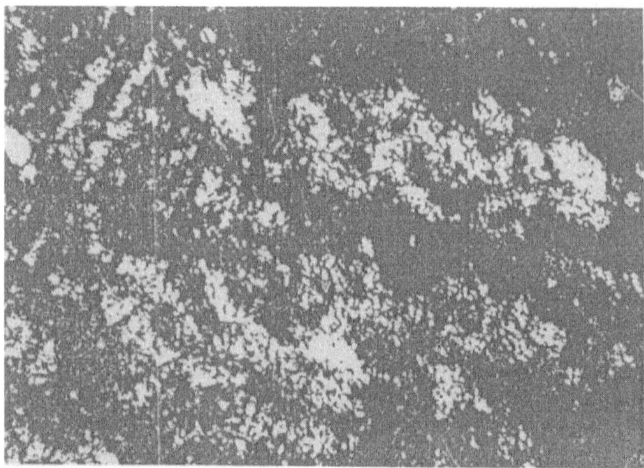

Fig. 1 Microscopic liquid crystalline domains in petroleum resins at room temperature in polarized light

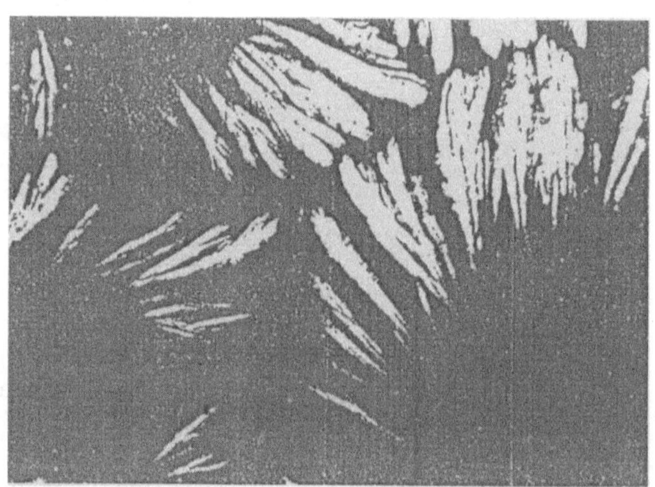

Fig. 2 Microscopic liquid crystals in petroleum resins at 90 °C in polarized light

Stability of petroleum emulsions

W/O petroleum emulsions are stable against coalescence but not against flocculation. Two flocculent drops touching each other do not coalesce spontaneously even not when mechanically deforming while giving up their spherical form.

Petroleum emulsions are stabilized by "thick films" according to Gibbs [5] which are composed of mixed surfactants and which possess a "Gibbs" elasticity. The more the interfacial activities of the components differ from each other, the higher is the elasticity module of the multi-component films. Rehbinder calls those thick interface films with visco-elastic qualities stabilising emulsions "structurally mechanic barrier".

Paczynska-Lahme [6] investigated in petroleum emulsions of both types, W/O and O/W, optical anisotrope interface films. Due to its double refracting, observed structure-formation in interface films of emulsions which are wholly or for the greater part stabilized by petroleum resins, increases continuously with raising temperature. Lyotropic, liquid–crystalline mesophases develop.

The structures depend on temperature, water and salt content. Figures 1 and 2 show petroleum resins first at room temperature with double-refracting, liquid–crystalline areas (Fig. 1) and second at 90 °C with a marked structure (Fig. 2).

Thus for the multi-component system petroleum resins have been proved as Friberg [7, 8] described for defined systems: Because of their visco-elasticity liquid–crystalline phases lead to an "enhanced stability of emulsions" (according to Friberg). All the investigated petroleum resins from very different petroleums show the formation of a liquid–crystalline phase. Hexagonal and lamellar lyotrope structures were observed.

It is very difficult to break recovered petroleum emulsions with a high ratio of petroleum resins to asphaltenes – this fact can be explained by the regular occurrence of liquid–crystalline structures in petroleum resins.

Paraffine crystals, oil-wetted clay particles as well as corrosion products from the recovery equipment can be incorporated into these "thick films" and stabilize these additionally.

Anion and cation active substances can be separated from these interface films through chromatographical

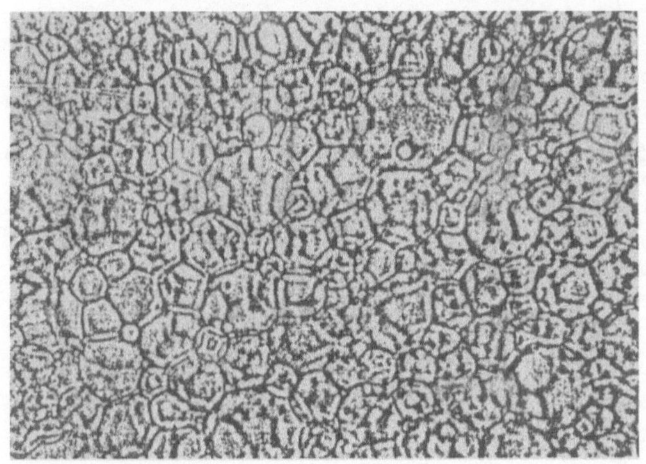

Fig. 3 Microscopic water drops in a petroleum polyhedron emulsion in polarized light

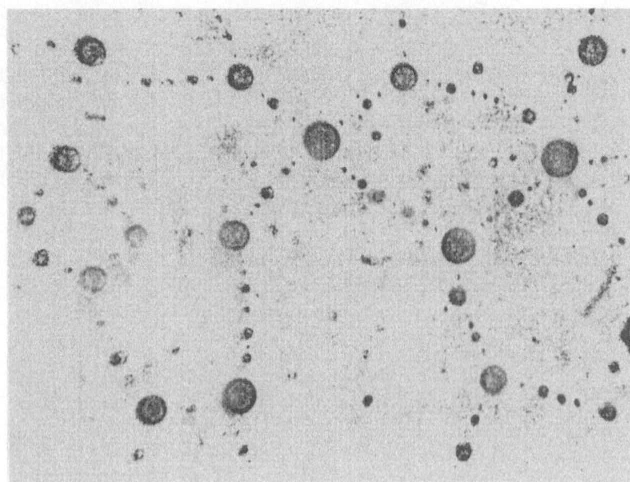

Fig. 4 Microscopic oil drops in water after phase inversion of a w/o petroleum emulsion

processes [9, 10]. These substances do not react with each other as they are incorporated in different micellas: the cation active substances are present in the petroleum resins, the anion active substances in the asphaltenes.

The W/O-petroleum emulsions investigated in this study exist in spherical and polyhedron emulsions, the latter being rich in water. Figure 3 shows a W/O-polyhedron emulsion of extremely high water content. Polyhedron structures predominantly occur in layers forming between a water and a petroleum phase in the process of technical demulsification of recovered petroleum. If a small quantity of water is added to such a W/O-emulsion or the system temperature is slightly raised, the emulsion changes into an O/W-emulsion. The oil films from the original external phase now form spherical drops. As shown in Fig. 4, their order still makes the original polyhedron structure apparent. Large drops form the corners of the structure. These are linked by a line of smaller drops. Phase inversion occurred because the W/O-emulsion preferred by the stabilizing interfaces could not be maintained due to high excess of water.

In many cases, multiple emulsions have also been found in recovered petroleum-salt water systems.

The stabilizing interface films in emulsions are subject to aging. They are composed by micellas of petroleum resins and asphaltenes. With these micellas, aging of the structure [11] can be perceived, combined with a coarsening of particles from sol to gel:

$$\text{sol} \rightarrow \text{structured sol} \rightarrow \text{coagel} \rightarrow \text{gel} .$$

With further aging it is even possible that the gel leaves the colloidal state and becomes coarsely disperse.

The advancing order results in increasing anisotropy and increasing elasticity of the interface films. This corresponds to the observation that the earlier a demulsifier additive is added, the more effective it is, i.e., an additive to the ring space is more effective than to the deposit tank.

With double breaking domains (Fig. 1), ultrasound treatment leads to further structural order even at room temperature.

Demulsification

Water and dissolved salt in the recovered W/O-emulsions have to be removed by demulsification of the emulsion and separation of the water. Petroleum is recovered with water content up to over 90%. In most cases, there are two demulsification processes for every emulsion: first with treatment on the oil field and then in the refinery before the distillation of the oil.

In most cases demulsifiers, i.e., surfactants are added to the demulsification process for the rewetting of the surfaces. They cause the inversion of the emulsion in case of excess concentration. One can observe that after addition of the demulsifier the drops lose their spherical form and coalesce spontaneously with touch.

A wide variety of different substances have already been used and even patented. However, there is no universal demulsifier for all kinds of petroleum. Today, propylene-/ethylene oxide-bloc polymers are often used. We effect many experiments with the "Dissolvan-types" of Hoechst AG [12].

References

1. van der Waarden M (1958) Kolloid-Z 156:116–122
2. Speight JG (1980) The Chemistry and Technology of Petroleum. Marcel Dekker New York, Basel, Chapt 7
3. Neumann HJ (1967) Habilitationsschrift Braunschweig
4. Neumann HJ (1970) Erdöl Kohle-Erdgas-Petrochem 23:496–499
5. Gibbs JW (1928) Collected works Vol 1 Yale Univ Press
6. Paczynska-Lahme B (1990) Progr Colloid Polym Sci 83:196–199
7. Friberg S (1971) J Coll Interf Sci 37:291
8. Friberg S (1971) J Coll Interf Sci 40:291–295
9. Haardt HJ (1973) Diss Braunschweig
10. Durmosch MR (1977) Diss Braunschweig
11. Neumann HJ, Rahimian I, Paczynska-Lahme B (1992) Bitumen 54:54–56
12. Hoechst AG (1987) The effective way of separating water from oil and oil from water

Progr Colloid Polym Sci (1996) 101:105–112

T. Förster
B. Guckenbiehl
H. Hensen
W. von Rybinski

Physico-chemical basics of microemulsions with alkyl polyglycosides

Dr. T. Förster (✉)
Dr. B. Guckenbiehl · Dr. H. Hensen
Dr. W. von Rybinski
Henkel KGaA
40191 Düsseldorf, FRG

Abstract Alkyl polyglycosides are a nonionic surfactant class based on sugar and fatty alcohol; they are being used to an ever-increasing extent in cleaning and personal care products because of their excellent environmental and skin compatibility. Although they are nonionic like the fatty alcohol ethoxylates, on account of the different interaction of the sugar residue with water, the phase behavior of the alkyl polyglycosides clearly differs from that of the fatty alcohol ethoxylates. This also affects the formation of microemulsions in oil–water–surfactant systems. Microemulsions form when the hydrophilic–lipophilic properties of the surfactant film at the oil water interface are balanced. Depending on the surfactant type, this hydrophilic–lipophilic balance can be adjusted by means of different formulation parameters: In the case of fatty alcohol ethoxylates this is possible via the temperature as a parameter which is easily changed; in the case of the alkyl polyglycosides it can only be achieved by mixing with hydrophobic cosurfactants. In this respect alkyl polyglycosides are similar to anionic surfactants. This leads to interesting applications since – in contrast to fatty alcohol ethoxylates – temperature-stable microemulsions on the basis of APG can be produced. Through addition of anionic surfactants and variation of the oil/water ratio, microemulsions with "tailor-made" application properties such as viscosity or interfacial tension can be obtained. Small-angle x-ray scattering experiments and conductivity tests provide an insight into the structure of these microemulsions.

Key words Microemulsion – interfacial tension – alkyl polyglycoside – phase behavior – viscosity – SAXS

Introduction

As an interesting base for pharmaceutical, cosmetic but also technical products, microemulsions have been examined intensively for a long time. In this context microemulsions from oil, water and ethoxylated nonionic surfactants have been of special interest; on the one hand, they are widely used in practice, and on the other hand, they can form well-defined ternary mixtures, which renders them useful for systematic experimental studies [1, 2]. One characteristic of these systems from oil, water and ethoxylated nonionic surfactant is their distinct temperature dependence which provides the basis for the known phase inversion temperature (PIT)-phenomenon [1–3].

For some years, alkyl polyglycosides (APG) have been available as a new class of nonionic surfactants on a large scale [4]. Unlike typical binary fatty alcohol ethoxylate water mixtures the examination of the phase behavior of binary APG water mixtures already showed that there is a difference in the interaction of the sugar head group of the APG with water as compared to the ethoxylate head group: Particularly in the case of short-chain APG simple phase behavior is observed which only shows a weak dependence upon temperature [5–8]. This low temperature dependence also becomes apparent in measurements of the interfacial tension [9].

Therefore, one does not expect temperature dependent phase inversion in APG-containing emulsions. A first investigation on microemulsification of cyclohexane with APG and a specific alkyl glyceryl ether shows that the emulsion type can be adjusted by means of the mixing ratio between the hydrophilic APG and the hydrophobic coemulsifier [10]. A microemulsion forms when an optimal emulsifier ratio is adjusted. This optimum emulsifier mixing ratio is hardly affected by temperature.

This work is intended to present a physicochemical description of APG-containing microemulsions with alkanes and to elucidate the properties of these microemulsions. Using the example of a multicomponent system, possible cosmetic applications will be shown.

Methods and materials

Materials

The following Henkel sales products were used as emulsifiers: C_{12-14}-APG (Plantaren 1200), sodium lauryl ether sulfate (Texapon NSO) and sorbitan monolaurate (Dehymuls SML). The emollients were dodecane (Fluka) and dioctyl cyclohexane (Cetiol S from Henkel). PH values were adjusted at 6.5 with citric acid.

Methods

Sample preparation

20 ml-samples were prepared by mixing the components at 50 °C for 30 min with a LM 34 stirrer (Pendraulik). Before further tests the samples were stored for at least 4 weeks at room temperature in order to guarantee an equilibrium phase adjustment.

Phase behavior

Liquid–crystalline surfactant phases, emulsions and microemulsions were identified visually and through polarization microscopy (Zeiss). The emulsion type (o/w, w/o or microemulsion) was determined by means of temperature dependent conductivity measurements (Radiometer) and turbidity measurements with a fiber optic photometer (Metrohm).

The structures of selected samples were determined by means of small-angle x-ray scattering (SAXS apparatus (Philips) with Kratky camera (Paar)).

Viscosity measurements

The viscosity behavior was studied using a shear rate controlled rotational rheometer (RFS II from Rheometrics) with thermostat and a plate–plate measuring system.

Interfacial tension measurement

The interfacial tension between the equilibrated aqueous and oily phases was measured with a spinning-drop tensiometer (Kruess).

Results and discussion

In principle, emulsions from oil, water and ethoxylated nonionic surfactant show a temperature-induced phase inversion under formation of a microemulsion [1]. The system from tetradecane, water and $C_{12}E_5$ inverts from an oil-in-water emulsion into a water-in-oil emulsion in the temperature range between 45 °C and 55 °C (Fig. 1). In the phase inversion range, a microemulsion forms, leading to the so-called Kahlweit fish in the phase diagram. At low emulsifier concentrations (below 15%) the microemulsion phase reaches the equilibrium with an oil and an aqueous phase, forming a so-called three-phase microemulsion (w + D + o) in the body of the Kahlweit fish. Emulsifier concentrations over 15% are sufficient to solubilize the entire oil and water quantity in the form of a single-phase microemulsion (D) or a lamellar phase (L_α).

The hydrophilic–lipophilic properties of the fatty alcohol ethoxylates are balanced in the PIT range, which is manifested in a distinct minimum of interfacial tension (Fig. 1) [11].

In principle a similar figure results for emulsions consisting of oil, water and an emulsifier mixture from APG with a hydrophobic coemulsifier when the mixing ratio of APG and hydrophobic coemulsifier is varied instead of the temperature as a formulation parameter. At the optimal mixing ratio of 1:1 the system from dodecane, water, C_{12-14}-APG and the hydrophobic coemulsifier sorbitan monolaurate (SML) forms microemulsions (Fig. 2). Higher SML proportions lead to w/o emulsions, higher APG proportions form o/w emulsions. Again, a Kahlweit fish

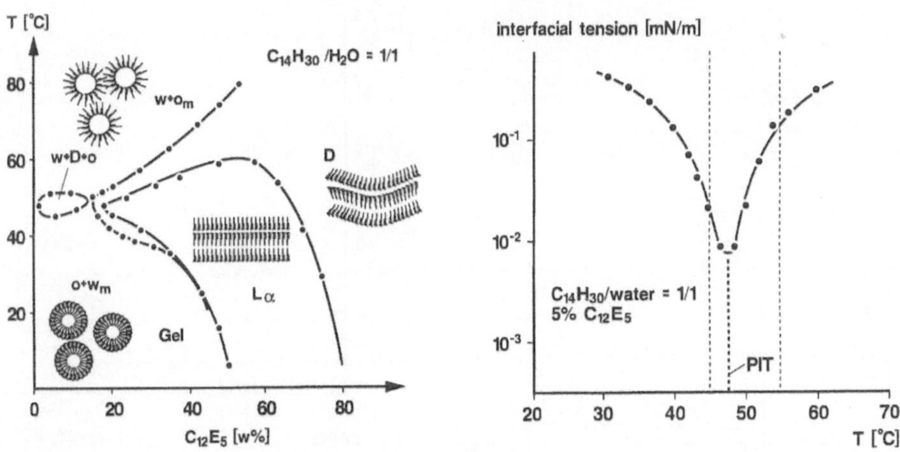

Fig. 1 Phase behavior and interfacial tension of tetradecane-water emulsions containing $C_{12}E_5$ (from (1) and (11))

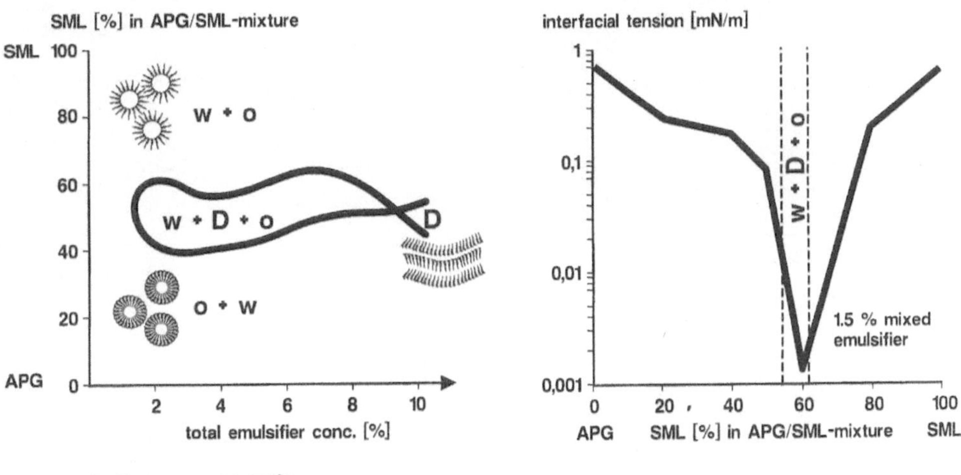

Fig. 2 Phase behavior and interfacial tension of dodecane-water emulsions containing APG/SML mixtures

develops in the phase diagram through variation of the total emulsifier concentration, with a three-phase microemulsion in the body and a single-phase microemulsion in the tail.

The similarity of the phase inversion range of both surfactant types is not only apparent in the phase behavior, but also in the course of the interfacial tension of the emulsifier mixture. The hydrophilic–lipophilic properties of the emulsifier mixture are balanced at the APG/SML ratio of 4:6 and the interfacial tension is minimal. Even the order of magnitude of the interfacial tension value (minimum below 10^{-2} mN/m) is comparable for both surfactant types (Fig. 2).

In the case of the APG-containing microemulsion, the high interfacial activity is due to the fact that the hydrophilic APG with the voluminous polyglycoside head group is mixed at the ideal ratio with the hydrophobic coemulsifier SML with small head group in the oil/water interface. In contrast to ethoxylated nonionic surfactants the hydration – and therefore the effective size of the head group – is hardly temperature dependent [5, 7, 10] so that no temperature dependence is expected for the phase

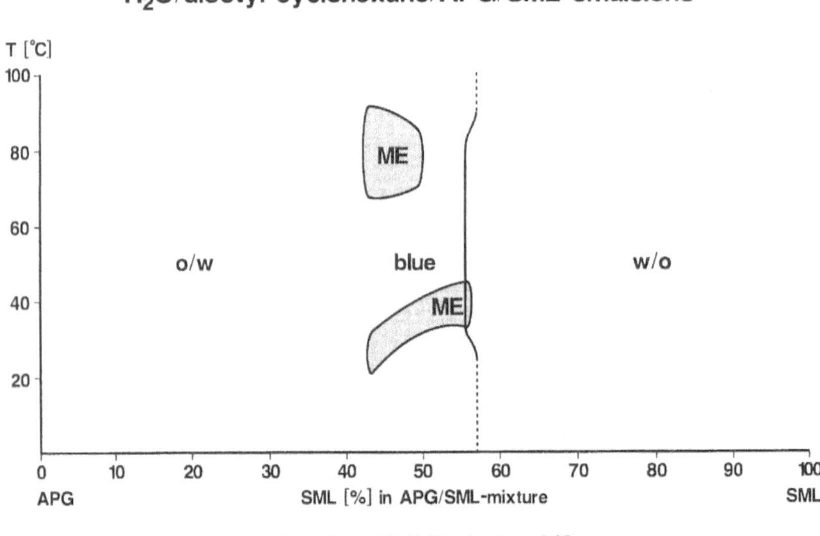

Fig. 3 Influence of temperature on emulsion systems containing dodecane, water, APG and SML

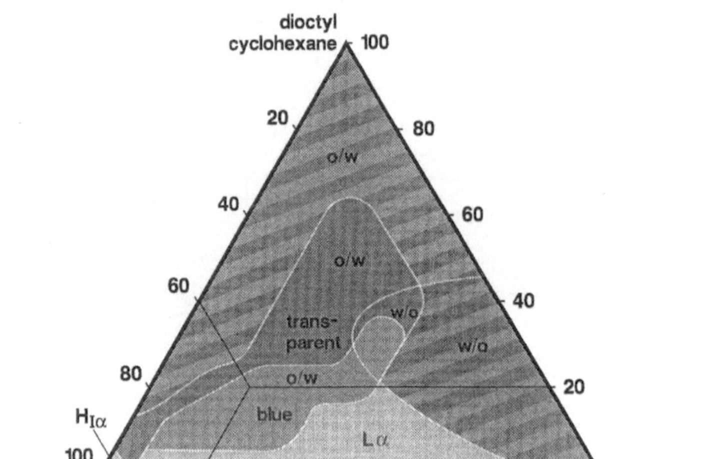

Fig. 4 Pseudo ternary phase diagram of APG, FAES, SML, dioctyl cyclohexane and water at 25 °C and a fixed water content of 60%

behavior of the emulsion (as is also true for the interfacial tension [9]). This assumption is true, as shown in Fig. 3 for the system dioctyl cyclohexane, water, C_{12-14}-APG and SML. Independent of the temperature the system forms transparent microemulsions or very fine-disperse blue emulsions (particle size around 100 nm) for APG/SML ratios between 45:55 and 55:45. The microemulsions and the blue emulsions are electrically conductive in the entire APG/SML mixing range, i.e., they are of the o/w type.

In order to be able to use APG microemulsions for technical or cosmetic applications tailor-made formulations are required which show specific application properties. Figure 4 shows the pseudo-ternary phase triangle of a five-component system from cosmetic raw materials for a constant water content of 60%. The emollient is dioctyl cyclohexane. A 5:3 mixture of APG and a fatty alcohol ether sulfate (FAES) serves as hydrophilic emulsifier. The anionic surfactant is added due to its high foaming power,

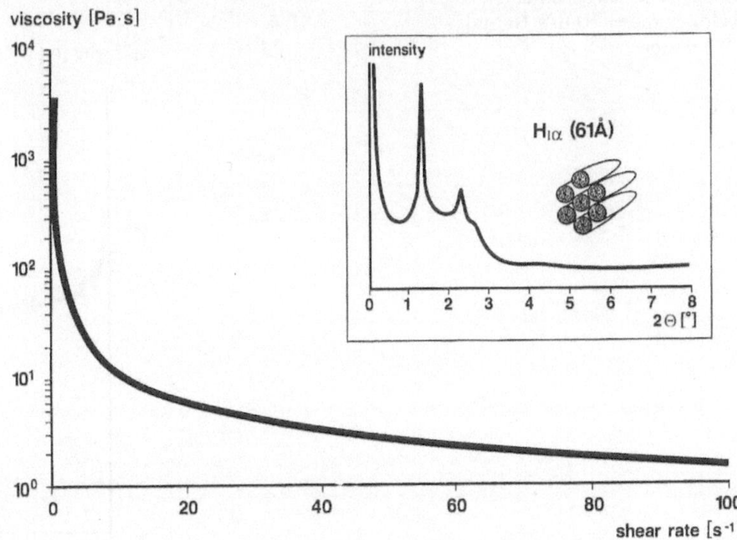

Fig. 5 Viscosity behavior and phase structure of APG/FAES/SML/dioctyl cyclohexane = 25/15/0/0 and 60% water

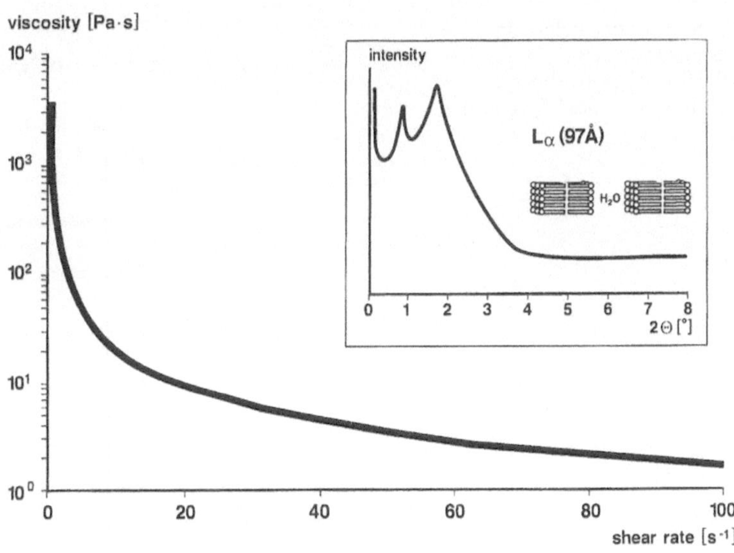

Fig. 6 Viscosity behavior and phase structure of APG/FAES/SML/dioctyl cyclohexane = 10/6/24/0 and 60% water

Table 1 Phase and viscosity behavior of selected systems (systems with 60% water content from Fig. 4 at 25 °C)

APG/FAES/SML/ dioctyl cyclohexane	Phase	Repeat distance/nm	Viscosity/Pa s at 1/s	10/s
25/15/0/0	$H_{I\alpha}$	6.1	113	11
22/14/4/0	L_α	7.0	2.0	2.0
10/6/24/0	L_α	9.7	201	19
10/6/16/8	L_α	11.9	151	12
10/6/8/16	bicont. ME		0.6	0.3
5/3/16/16	w/o ME		2.6	2.6

as common in a large variety of body cleansing agents. Therefore, the marked formulation in the blue o/w emulsion range contains 15% APG, 9% FAES, 8% SML, 8% dioctyl cyclohexane and 60% water.

Starting from an oil- and coemulsifier-free system a 40% APG/FAES mixture forms a viscous hexagonal liquid crystal ($H_{I\alpha}$) in water. The viscosity curve in Fig. 5 indicates the shear-thinning behavior for this hexagonal phase.

Only a small fraction of the APG/FAES mixture has to be exchanged for the hydrophobic cosurfactant SML in order to obtain a low-viscous lamellar phase (Table 1 and

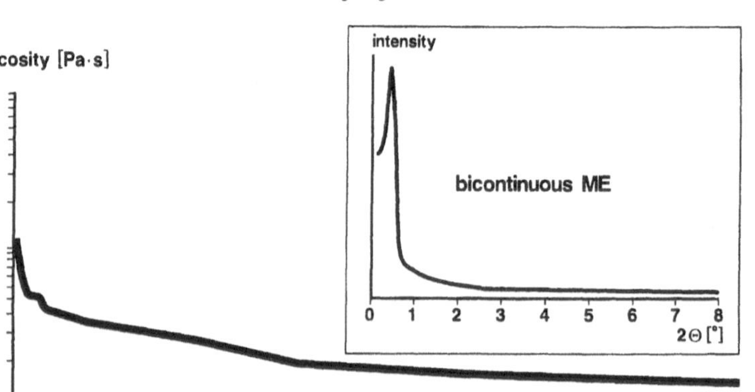

Fig. 7 Viscosity behavior and phase structure of APG/FAES/SML/dioctyl cyclohexane = 10/6/8/16 and 60% water

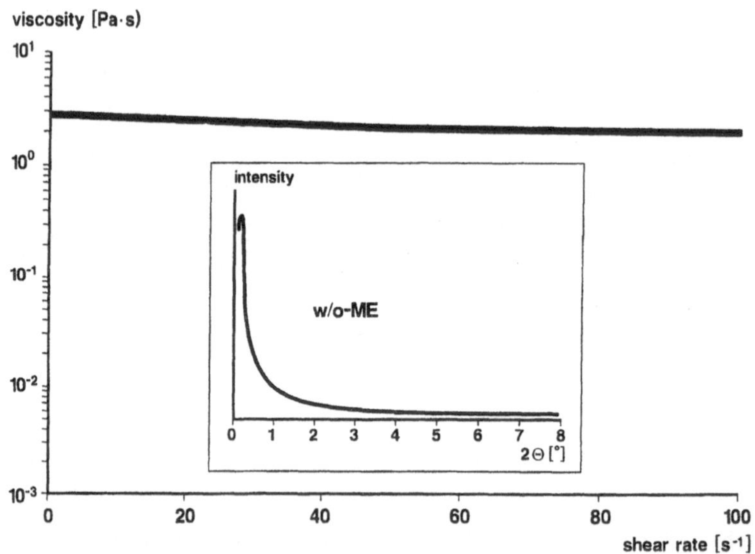

Fig. 8 Viscosity behavior and phase structure of APG/FAES/SML/dioctyl cyclohexane = 5/3/16/16 and 60% water

Fig. 4). An increase of the SML proportion maintains the lamellar phase but leads to a clearly increased viscosity which is higher than that of the hexagonal phase. Like the hexagonal phase the lamellar phases show also shear thinning (Fig. 6).

The two liquid crystals show very different reactions to oil addition: While the hexagonal liquid crystal can only take up very low oil quantities, the lamellar phase extends far into the direction of the oil corner (Fig. 4). The ability of the lamellar liquid crystal to take up oil increases clearly

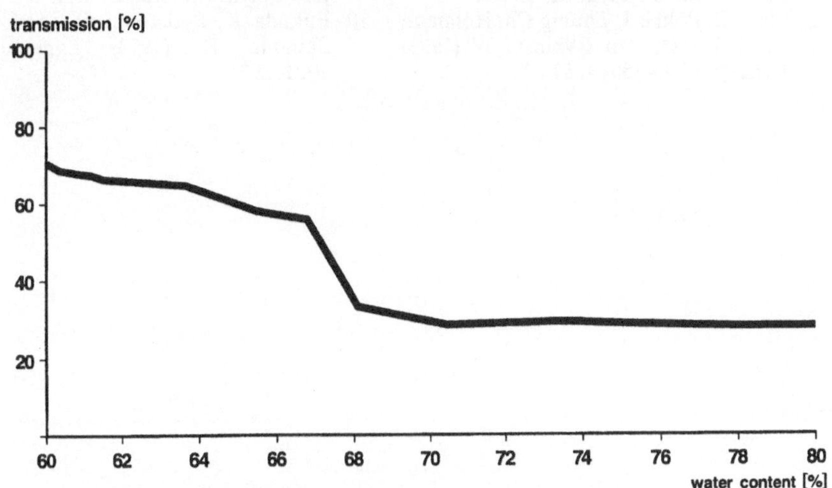

Fig. 9 Dilution of a microemulsion containing APG/FAES/SML/dioctyl cyclohexane = 5/3/8/24 with water

with increasing SML proportion. Oil solubilization leads to a larger repeat distance between the lamellar layers (from 9.7 nm to 11.9 nm in the case of incorporation of 20% dioctyl cyclohexane). Therefore, the oil molecules are stored in the hydrophobic interior of the lamellar layers and not incorporated into the palisade layer parallel to the emulsifier molecules.

In the case of higher oil fractions, blue emulsions or transparent microemulsions are obtained; in the case of high SML proportions they are of w/o type (over 50% in the emulsifier mixture). At the optimal mixing ratio between hydrophilic APG/FAES-compound and hydrophobic SML the transparent microemulsion range extends to very high oil contents of the emulsion: a maximum of approx. 60% dioctyl cyclohexane can be microemulsified. The microemulsions are low-viscous and show almost Newtonian flow behavior; the w/o microemulsions have a slightly higher viscosity than the o/w microemulsions (Figs. 7 and 8). The SAXS spectra both of the w/o- and the o/w microemulsions show peaks at small q-values (0.33 and 0.18 nm^{-1}, respectively) and a q^{-4}-dependence at large scattering angles. This behavior, which is characteristic of microemulsions [12], can be explained with the presence of a bicontinuous disordered structure formed by water and oil domains [13] with a length scale of a few 10 nm.

Microemulsions with a high APG/FAES proportion are potentially suitable for cosmetic products: The APG/FAES-compound contributes to cleaning performance and foam while the oil acts as a refatting care component. In order to achieve a refatting effect with a microemulsion, the oil must be released, i.e., the microemulsion must separate during application. For a special model formulation the sharp increase of turbidity on dilution with water (Fig. 9) indicates the breakdown of the microemulsion. Thus after cleansing the consumer will perceive a slight caring oil film on his skin. In summary, it can be stated that on account of their good skin compatibility, APG microemulsions represent an interesting new application medium for cosmetics.

Acknowledgments The authors sincerely thank Mr. Marcus Claas and Mr. Jürgen Stodt for preparation of samples and thorough execution of the measurements.

References

1. Kahlweit M, Strey R (1985) Angew Chem 97:655–669
2. Shinoda K, Kunieda H (1983) In: Becker P (ed) Encyclopedia of Emulsion Technology, Vol 1. Marcel Dekker, New York, pp 337–367
3. Förster T, von Rybinski W, Wadle A (1995) Advances in Colloid and Interface Sci 58:119–149
4. Andree H, Middelhauve B (1991) Tenside Surf Det 28:413
5. Balzer D (1991) Tenside Surf Det 28:419–427
6. Warr GG, Drummond CJ, Grieser F, Ninham BW, Evans DF (1986) J Phys Chem 90:4581–4586

7. Hofmann R, Nickel D, von Rybinski W, Platz G, Pölike J, Thunig Ch (1993) Progr Colloid Polym Sci 93:320
8. Platz G, Pölike J, Thunig Ch, Hofmann R, Nickel D, von Rybinski W (1995) Langmuir 11:4250–4255
9. Kutschmann EM, Findenegg GH, Nickel D, von Rybinski W (1995) Colloid Polym Sci 273:565–571
10. Fukuda K, Söderman O, Lindman B, Shinoda K (1993) Langmuir 9: 2921–2925
11. Aveyard R, Binks BP, Fletcher PDI (1989) Langmuir 5:1210–1217
12. Lichterfeld F, Schmeling T, Strey R (1986) J Phys Chem 90:5762–5766
13. Chen SH, Chang SL, Strey R (1990) Progr Colloid Polym Sci 81:30–35

Progr Colloid Polym Sci (1996) 101:113–115
© Steinkopff Verlag 1996 COLLOIDS

H.-D. Dörfler
A. Große

Applications of microemulsions with different compositions in detergents technique

Prof. Dr. H.-D. Dörfler (✉) · A. Große
TU Dresden
Fachrichtung Chemie
Lehrstuhl für Kolloidchemie
Mommsenstraße 13
01062 Dresden, FRG

Abstract Basing on the knowledge of their phase diagrams, model textile detergency tests with selected cleaning media of such compositions that form at the given washing temperatures homogeneous, bicontinuous microemulsions were performed. These cleaning media consisted of quaternary systems of the type water/Halpaclean/Marlipal/*n*-pentanol. As model test fabric a polyester-cotton blend (65:35) soiled with 6 mg pigments and 42 mg wool fat and sebum per gram fabric was used. The composition of the cleaning media was varied by changing the ethoxylation number of the Marlipals or the concentration of *n*-pentanol in the quaternary system.

The results of the detergency tests with microemulsions of different compositions showed that the detergency of these media especially at low temperatures $T < 313$ K is notably higher than that of a standard detergent solution. The optimum detergency of microemulsion systems corresponds to a minimum of the interfacial tension γ between the microemulsion and the water or oil excess phase in the heterogeneous regions of the phase diagram.

Key words Bicontinuous microemulsions – textile detergency – quaternary system – water/Marlipal/Halpaclean/*n*-pentanol

Introduction

It is well known that microemulsions exist in defined phase regions of multicomponent systems consisting of water/hydrocarbon/surfactant/cosurfactant. They are distinguished by ultralow interfacial tensions against both hydrophilic and lipophilic substances and a high solubilization power [1]. Because of these properties, microemulsions are interesting media also for the application in textile detergency [2, 3].

The aim of the present work was to compare the detergency of select microemulsions of the ternary system water/Halpaclean/Marlipal/*n*-pentanol for oil- and pigment-soiled test fabrics with the detergency of a standard detergent solution.

Halpaclean (Haltermann GmbH, Hamburg) is a hydrocarbon mixture which contains > 95 wt-% *n*-undecane, Marlipal (Hüls AG, Marl) are commercial grade alkyl polyglycol ether mixtures C_iE_j with hydrocarbon chain lengths between $i = 12$ and $i = 14$ and with mean ethoxylation numbers $(j) = 4, 5, 6, 7,$ and 8.

The standard detergent solution consisted of the following composition: 19.25 g basic powder; 5 g perborate-tetrahydrate; 0.75 g activator tetraacetylethylene diamine per 1 liter of distilled water. The detergency results of standard detergent solution were compared with the results using microemulsions. The composition of the cleaning media was varied either by changing the mean ethoxylation number of the Marlipal or by adding defined amounts of *n*-pentanol. Further experimental details are given in [4, 5].

Table 1 Composition of the ternary (samples 1–3) and quaternary systems (samples 4–6) consisting of water/Halpaclean/Marlipal/*n*-pentanol applied in model detergency tests. Symbols $\{j\}$ mean ethoxylation number of the Marlipal; ΔT temperature interval in which homogeneous microemulsions are formed

Sample no.	Water [wt-%]	Halpaclean [wt-%]	Marlipal [wt-%]	$\{j\}$	*n*-pentanol [wt-%]	ΔT [K]
1	45.0	45.0	10	4	0.0	306–312
2	44.5	44.5	11	5	0.0	327–332
3	43.5	43.5	13	6	0.0	337–340
4	42.7	42.7	12.7	6	1.9	324–328
5	41.8	41.8	12.5	6	3.9	308–313
6	41.0	41.0	12.3	6	5.7	297–302

As test fabric the polyester-cotton blend wfk 20 C (wfk-Testgewebe GmbH, Brüggen), soiled with 6 mg pigments and 42 mg wool fat and sebum per gram fabric was applied.

Results

Starting from the phase diagrams of the four-component systems water/Halpaclean/Marlipal/*n*-pentanol described by us in [5], selected samples from this system were used in model detergency tests. The composition of these samples is given in Table 1. They consist of equal parts water and Halpaclean. Thus the microemulsions formed by the samples in the temperature intervals ΔT (see Table 1) are bicontinuous and show in the heterogeneous regions of the phase diagram minimum interfacial tensions ($\gamma \approx 10^{-3}\,\mathrm{mN\,m^{-1}}$) against the water and oil excess phases [1]. The washing temperatures were chosen in the middle of these temperature intervals, i.e., they correspond approximately to the phase inversion temperature of the quaternary systems water/Halpaclean/Marlipal/*n*-pentanol (samples 4–6).

In the case of samples 1–3 we worked in the ternary system without *n*-pentanol. The temperature interval ΔT in which homogeneous microemulsions are formed, was tuned by variation of the ethoxylation number of the Marlipal for the samples 1–3 and by addition of different amounts of *n*-pentanol for the samples 3–6 (see Table 1).

In model washing experiments the detergency of the microemulsions sample 1 at $T = 308$ K, sample 2 at $T = 323$ K, sample 3 at $T = 338$ K, sample 4 at $T = 325$ K, sample 5 at $T = 311$ K and sample 6 at $T = 298$ K, i.e., at temperatures within the limits of ΔT of each sample was evaluated and compared with the detergency of the standard detergent solution at the same temperature.

In Figs. 1 and 2 is summarized the results of our washing experiments. The reflectance differences ΔR of the test fabrics in Fig. 1 after being cleaned by the micro-emulsions samples 1–3, and in Fig. 2 the treated with

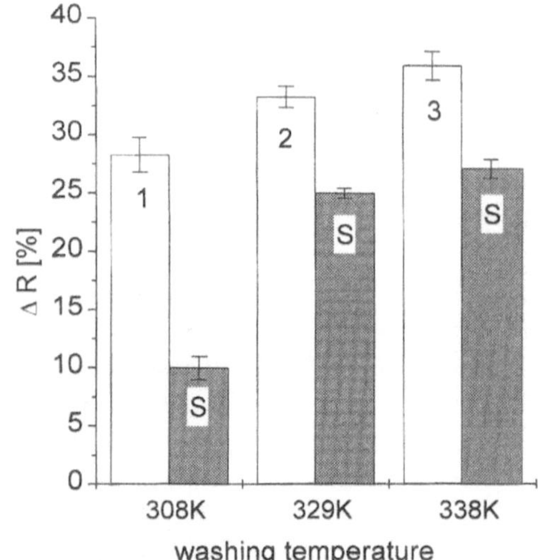

Fig. 1 Reflectance differences ΔR of polyester-cotton blend fabric wfk 20 C after having been cleaned with the microemulsions formed by sample 1 at $T = 308$ K (1) sample 2 at $T = 329$ K (2), and sample 3 at $T = 338$ K (3) – samples according to Table 1 – in comparison to the reflectance differences after cleaning with the standard detergent solution (S) at the same temperatures

microemulsions of samples 3–6 leads to the following results: For comparison, the reflectance differences gained by washing with the standard detergent solution (S) at the same temperatures are plotted. Both experiments show the same tendency: Already at the lowest washing temperatures ($T = 298$ K and $T = 308$ K) the soil removal by the microemulsions formed by the samples 1 and 6, resp. is significantly higher than by the standard detergent solution at high solution at high temperatures ($T = 338$ K). By this comparison it must be mentioned that the standard detergent solution contains additionally builders, bleaching and antiredeposition agents and optical brighteners. The use of the samples 2 and 3 or 5 and 4 (see Table 1) at higher washing temperatures improves the detergency of microemulsions slightly while the detergency of the standard detergent solution increases notably. This

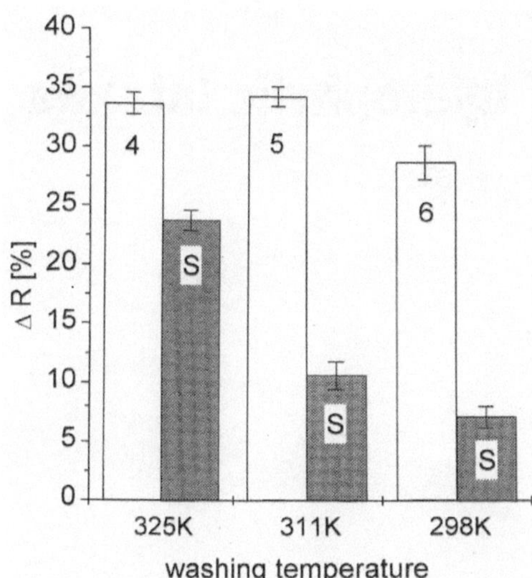

Fig. 2 Reflectance differences ΔR of the polyester-cotton blend fabric wfk 20 C after having been cleaned with the microemulsions formed by sample 6 at $T = 298$ K (6), sample 5 at $T = 311$ K (5), sample 4 at $T = 325$ K (4) – samples according to Table 1 – in comparison to the reflectance differences after cleaning with the standard detergent solution (S) at the same temperatures

corresponds to the fact that the interfacial tensions between the microemulsions and the water or oil excess phase in the heterogeneous regions of the phase diagram [5] is of the same order of $\gamma \approx 10^{-3}$ mN m^{-1} for all samples.

A comparison between Figs. 1 and 2 shows that the samples 5 and 6 containing the Marlipal $C_{12/14}E_6$ and n-pentanol exhibit at about the same washing temperature the same detergency in the limits of experimental error as the pentanol-free system sample 1 with the lower ethoxylated Marlipal $C_{12/14}E_4$. Thus it does not influence the detergency of the samples investigated here whether the composition of the microemulsions at a given washing temperature is varied by decreasing the ethoxylation number of the Marlipal or by addition of n-pentanol. If a recipe according to sample 1 with a low ethoxylated Marlipal $(\{j\}) = 4$ and without or with cosurfactant concentration $c_{co} < 1$ wt-% n-pentanol is used, however, the quantity of Marlipal and n-pentanol needed to prepare a homogenous microemulsion is lower while the detergency of the microemulsion remains constant.

Further model detergency tests showed that the detergency of a microemulsion is nearly independent on the liquor ratio. If only 2 g of the microemulsions from sample 1 where applied per gram of the test fabric, i.e., if the fabric was completely wetted by the cleaning medium, its maximum detergency was nearly reached. The existence regions of the microemulsions and the minimum concentration of Marlipal needed to prepare homogeneous microemulsions were not drastically affected by the cleaning process. This allows the conclusion that microemulsions can be used in very small quantities and can be re-used several times.

References

1. Kahlweit M, Strey R (1985) Angew Chem 97:655
2. Schwuger MJ, Schomäcker R, Stickdorn K (1995) Chem Rev 95:849
3. Solans C, Garcia Dominguez J, Friberg SE (1985) J Disp Sci Technol 6:523
4. Dörfler HD, Große A (1996) Tenside Surf Det; in press
5. Dörfler HD, Große A, Krüßmann H (1996) Tenside Surf Det; in press

Progr Colloid Polym Sci (1996) 101:116–119
© Steinkopff Verlag 1996

R. Zimehl
J. Priewe

Colloid stability of hydrophylic latexes

Dr. R. Zimehl (✉) · J. Priewe
Institute of Inorganic Chemistry
University of Kiel
Olshausenstraße 40-60
24098 Kiel, FRG

Abstract Results of coagulation experiments on negatively charged polystyrene latexes are reported. The different latex dispersions were stabilized by electrostatic, electrosteric and steric mechanisms. For the electrostatic and electrosteric stabilized dispersion the effectiveness of monovalent cations as coagulants increases according to the lyotropic sequence. For sterically stabilized latex dispersions monovalent cations as coagulants proved to be ineffective. With sodium tetraphenyl borate a narrow instability region at low electrolyte concentration was detected. The specific ion effects were evaluated and interpreted in the light of hydration regulation. The hydration forces may be modified or regulated by exchanging ions of different hydration on the surfaces of approaching particles.

Key words Latexes – stability – adsorption – ion effects – coagulation

Introduction

The stability of a colloidal dispersion is dominated by a complex interference of van-der-Waals forces, electrostatic interactions and osmotic forces (sterically stabilized systems). For technical application dispersions usually should have high solid volume fraction and the primary particles will approach close to one another. At these small particle separations the repulsive interactions must be dominant to ensure colloidal stability. In practice this is often achieved by steric or electrosteric stabilization.

To investigate the stability of colloidal systems polystyrene latexes have been used as almost ideal model dispersions. Especially amphoteric latexes are used to simulate the properties of dispersed pigment and oxide particles. The amphoteric polystyrene latex dispersions exhibit an excellent colloid stability, but a hydrated surface layer of polyelectrolyte chains renders the particle nature from hydrophobic to hydrophilic. The stability of these hydrophilic latexes can no longer be described by the classical electrostatic theory (DLVO theory). Additional non-DLVO forces have to be taken into account to describe the stability of hydrophilic particles in a more appropriate way.

Materials and methods

Chemicals

Ion exchanged water was used throughout all experiments. Sodium tetraphenyl borate was a Fluka product of 99% purity. All other electrolyte were of A.R. quality, α-Methacryloyl-ω-methoxy-poly(ethylenglycol) (Bisomer) was a kindly gift from BP-chemicals and was used as received.

Latex dispersions

Electrostatically stabilized polystyrene latex dispersions were prepared by emulsifier free emulsion polymerization of styrene with potassium peroxo disulfate [1, 2].

The synthesis of electrosterically stabilized dispersions by emulsion copolymerization of styrene with potassium styrene sulfonate is given in detail in other publications [2–4].

Sterically stabilized latices were synthesized by copolymerization of a styrene/divinylbenzene mixture (1% DVB by weight) with functionalized PEO of different chain length (Bisomer) [5, 6].

Coagulation experiments

Stability measurements were performed by different techniques which have been already described in literature [1, 2, 7]:

Test-tube experiments: A series of dispersions containing increasing amounts of salts were prepared by adding 1 ml of the salt solution to 1 ml of the latex dispersion. The electrolyte concentration at which a floc phase is formed is determined by visual inspection after a sufficient equilibration time (usually 24–48 h).

Photometric experiments: The optical density D of dispersions at different salt concentration c is recorded as a function of time t. From dD/dc constant time the critical coagulation concentration is derived.

Quasi elastic light scattering: Photon correlation spectroscopy (Malvern Zetasizer 4) was used to determine the particle size distribution of some coagulating dispersions.

Results

A very simple but quite effective method for determining the stability of a colloidal system is the so-called test tube experiment. The critical electrolyte concentration at which a floc phase is formed is displayed in Fig. 1 for electrostatically stabilized latex dispersions. It can be seen that the valency of the gegenion mainly controls the dispersion stability. Closer inspection of the coagulation process for some monovalent gegenions with light scattering methods show small differences in the critical coagulation concentration.

Figure 2 exemplifies the influence of ionic comonomers in emulsion copolymerization on dispersion stability. The mean critical coagulation concentration of the monovalent gegenions taken from Fig. 1 for the electrostatically stabilized polystyrene dispersions is included in Fig. 2 as a dashed line. The coagulation behavior of the electrosterically stabilized dispersions is superficially similar to that of the conventional charged latexes but the relatively high critical coagulation concentrations could not be accounted for solely in terms of an enhancement of electrostatic repulsion.

The salt stability of our dispersions could be much more enhanced by introducing nonionic stabilizers such as polyethylene oxide chains or dextran molecules to the

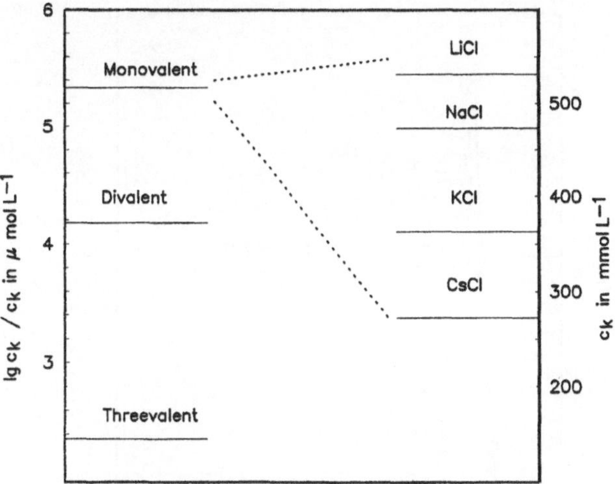

Fig. 1 Critical coagulation concentration of anionic polystyrene latexes, prepared in the presence of potassium peroxo disulfate

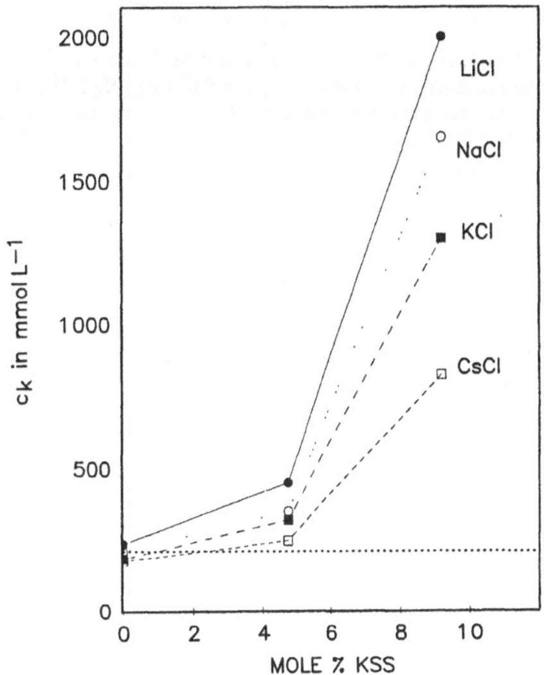

Fig. 2 Critical coagulation concentration of electrosterically stabilized latexes, prepared in the presence of different amounts of potassium styrene sulfonate ("KSS")

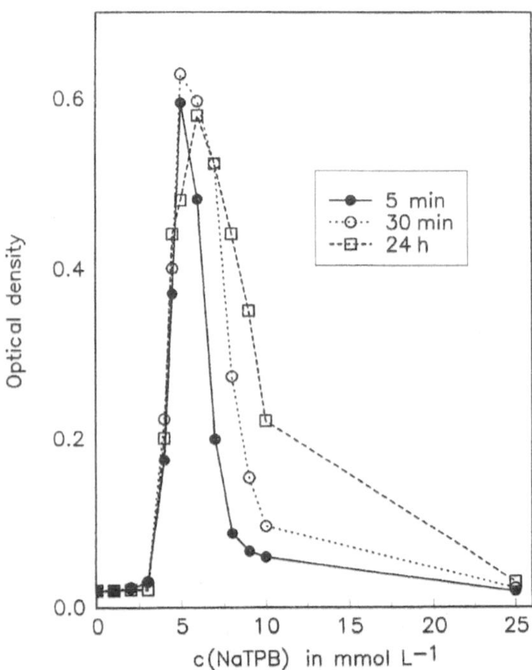

Fig. 3A Optical density plots for polyethylene oxide stabilized dispersions in the presence of sodium tetraphenyl borate (NaTPB); the photometric measurements were performed at different times after mixing

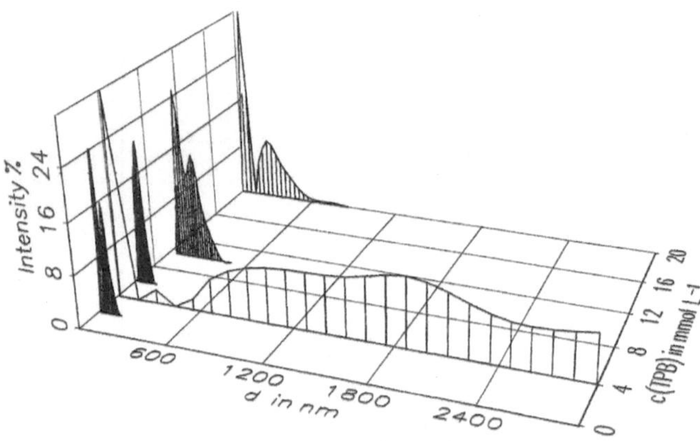

Fig. 3B Particle size distribution of the same dispersion as in Fig. 3a at different sodium tetraphenyl borate concentrations; measurements were performed 30 min after mixing

particle surface. It was nearly impossible to coagulate the sterically stabilized latex dispersions by electrolyte addition. The polymer particles must be forced to separate from the dispersion by increased temperature and very high salt concentration. The process is rather a phase separation than flocculation or coagulation. Surprisingly, we found that some of the sterically stabilized latex dispersions could readily be coagulated in the presence of sodium tetraphenyl borate. Small instability regions were detected by light scattering methods at tetraphenylborate concentrations between 5 to 10 $mmol \cdot L^{-1}$ (Fig. 3a) and a domain of aggregated particles was unequivocally indicated as a diffuse particle size distribution by photon correlation spectroscopy (Fig. 3b).

Discussion

We think that the differentia in coagulation of our systems can be readily interpreted in terms of an exchange process at the solid liquid interface.

Let us consider a latex dispersion stabilized by sulfate groups at the surface of the polymer particles. If some electrolyte, e.g. caesium chloride, is added to the dispersion, any ion in the vicinity of the sulfate group will be displaced by caesium ions. Some of the ions may bind directly to the sulfate groups to form an inactive ion pair, but most of them remain in the diffuse double layer around the isolated particles. Caesium is weakly hydrated and the critical coagulation concentration is indeed explicable by the DLVO theory. However, by adding the same amount of lithium chloride to the dispersion the particles will not aggregate to larger flocs, because of stronger hydration forces arising from the binding of more hydrated lithium ions.

By copolymerization of styrene with ionic comonomers such as styrene sulfonate core shell particles are created. Highly swollen polyelectrolyte layers at the polymer particle surface causes an additional repulsive term. The stability of the dispersions is enhanced and the specific ion effects are still operative.

Even across uncharged lipid bilayers repulsive forces have been measured [8]. Thermally excited hydrophilic polymer chains and headgroups such as strongly hydrated sugar molecules protruding from the surfaces overlap as they approach each other. The repulsive force between them is then essentially entropic. On the other hand, it was confirmed through the examination of the electrophoretic mobility of polystyrene latexes bearing disaccharide surface groups that the presence of hydrated oligosaccharide chains can affect the structure of the double layer around the latex particles [9]. In this respect the behavior of our sterically stabilized dispersions in the presence of tetraphenyl borate ions is not too surprising. Polyethylene oxide interacts strongly with water molecules and highly swollen polymeric layers at the surface of the polystyrene core repel each other. Hydrated lithium or caesium ions cannot alter the hydration layer repulsion between latex particles but tetraphenyl borate ions might interact

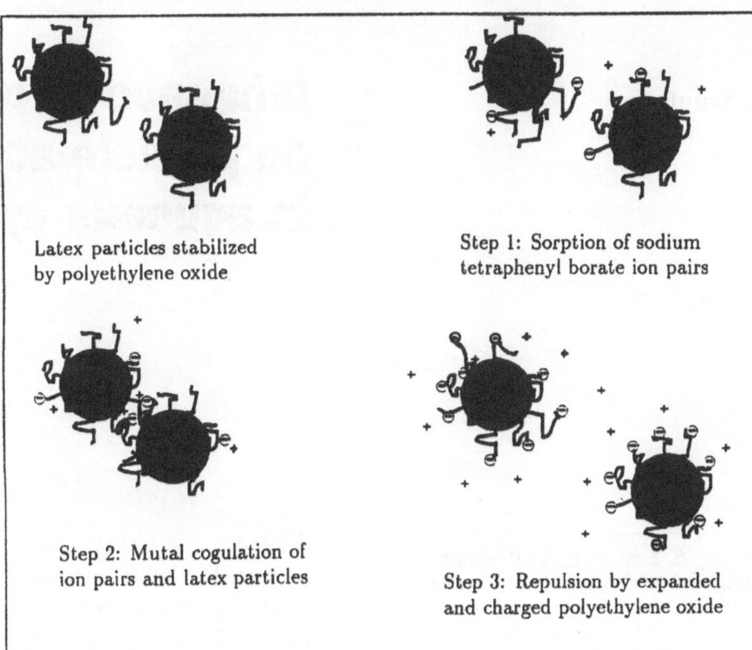

Fig. 4 Schematic presentation of the coagulation process of sterically stabilized polystyrene/co-polyethylene oxide latex particles

specifically with the hydration layer of the polyethylene oxide modified polystyrene particles and cause mutual coagulation.

In short, we suggest the following model for the highly specific interaction of sodium tetraphenyl borate with our polyethylene oxide stabilized polystyrene particles (Fig. 4):

Sodium tetraphenyl borate ion pairs are preferentially sorbed at the hydrated polyethylene oxide surface layer. At a concentration of about 4 mmol L^{-1} mutual coagulation of sodium tetraphenyl borate ion pairs and latex particles brings about a very shallow aggregation domain. At somewhat larger tetraphenyl borate concentrations the polyethylene oxide chains are increasingly charged by tetraphenyl borate anions. The polyethylene oxide layer expands and the sodium ions form a diffuse layer of countercharges. The dispersion character now resembles that of an electrostatically stabilized latex [6].

Conclusions

From the previous discussions we can infer that the stability of hydrophobic and even more of hydrophilic polystyrene latex dispersions is not of a simple character. Between two approaching hydrophilic particles there is an exponentially repulsive hydration or structural force. The main reason for the additional repulsion is the strong interaction of stabilizing polymer chains with water: hydrophilic polymer segments immobilize a large amount of water to form a gel-like stabilizing shell at the particle water interface. The strength of the repulsive force depends on the energy needed to disrupt the hydrogen-bonding network and to dehydrate the particle surfaces as they approach each other.

Acknowledgement We are grateful to Prof. G. Lagaly for helpful discussions.

References

1. Zimehl R, Lagaly G (1986) Progr Colloid Polym Sci 72:28–36
2. Zimehl R, Lagaly G (1987) Colloids and Surfaces 22:225–236
3. Zimehl R, Lagaly G, Ahrens J (1990) Colloid Polym Sci 268:924–933
4. Zimehl R, Wieboldt J (1989) TIZ International Powder Magazine 113:629
5. Ottewill RH, Satgurunathan R (1987) Colloid Polym Sci 265:845–853
6. Priewe J, Lagaly G, Zimehl R, to be published
7. Leaver J, Brooksbank DV, Horne DS (1994) J Colloid Interface Sci 162:463–469
8. McInthosh TJ, Simon SA (1986) Biochemistry 25:4058–4066
9. Charreyre MT, Boullanger P, Delair Th, Mandrand B, Pichot C (1993) Colloid Polym Sci 271:668–679

Progr Colloid Polym Sci (1996) 101:120–124
© Steinkopff Verlag 1996

S. Schulz
R. Gimbel

Influence of polymeric auxiliaries on particle adhesion in aqueous systems

Dr.-Ing. S. Schulz (✉) · R. Gimbel
Gerhard-Mercator-Universität-GHS Duisburg
FB 7/FG 15 Wassertechnik
Bismarckstraße 90
47057 Duisburg, Germany

Abstract Turbidity is usually removed in water treatment by flocculation (followed by sedimentation or flotation) and/or deep bed filtration. To improve the removal efficiency polymeric auxiliaries are often used. It has been examined how these polymers effect the adhesion behavior of particles. The tested polymers differ in charge character, charge density and molecular weight. Experimental results point out that the adsorbability as well as the strengthening of adhesion force of polymer are considerably influenced by its electrostatic interaction with the particle surface. But there is no direct interdependence between adsorbability and strengthening of adhesion. Experiments show that besides the influence of the polymer properties there is a strong dependence on the particle contact time. The influence of the contact time is attributed to rearrangement processes of the polymer in the particle contact area and can be described as a first order reaction. The average particle adhesion force at the respective contact time is used as a measure for the actual polymer configuration in the adhesion area. The reaction velocity for the rearrangement is dependent on various parameters. It increases with rising polymer loading and decreasing molecular weight of polymer. In addition, the reaction takes place five-fold quicker in the case of preloading only one adhesion partner before contact instead of preloading both adhesion partners. Furthermore, the rearrangement of the polymer is dependent on the kind of polymer. Stronger interactions between polymer and adhesion partner cause stronger and longer lasting rearrangements than do weaker interactions. For the case of deep bed filtration it can be demonstrated that consequent transfer of the knowledge which has been established by fundamental examinations can optimize the use of polymer.

Key words Particle adhesion – polymeric auxiliaries – water technology – adsorbability – time dependence

Introduction

Flocculation in combination with sedimentation or flotation and/or deep bed filtration are the generally used methods in water technology for the removal of turbidity. To improve the separation efficiency polymeric auxiliaries are often added. So far the dose is mostly determined empirically. For a purposeful use of polymers knowledge about their mode of functioning is needed. For the theoretical description of a particle aggregation process or the particle deposition in a deep bed filter this process is usually divided in two successive steps: at first the transport, which brings the particles into contact with each other or with the filter grain surface, and second the following adhesion, which determines the stability of the particle deposition. Polymers can improve both: transport and adhesion. Within the presented investigations especially the influence on the particle adhesion is examined. Parameters of investigation are the polymer properties (charge character, charge density and molecular weight), the adsorbability of the polymer onto the particles and the particle contact time.

Method of examination

The adhesion process is studied on the model system micro glass beads (also called Ballotini) adhering to a smooth polished quartz glass plate. The surrounding medium is model water. The model water is a synthetic tap water made of bidistilled water with different salts. Figure 1 shows schematically the procedure of experimental realization. In this case only the quartz glass plate is preloaded with polymer.

In principle, the polymer may be adsorbed onto one or both adhesion partners. The resulting polymer loading is estimated by mass transfer equations. Subsequent to the adsorption the glass microspheres are brought into contact with the quartz glass plate by sedimentation. After a fixed contact time the particles are exposed to defined, normally acting separation forces (gravitational and/or centrifugal separation forces). The same segment of the quartz plate is photographed under a microscope before and after applying the separation force. By comparison of the two photographs the fraction of still adhering particles is determined. To quantify the results of various experiments an average adhesion force $F_{H50\%}$ is defined.

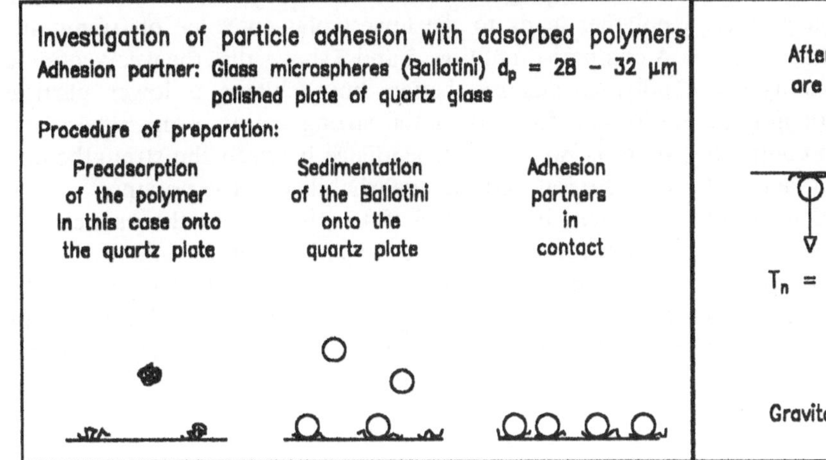

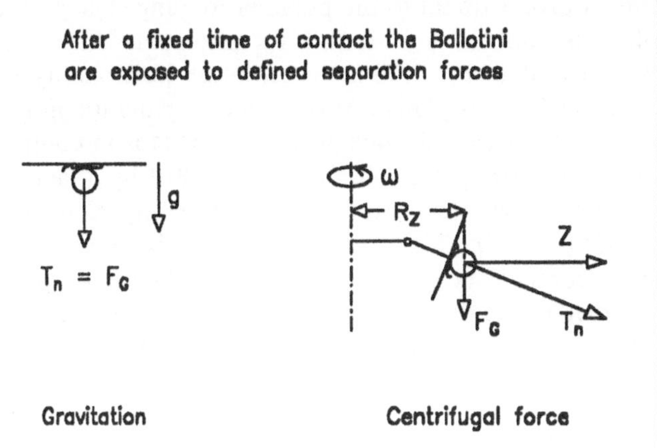

Fig. 1 Schematic representation of the experimental procedure [3]

Table 1 Characteristics of all polymers used [3]

Polymer name	Ionic character		Synthesis Copolymer weight%	mole%	Molecular weight determined by viscosimetry g/mole
Praestol 2530	anionic		31.4	25.7	$5.4 \cdot 10^7$
A 5709	anionic		30	24.5	$5.4 \cdot 10^6$
A 5708	anionic	acrylate	30	24.5	$1.7 \cdot 10^6$
A 5774	anionic		30	24.5	$1.2 \cdot 10^6$
Praestol 2500	weak anionic		1.9	1.4	$2.6 \cdot 10^7$
A 5104	weak cationic	tri-	5	1.8	$9.4 \cdot 10^6$
A 5013	cationic	metyl-	30–35	13–16	$2.8 \cdot 10^7$
A 6403	cationic	amino-	30–35	13–16	$1.3 \cdot 10^7$
A 6402	cationic	propyl-	30–35	13–16	$8.5 \cdot 10^6$
A 6436	cationic	acryl-	30–35	13–16	$6.2 \cdot 10^6$
A 6510	cationic	amide	30–35	13–16	$1.1 \cdot 10^6$

The average adhesion force is set equal to the separation force where 50% of the particles are still adhering.

Polymers

Table 1 gives a summary of the used polymers. These polymers are based on polyacrylamide and purchased from the Stockhausen company. They differ in charge character (anionic and cationic), charge density (weak and strong cationic respectively weak and strong anionic) and in molecular weight (in the range from $1.1 \cdot 10^6$ to $5.4 \cdot 10^7$ g mole).

Results of experiments

Figure 2 shows the influence of polymer charge character and charge density on the strengthening of particle adhesion. For all polymers the measured adhesion force is almost proportional to the polymer loading. The greatest effect on adhesion is found by adsorbing the strong cationic polymer A 5013 (constant of proportionality K is about 41.4 nN/(mg/m^2)). At the same polymer loading the effect of the weak cationic polymer A 5104 is in comparison to the strong cationic polymer A 5013 by a factor of about 6.3 lower (constant of proportionality K is about 6.6 nN/(mg/m^2). This corresponds to the six- to seven-fold lower content of cationic component of the weak cationic polymer in comparison to the strong cationic one. This means that the effect of strengthening the particle adhesion force by polymer is almost proportional to its content of opposed ions of adhesion partners. The weak anionic polymer Praestol 2500 shows in comparison to cationic polymers a lower effect on adhesion. Summarizing these observations it can be pointed out that strengthening of adhesion is decisively dependent on electrostatic interaction between polymer and adhering surfaces.

Furthermore, it was investigated whether there is a relationship between the adsorbability of polymer onto the adhering surfaces and the strengthening of adhesion. The adsorption experiments are carried out with quartz powder. Quartz powder has almost the same surface properties as the adhesion partners. For determining the residual polymer concentration the method of Burkert [1] is used, which has been improved by Sehn [2]. Figure 3 shows the obtained adsorption isotherms. The weak cationic polymer A 5104 reaches the highest adsorbability with a plateau value of about 4.2 mg/m^2 (Here with the plateau value the turning point of the isotherm is designated). The strong cationic polymer is less adsorbable with a plateau value of about 2.75 mg/m^2. The adsorbability of the weak anionic polymer leads to the lowest plateau value of 0.6 mg/m^2. A comparison of Figs. 2 and 3 shows that the weak cationic polymer adsorbs better, but achieves a lower particle adhesion force than the strong cationic one. Therefore, there is no direct dependence between the strengthening of adhesion and the adsorbability of the polymer.

Later, it was checked to what extent the particle adhesion force is influenced by the contact time of the adhesion partners. The results for adsorbed layers of the strong cationic polymer A 5013 are presented in Fig. 4 for two different cases. In the first case only the microglass

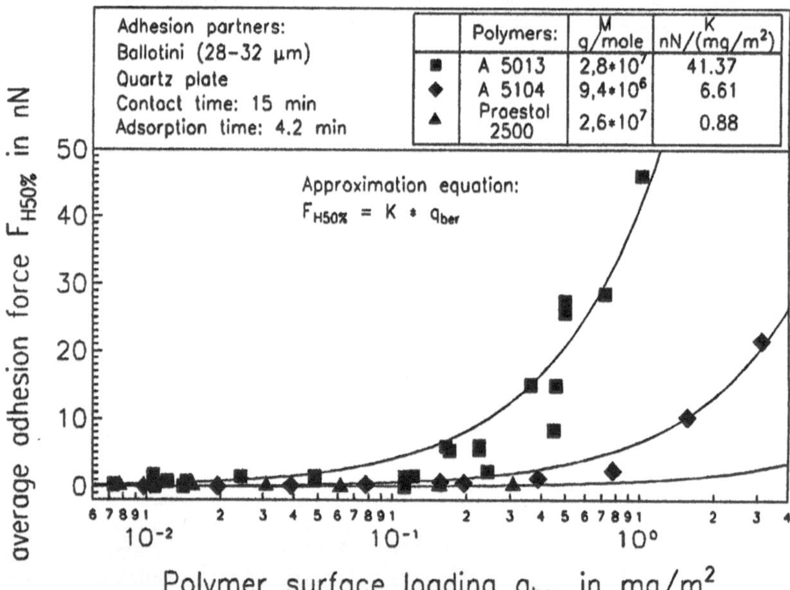

Fig. 2 Average particle adhesion force $F_{H50\%}$ in dependence of polymer loading for the strong cationic polymer A 5013, the weak cationic A 5104 und the weak anionic Praestol 2500[3]

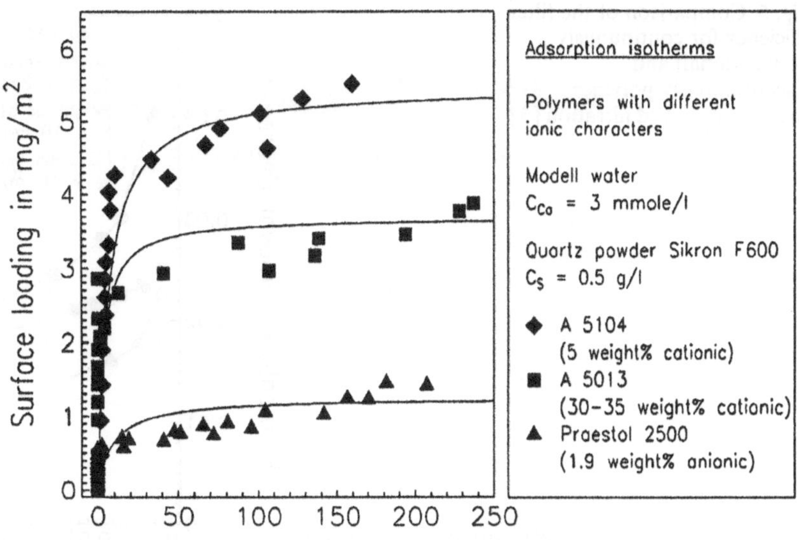

Fig. 3 Adsorption isotherms of the weak cationic polymer A 5104, the strong cationic A 5013 und the weak anionic Praestol 2500 onto quartz powder [3]

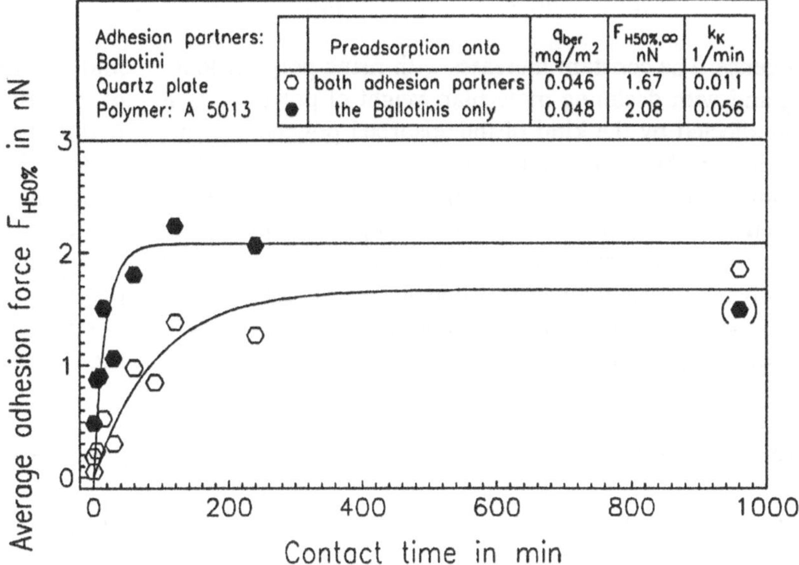

Fig. 4 Influence of contact time on the average adhesion force $F_{H50\%}$ by preadsorbing the strong cationic polymer A 5013 onto only one and onto both adhesion partners [3]

beads are preloaded and in the second case the polymer is equally distributed on both adhesion partners. In both cases the total amount of the polymer adsorbed in the contact area is the same. For both kinds of preloading a rise in the average particle adhesion force is observed with increasing contact time. The increase of particle adhesion force is attributed to time-dependent rearrangement processes of the polymer in the contact area.

The adhesion force in dependence of contact time is regarded as a measure of the respective polymer configuration in the contact area. Thus the polymer rearrangement can be described by the following equation of a first order reaction:

$$\frac{dF_{H50\%}(t_K)}{dt_K} = -k_K \cdot (F_{H50\%}(t_K) - F_{H50\%,\infty}) \tag{1.1}$$

$F_{H50\%}(t_K)$ average adhesion force at contact time t_K; $F_{H50\%,\infty}$ average adhesion force in equilibrium after long contact times ($t_K \to \infty$); k_K velocity constant of polymer rearrangements during particle contact.

By integration one obtains the following equation for describing the average adhesion force respectively the polymer configuration in dependence of the contact time:

$$F_{H50\%}(t_K) = F_{H50\%,\infty} \cdot (1 - e^{-k_K \cdot t_K}) \,. \tag{1.2}$$

Further investigations have shown that the constant of reaction velocity k_K is dependent on various parameters of

Introduction

The textile finishing industry is a high water-consuming industry. For producing one ton of textiles about 100–250 m^3 water is used. Within the main production steps like pretreatment, dyeing and finishing the pretreatment of textiles is very important for producing high qualitative endproducts [1–3]. Various natural and synthetic fiber admixtures (natural polymers, sizing agents) must be removed from the fiber surface by desizing, bleaching and washing processes. This purification step prepares the textile for the next production steps like dyeing and finishing. In addition to the mentioned substances a special class of auxiliary agents is to be considered. This group contains fiber preparations, which are special formulations of various lipophilic substances like mineral oils (light white oil), natural oils (fatty acid-triglyceryds), fatty acid esters (butyl stearate, tridecyl stearate) and ethoxylated substrates like polyoxyethylene esters. These auxiliary agents are mostly used as lubricants to obtain optimal friction behavior during the production of a fabric [4]. With regard to the removal of these lubricants problems may occur like impeded emulsification and possible oil redeposition on the fiber surface. Furthermore, conventional washing processes affect the environment, because in addition to the lubricants eliminated ecologically incompatible detergent components (surfactants, builders, complexation agents) are used in the liquor-bath. Economic aspects like increasing costs for the disposal of wastewater have to be considered too [5]. For these reasons new technologies must be developed to facilitate effective and ecologically compatible washing processes in the textile industry. The use of biotechnology offers interesting innovative attempts for solving these problems.

During the growth of microorganisms on lipophilic substrates like hydrocarbons, certain microbial cultures produce a variety of surface active substances, so called biosurfactants (Fig. 1). Such natural systems have a potential for ecologically washing textile, where fiber preparations such as lubricants can be degraded microbially and further mineralized to CO_2 and H_2O.

Biosurfactants have the function to increase the bioavailability of water-insoluble substrates. By producing a wide spectrum of natural surfactants the passive diffusion of the hydrocarbon substrates into the cell is facilitated. In the presence of natural surfactants liquid lipophilic substrates are emulsified, incorporated into micelles and spread over the cell wall. On the other hand, different kinds of biosurfactants are components of the cell-wall, where the lipophilic lipid moiety is oriented to the medium. When such a microorganism comes into contact with an oil droplet, the interfacial tension between the cell wall and the oil phase is decreased. A monomolecular oil layer around the cell is formed, and a passive diffusion of the oil into the cell takes place [6].

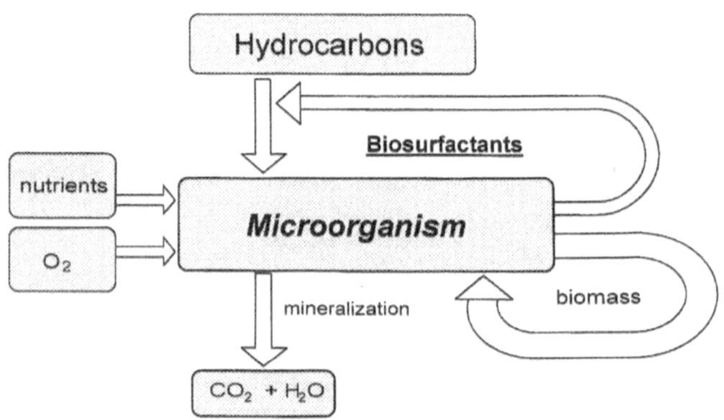

Fig. 1 Scheme to illustrate the biodegradation of hydrocarbons by microorganism and simultaneous synthesis of biosurfactants

Microbial surfactants are commonly differentiated on the basis of their biochemical nature and the microbial species synthesizing them. The most important types of surfactants and the producing microbial species are listed in Table 1.

Major classes of biosurfactants includes glycolipids, lipopeptides, phospholipids, fatty acids, polymeric surfactants and other substances [13]. The most commonly isolated and studied group are the glycolipids, constituted of carbohydrates in combination with long-chain aliphatic- or hydroxy aliphatic acids [13]. Furthermore, biosurfactants can be classified into anionic and nonionic species as well as into extracellular and cell wall bounded substrates. A large variety of biosurfactants has been reported [14–19]. The type, quantity and physico-chemical properties of microbial surfactants are influenced by the nature of the carbon substrate and the concentration of mineral nutrients in the medium. Furthermore, culture conditions such as pH, temperature, agitation and dilution rate under continuous conditions influence the chemical constitution of biosurfactants.

Actually biosurfactants become more interesting for applications, because they present a much broader range of surfactant types and properties than the available synthetic surfactants. They are usually biodegradable thus reducing the potential of environmental pollution. Furthermore biosurfactants can be synthesized from renewable substrates. With regard to textile processing the use of biosurfactants offers interesting ecological applications. The functions of biosurfactants most likely to be needed in textile finishing are emulsification, solubilization, dispersing, wetting and detergency. This paper reports the potential application of microbial surfactants in textile washing processes for removing various lipophilic preparations from fiber surfaces.

Table 1 Major types of biosurfactants produced by microorganisms and their properties [7], (σ_s: lowest surface tension obtainable, CMC: critical micelle concentration, n.a.: not available)

Group	Components	Charge	Type	σ_s [mN/m]	CMC [mg/l]	Microorganisms	Ref.
Glycolipids	Trehaloselipids	nonionic/ anionic	cell wall bounded/ extra-cellular	32–36	4	**Rhodococcus erythropolis**	8
	Rhamnolipids	anionic	extra-cellular	25–30	0.1–10	**Pseudomonas sps.**	9
	Sophoroselipids	nonionic	extra-cellular	33	82	**Torulopsis bombicola**	10
Lipopeptides	Surfactin	anionic	extra-cellular	27–32	23–160	**Bacillus subtilis**	11
Pospholipids	Lecithin	amphoteric	cell wall bounded	n.a.	n.a.	**Corynebact. alkanolytikum**	
Fatty acids	Corynomycolic-acid	anionic	extra-cellular	30	150	**Corynebact. lepus**	12

Experimental section

Organisms and materials

Rhodococcus erythropolis (DSM 311), *Rhodococcus globerulus* (DSM 43290) and *Candida antarctica* (DSM 70725) were selected as suitable microorganisms to synthesize biosurfactants. These organisms were obtained from DSM (Deutsche Sammlung von Mikroorganismen, Braunschweig, Germany). The substrates for cultivation and fermentation media, the various oils and other chemicals were reagent grade and obtained from Fluka (Neu Ulm, Germany), Sigma (Deisenhofen, Germany) and Difco (Detroit, USA). Various commercial fiber preparations were used as received.

Cultivation

Rhodococcus erythropolis and *Rhodococcus globerulus* were cultivated in Erlenmeyer flasks at pH 7.2 and 30 °C in a medium with glucose (4.0 g/l), yeast extract (4.0 g/l), malt extract (10.0 g/l) and $CaCO_3$ (2.0 g/l) as ingredients. Cultures were shaken using a rotary shaker (Certomat U, Braun Melsungen, Germany) at 200 min^{-1}. *Candida antarctica* was cultivated at pH 7.0 and 30 °C in a medium containing glucose (10.0 g/l), bactopeptone (5.0 g/l), yeast extract (3.0 g/l) and malt extract (3.0 g/l). A scale-up fermentation of *Candida antarctica* using soybean oil as carbon source was made as described by Kitamoto et al. [20].

Screening

The screening studies with *Rhodococcus strains* were made in a medium containing yeast extract (1 g/l), $(NH_4)_2HPO_4$ (1.5 g/l), K_2HPO_4 (1.26 g/l), KH_2PO_4 (0.5 g/l), Na_2HPO_4 $*2H_2O$ (0.5 g/l) and $MgSO_4*7H_2O$ (0.1 g/l). Furthermore, a solution of trace elements as described by Rapp et al. [8] was added to this medium (1 ml/l). The pH-value was adjusted to 7.0 with 1M NaOH and a selected oil (20 g/l) was added as carbon source. In 500 ml Erlenmeyer flasks 100 ml media were inoculated with *Rhodococcus* cultures. The cultures were shaken with 200 min^{-1} rotation frequency at 30 °C. The cell growth was monitored semi-quantitively by visual inspection of the culture broth.

Fermentation

Scale-up fermentations were carried out at 30 °C in 500 ml fermenter flasks. 400 ml of a medium containing 20 ml/l lipophilic carbon source was inoculated with 4 ml of a *Rhodococcus* culture. The culture broth was aerated with 190 l/h flow rate. The pH-value was measured by a combined pH glass electrode. The O_2-concentration was determined with a polarographic electrode (EO 96, WTW). The concentration of substrate was measured with a Hewlett Packard gas chromatograph (HP 5890 Series II) equipped with a flame ionization detector. A HP 5 capillary column (crosslinked 5% Ph Me Silicone, 25 m × 0.32 mm × 0.52 μm film thickness) was used. The carrier gas was N_2 with a constant flow rate of 1 ml min^{-1}. The biosynthesis of natural surfactants was monitored by measurement of the surface tension, where small volumes of the aqueous phase were separated by sterile-filtration. Measurements were carried out by the Wilhelmy method using a Krüss Processor-Tensiometer (K 12).

Washing

70 ml of the aqueous phase were separated from the culture broth by sterile-filtration after 50 h of fermentation time. Poly(ethylene terephthalate) (PET) textiles loaded (2%) with hydrophobic substrates (Coconut oil) were used

for determining the washing efficiency of the aqueous phase. The textiles were treated for 60 min at 40 °C with the aqueous culture broth containing biosurfactants. Furthermore, a medium free of carbon source was used as washing agent in control experiments. The material-liquor ratio was 1:35. The remaining residual oil was extracted using cyclohexane from the textile after washing and the amount was determined by gaschromatography as described above.

Results and discussion

Surfactants must have a high aqueous solubility, a low CMC as well as optimal emulsification and wetting properties to develop good detergent activities. Figure 2 shows the dependence of the surface tension from the concentration of an isolated biosurfactant produced from *Candida antarctica* growing onto soya oil as carbon source. Spectroscopic and chromatographic characterization of the isolated biosurfactant indicate a chemical constitution of a glycolipid. We assume that under the experimental conditions a mannosylerythritol-lipid is synthesized by *Candida antarctica* as it is described by Kitamoto et al. [20]. As indicated in Fig. 2 the surface tension shows a strong dependence on the concentration of the biosurfactant.

Even a small biosurfactant concentration of about 10 mg/l can reduce the surface tension down to 30 mN/m in contrast to synthetic surfactants, where higher concentrations (≈100 mg/l) are necessary for reducing the surface tension to this value.

In further investigations different *Rhodococcus strains* were used to investigate their degradation properties for various lipophilic fiber preparations. *Rhodococcus strains*

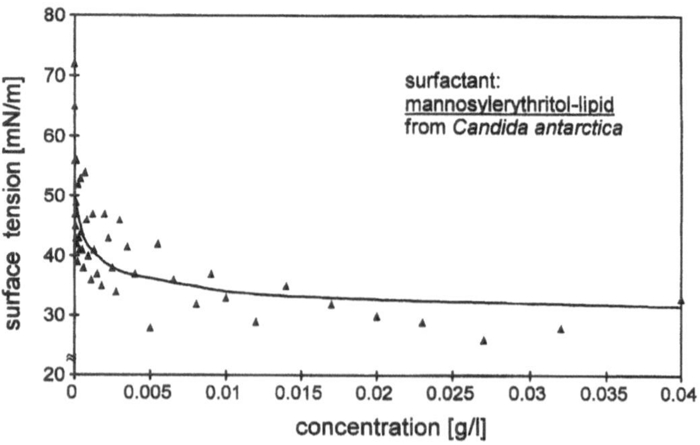

Fig. 2 Dependence of surface tension on the concentration of a biosurfactant isolated from *Candida antarctica* culture broth

are known for degrading hydrocarbons and simultaneously producing free anionic and cell-wall bounded biosurfactants like trehalosetetraesters and trehalosecorynomycolates with good emulsifying properties (Table 1) [21, 22]. In screening studies the degrading ability of *Rhodococcus strains* was tested for selected pure lipophilic substrates as well as for some commercial formulations. Two kinds of microbes (*Rhodococcus globerulus, Rhodococcus erythropolis*) were used for these experiments. Qualitative results of this screening program are reported in Table 2.

Good results were obtained with *Rhodococcus globerulus* which is especially able to decompose various pure substrates like triglycerides (natural oils), fattyacid esters and polyoxyethylene esters with good success. Using different mineral oil fractions like heavy and light white oil only the last fraction can be used as carbon source by

Table 2 Biodegradation of textile relevant lipophilic substrates by *Rhodococcus strains*, (+ + very good degradability, + good degradability, – poor or no biodegradability, n.d.: not determined)

Substrates	Components	**Rhodococcus globerulus** DSM 43290	**Rhodococcus erythropolis** DSM 311
Natural oils	Coconut oil	+ +	+ +
	Peanut oil	+	+
	Olive oil	+	+
Mineral oils	Light white oil	+	–
	Heavy white oil	–	–
Hydrocarbons	Hexadecane	+ +	+ +
Fatty acid esters	Stearic Acid Butyl Ester	+ +	n.d.
	Oleic Acid Propyl Ester	+	+
	Sebacic Acid Di(2-Ethylhexyl)Ester	+	–
	Sebacic Acid Dibutyl Ester	–	–
Polyoxyethylene ester	Polyoxyethylene 8-Stearate	+	+
Commercial products	spinning oil (modified fatty acid ester)	+	n.d.
	finish (fatty acid ester)	–	n.d.
	warping oil (hydrocarbons)	–	–
	winding oil (fatty acid ester/fatty-alcoholethoxylates)	–	n.d.

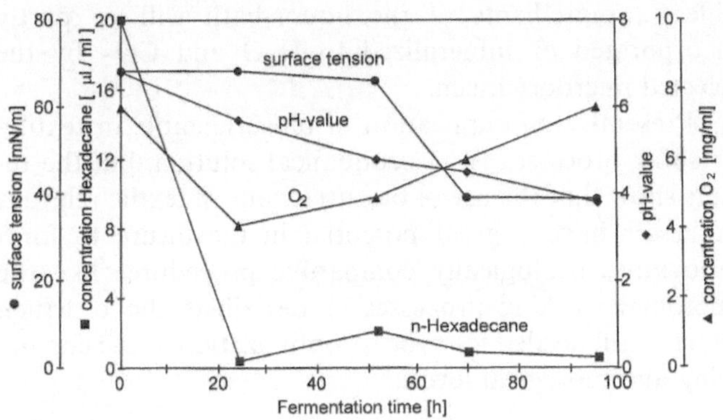

Fig. 3 pH, O_2, surface tension and *n*-hexadecane concentration during fermentation of *Rhodococcus erythropolis*

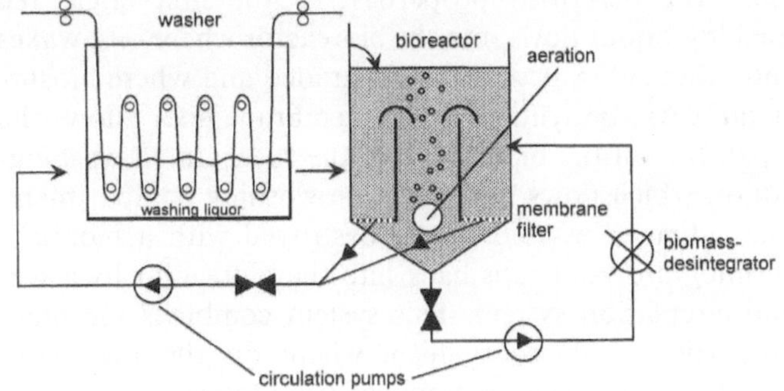

Fig. 4 Proposed scheme of a microbial washing-system for continuous textile washing [19, 24]

Rhodococcus globerulus under these experimental conditions. An explanation is the high hydrocarbon chain length in the heavy white oil fraction. With regard to this it is described in the literature that any microorganism capable of growing on *n*-alkanes usually consumes C_{11}- to C_{14}-*n*-alkanes most rapidly; C_{15}-C_{18} *n*-alkanes moderately, and C_{23}-C_{24} alkanes are assimilated more slowly than others. The C_6–C_9 alkanes are not only scarcely utilized but are also frequently toxic [23].

Testing various commercial fiber preparations which are used in textile industry only spinning oil can be degraded successfully by the selected microorganism. Other commercial products like finishing agent (fatty acid ester), warping oil (hydrocarbons) and winding oil (fatty acid ester) are unaffected. These fiber preparations are possibly stabilized against microbial degradation with biocides. Using spinning preparations in emulsions biocides like formaldehyde and heterocyclic compounds (isothiazolinone, imidazole) are sometimes used for microbial stabilization [4].

For proving the detergency activity of natural biosurfactants like trehalosetetraester and trehalosedicorynomycolate a fermentation was carried out with **Rhodococcus erythropolis** (DSM 311) onto *n*-hexadecane as carbon source. Figure 3 shows the surface tension, the O_2- and *n*-hexadecane concentration and pH changes as a function of the fermentation time.

A nearly complete alkane degradation is possible within 20 h under these fermentation conditions. Due to the microbial activity the O_2-consumption reaches a maximum within the same time range. The production of metabolic acidic components cause a lowering of pH-value. The surface tension which was measured in the aqueous phase decreases just after 50 h from initially 70 mN/m down to 40 mN/m. This delay in surface tension decrease is explained by the need of sufficient time for extraction of the biosurfactants from the oily phase into the aqueous phase.

Table 3 Washing of lubricants from polyester textiles with biosurfactants from *Rhodococcus erythropolis*, (material to liquor ratio: 1:35, coconut oil as lubricant, 2% loaded)

Washing bath	Washing time [min]	Temperature [°C]	Amount of oil washed out [%]
nutrient medium	60	40	14
aqueous culture broth	60	40	23

A washing test was made with the separated aqueous phase of the culture broth to investigate detergency properties of the glycolipids. As can be seen in Table 3 the amount of the oil removed from the fabric increases by using the biosurfactant containing liquor compared with the control experiment where a surfactant-free nutrient medium is used as washing agent.

Based on this preliminary findings further work has to be done for optimizing such natural washing systems. Parameters like washing temperature, treatment time as well as the emulsification properties have to be investigated for different microbial cultures and their resulting biosurfactants. Furthermore, it must be considered that the fermentation-system used in this work is not optimized because the selected microorganism are not adapted to the fermentation conditions choosen.

In principal it will be possible to form continuous washing technologies where textile washing-machines are combined with bioreactor-systems where suitable microorganisms are adapted [24]. Figure 4 shows a possible scheme of a microbial washing-system for continuous textile washing.

The system contains two coupled chambers. The first is a conventional continuous textile washing machine and the second one is an aerated bioreactor. The reactor system contains suitable microorganism which has to fulfill the described properties. In the first circle the washing-liquor flows into the bioreactor where oils, waxes and other substances will be degraded and where biosurfactants will be synthesized. A membrane filter allows the separation of the biomass from the regenerated washing-liquor, which flows back into the washing compartment. The active excess biomass is destroyed with a biomass-desintegrator and runs back into the bioreactor by a second circulation system. This system combines the main properties of microorganism where on the one hand lipophilic substrates are removed from the fiber surface by microbial degradation. In addition, microbial surfactants are produced simultaneously by the cell metabolism which causes an emulsifying of the preparations and supports the washing process. These processes can occur completely self-sufficiently because the excess biomass as well as the different ingredients of the liquor-bath will be partly incorporated or mineralized to H_2O and CO_2 by the selected microorganism.

Presently the application of biosurfactants in textile-finishing processes is no economical solution, but the results show that the use of biosurfactants in textile relevant processes have a great potential in the future to form innovative, ecologically compatible procedures like the microbial washing processes as described above. Much effort is still needed for process optimization at the engineering and biological levels.

Acknowledgments We are grateful to the Forschungskuratorium Gesamttextil for their financial support for this research projekt (Aif-No. 9111). This support was granted from resources of the Federal Ministry of Economic Research Organisation (Arbeitsgemeinschaft Industrieller Forschungsvereinigungen, AIF).

References

1. Frahne D (1992) Textilveredlung 73: 384–388
2. Dünser H (1992) Melliand Textilber 73:280–284
3. Schollmeyer E (1994) Textilveredlung 29:318–322
4. Schönberger H (1994) Umweltbundesamt, Texte 3: agenda 2
5. Hemmpel W-H (1992) Textilveredlung 27:159–163
6. Syldatk C, Wagner F (1987) In: Kosaric N (ed) Biosurfactants and Biotechnology, Marcel Dekker, INC New York, pp 89–120
7. Desai AJ, Patel R, Desai J (1994) J Sci Ind Res 53:619–629
8. Rapp P, Bock H, Wray V, Wagner F (1979) J Gen Microbiol 115:491–503
9. Hisatsuka K, Nakaharat, Sano N, Yamada K (1981) Agric Biol Chem 35:868
10. Cooper DG, Paddock DA (1984) Appl Environ Microbiol 47:173–176
11. Cooper DG, Zajik JE, Gerson DF (1978) Appl Environm Microbiol 37:4–10
12. Arima K, Kakinuma A, Tamura G (1968) Biochem Biophys Res Commun 31:488
13. Desai AJ, Desai J (1993) In: Kosaric N (ed) Biosurfactants, Surfactants science series Vol 48. Marcel Dekker, INC New York, pp 66–92
14. Robb ID (1990) Spec Publ-R Soc Chem (Ind Appl Surf II) 77:22–35
15. Syldatk C, Matulovic U, Wagner F (1984) Biotech- Forum 1:58–66
16. Kosaric N (ed) (1993) Biosurfactants, Surfactants science series Vol 48. Marcel Dekker, INC New York
17. Fiechter A (1992) Trends Biotechnol 10:208–217
18. Georgiou G, Lin S-C, Sharma MM (1992) Biotechnology 10:60–65
19. Kesting W, Bach E, Tummuscheit M, Schollmeyer E (1994) Tenside Surf Det 31:362–371
20. Kitamoto D, Haneishi K, Nakahara T, Tabuchi T (1990) Agric Biol Chem 54:37–40
21. Kretschmer A, Bock H, Wagner F (1982) Appl Environ Microbiol 44: 864–870
22. Wagner F (1987) Proc-World Conf Biotechnol (Fats Oils Ind) Meeting Date 1987, ed by Applewhite TH, AOCS: Champaign I11:189–194
23. Kosaric N, Chio HY, Blaszczyk R (1990) Tenside Surf Det 27:294–297
24. Kunz P (1991) Abwassertechnik 2: 54–57

Progr Colloid Polym Sci (1996) 101:131–134

S. Haas
H. Hoffmann

Interfacial tension of double-chain cationic surfactants

S. Haas (✉) · H. Hoffmann
Universität Bayreuth
Lehrstuhl für Physikalische Chemie
Universitätsstraße 30
95440 Bayreuth, FRG

Abstract The paper deals with surface and interfacial tension measurements for the cationic, double-chain *N*-alkyl-*N*-alkyl′-*N*,*N*-dimethylammonium bromides. The surfactants C_xC_yDMABr, with the variable alkylchains C_x and C_y, are synthesized by a quaternization reaction of the alkyldimethylamine C_xDMA with the alkylbromide C_yBr in acetonitrile and purified by recrystallization in ether/methanol. The CMC for these surfactants was determined by surface tension measurements with a ring-tensiometer and by interfacial tension measurements with a spinning drop tensiometer.

The surface tension measurements show that the CMC of the surfactants decreases with increasing chain length. For surfactant concentrations $c_S >$ CMC the surface tension decreases further and shows a second break (CMC_{II}). The CMC_{II} is an indication of a change in micelle shape from globular to rodlike micelles.

The interfacial tension measurements show that the values of the interfacial tension at the CMC decreases with the increase of the chain length. The same effect as for the CMC could be found for the values of the interfacial tension of a 10 mM surfactant solution.

The plots of the interfacial tension versus the logarithm of surfactant concentration showed more than two breaks for some of these surfactants.

Key words Double-chain cationic surfactants – surface tension – interfacial tension

Introduction

The formation of microemulsion phases in water–surfactant–oil mixtures is of great interest in industry and research [1]. One requirement to form such microemulsions is the ultralow interfacial tension ($\gamma < 1 \cdot 10^{-2}$ mN/m) between the surfactant solution and the oil phase [2–4]. Therefore a lot of systems have been investigated to reach such ultralow interfacial tensions. It is well established now that the interfacial tension of surfactant solutions against oil is dependent on the cosurfactant concentration, the concentration of added salt and the chain length of the used oil [5–11].

The aim of our work was to find surfactants in the class of the double-chain cationic *N*-alkyl-*N*-alkyl′-*N*,*N*-dimethylammonium bromides, which show ultralow interfacial tensions against *n*-decane at low surfactant concentrations and without any additives.

Experimental section and methods

Materials

The surfactants used in our work were synthesized by a quarternization reaction of the alkyldimethylamine

C_xDMA with the corresponding alkylbromide C_yBr in acetonitrile at 80 °C.

$$C_xDMA + C_yBr \xrightarrow{CH_3CN, 80\,^\circ C} C_xC_yDMABr$$

(0.1 mol) (0.11 mol)

After the reaction (20 min to 4 h, dependent on the chain-length of the materials) the acetonitrile is removed under vacuum from the slightly yellow solution and the remaining raw product is purified twice by recrystallization in ether/methanol and is then dried in vacuo.

The used amines were a gift of Hoechst Gendorf (purity after gc > 95%) and the alkylbromides in p.a. quality were purchased from Merck.

The *n*-decane for the interfacial tension measurements was from Merck, p.a. quality, and not further purified before use.

Methods

The critical micelle concentration of the C_xC_yDMABr surfactants was determined by surface tension measurements with a ring tensiometer from Lauda.

The interfacial tensions of the surfactant solutions against *n*-decane were carried out with a spinning-drop tensiometer SITE 04 from Krüss.

Experimental results and discussion

According to the description in the experimental section, the following 13 surfactants were synthesized with a yield of 75 to 95%.

For all of those surfactants the surface tension was measured in dependence of the surfactant concentration. Figures 1 and 2 show the plots of the surface tension versus the logarithm of the molar surfactant concentrations of some of the surfactants. The measured values of surface tension were all corrected after Harkins and Jordan [12].

From these curves the CMC can be determined and the head group area can be calculated from the slope of the line through the values before the CMC according to the Gibb's equation for ionic surfactants:

$$A_0 = \frac{1}{N_L \cdot \Gamma_0}$$

$$\Gamma_0 = \frac{1}{2 \cdot R \cdot T} \cdot \frac{d\sigma}{d \ln c} = \frac{1}{2 \cdot R \cdot T} \cdot \frac{d\sigma}{2.303 \cdot d \log c} \quad (1)$$

For higher surfactant concentrations a second CMC can be determined for nearly all surfactants. This second CMC, short CMC_{II} is due to a micelle shape conversion from globular to rodlike micelles. This change could be seen in electric birefrigence measurements, where signals were found for concentrations $c > CMC_{II}$. The values for the CMC_I, the head group area per molecule at the air/water interface and the CMC_{II} values are listed in Table 2.

Table 1 Synthesized surfactants

	$C_{14}C_8DMABr$	$C_{16}C_8DMABr$
	$C_{14}C_9DMABr$	$C_{16}C_9DMABr$
	$C_{14}C_{10}DMABr$	$C_{16}C_{10}DMABr$
$C_{12}C_{11}DMABr$	$C_{14}C_{11}DMABr$	$C_{16}C_{11}DMABr$
	$C_{14}C_{12}DMABr$	$C_{16}C_{12}DMABr$
	$C_{14}C_{14}DMABr$	$C_{16}C_{14}DMABr$

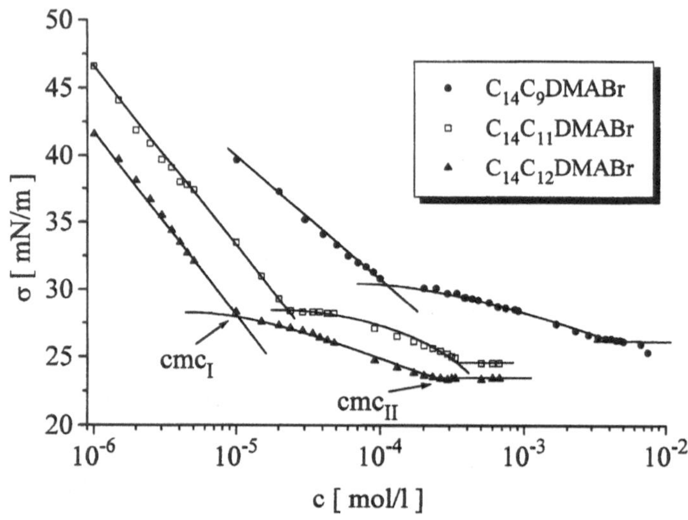

Fig. 1 Surface tension of $C_{14}C_yDMABr$ at 25 °C

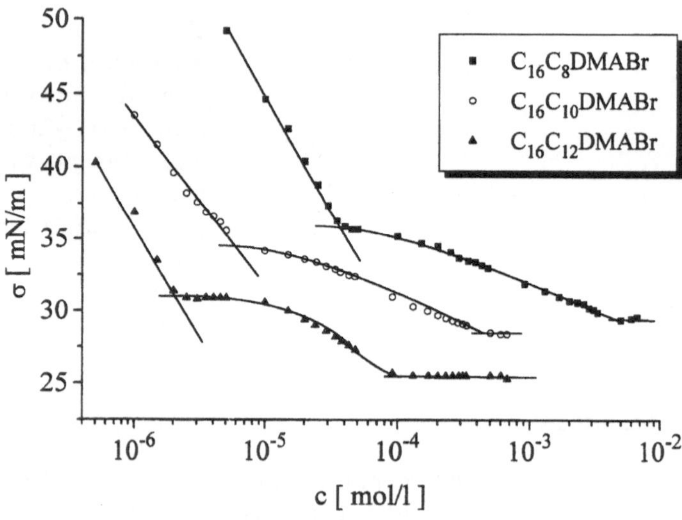

Fig. 2 Surface tension of $C_{16}C_yDMABr$ at 25 °C

Table 2 CMC_I-, CMC_{II}- and head group area values for C_xC_yD-MABr

Surfactant	CMC_I [mol/l]	CMC_{II} [mol/l]	A_0 [Å^2]
$C_{12}C_{11}DMABr$	$2\cdot10^{-4}$	–	115
$C_{14}C_8DMABr$	$2.6\cdot10^{-4}$	–	246
$C_{14}C_9DMABr$	$1.2\cdot10^{-4}$	$6.7\cdot10^{-3}$	248
$C_{14}C_{10}DMABr$	$6\cdot10^{-5}$	$1.2\cdot10^{-3}$	146
$C_{14}C_{11}DMABr$	$2.2\cdot10^{-5}$	$5\cdot10^{-4}$	135
$C_{14}C_{12}DMABr$	$8.5\cdot10^{-6}$	$1.8\cdot10^{-4}$	122
$C_{14}C_{14}DMABr$	$5\cdot10^{-6}$	$8\cdot10^{-5}$	148
$C_{16}C_8DMABr$	$4\cdot10^{-5}$	$3.8\cdot10^{-3}$	113
$C_{16}C_9DMABr$	$1.6\cdot10^{-5}$	$1.8\cdot10^{-3}$	195
$C_{16}C_{10}DMABr$	$6.2\cdot10^{-6}$	$4.8\cdot10^{-4}$	157
$C_{16}C_{11}DMABr$	$4.1\cdot10^{-6}$	$1.6\cdot10^{-4}$	90
$C_{16}C_{12}DMABr$	$2.1\cdot10^{-6}$	$9.1\cdot10^{-5}$	127
$C_{16}C_{14}DMABr$	$1.6\cdot10^{-6}$	$2\cdot10^{-5}$	111

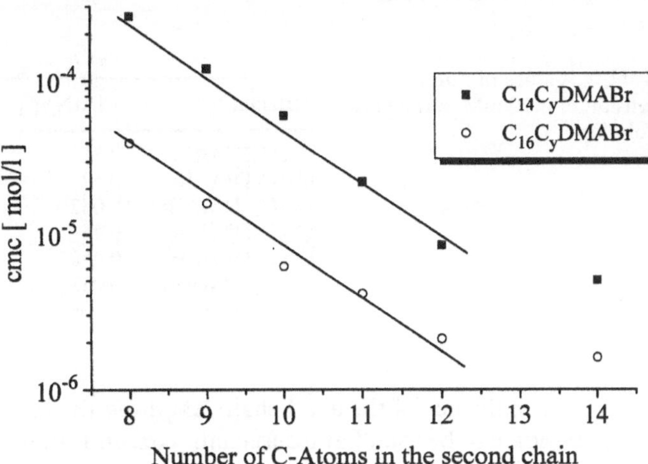

Fig. 3 Dependence of the CMC_I on the length of the alkyl chains

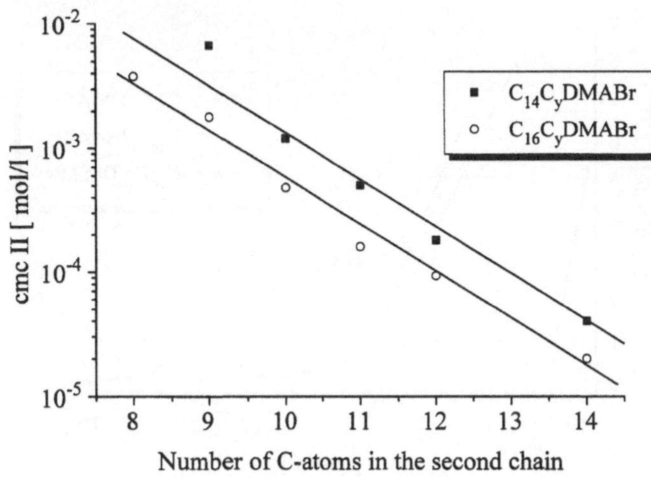

Fig. 4 Dependence of the CMC_{II} on the length of the alkyl chains

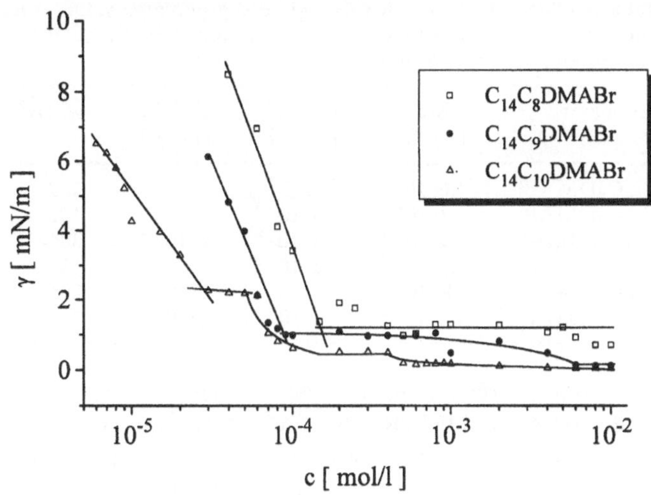

Fig. 5 Interfacial tension of $C_{14}C_yDMABr$ against decane at 25 °C

The CMC values decrease by a factor between 2 and 3, when the number of the CH_2-groups in the second alkyl-chain decreases by one CH_2-group. This can be seen from Fig. 3. The increase of the C_x-chain by the same number of methylene groups decreases the CMC more than the increase of the C_y-chain. No systematic effect can be found for the head group area per molecule at the air/water interface: the values vary from 100 to 200 Å^2 but show no dependence on the length of the two alkylchains. Therefore it is not possible to say in which way the surfactant molecules are present at the air/water interface. As Fig. 4 shows, the same effect as for the CMC_I can be found for the CMC_{II}. The CMC_{II} is lowered by adding methylene groups to both alkylchains, but the effect is stronger for the C_x-chain.

For the surfactants $C_{14}C_yDMABr$ and $C_{16}C_yDMABr$ with $y = 8, 9$ and 10 the interfacial tension against *n*-decane was measured for concentrations from below the CMC from surface tension measurements up to 10 mM. The plots of the interfacial tension values versus the logarithm of surfactant concentration are shown in the Figs. 5 and 6.

The interfacial tension results can be analyzed in the same way as the surface tension results and the CMC and the head group area per molecule at the water/oil interface were determined. As one can see from the plots, the values of the interfacial tension are not constant for $c > CMC$ and for most surfactants one or two more breaks in interfacial tension are noticed. The values for the CMC, for the head group area per molecule at the water/oil interface and for the concentrations of the breaks in interfacial tension are listed in Table 3.

The data show that the CMC determined from the interfacial tension measurements depend little on the length of the first alkylchain but depend strongly on the length of the second chain. For the $C_{14}C_yDMABr$ surfactants, the CMC from surface tension measurements

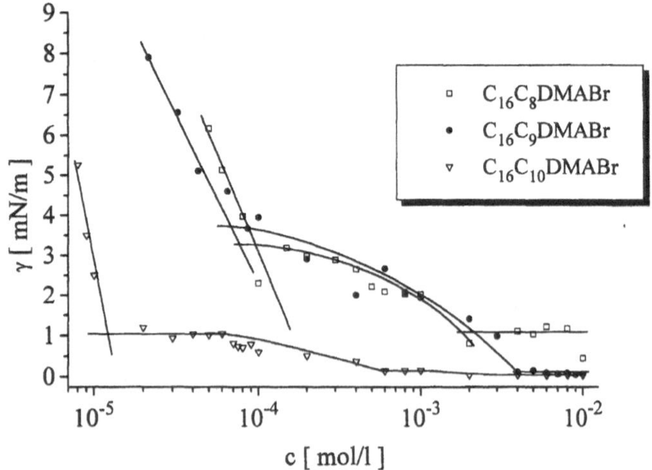

Fig. 6 Interfacial tension of $C_{16}C_yDMABr$ against decane at 25 °C

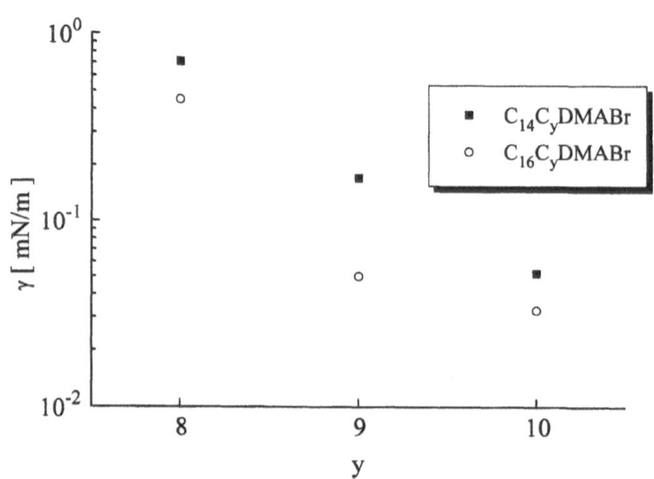

Fig. 7 Dependence of the interfacial tension of a 10 mM solution on the length of the alkyl chains

Table 3 CMC_I-, CMC_{II}-, CMC_{III}- and head group area values from interfacial tension measurements against *n*-decane for C_xC_yDMABr ($x = 14$ and 16; $y = 8$, 9 and 10)

surfactant	CMC_I [mol/l]	CMC_{II} [mol/l]	CMC_{III} [mol/l]	A_0[Å^2]
$C_{14}C_8DMABr$	$1.6 \cdot 10^{-4}$	–	–	107
$C_{14}C_9DMABr$	$7 \cdot 10^{-5}$	$6 \cdot 10^{-3}$	–	146
$C_{14}C_{10}DMABr$	$3 \cdot 10^{-5}$	$1.1 \cdot 10^{-4}$	$4 \cdot 10^{-3}$	304
$C_{16}C_8DMABr$	$1.1 \cdot 10^{-4}$	$2 \cdot 10^{-3}$	–	155
$C_{16}C_9DMABr$	$8 \cdot 10^{-5}$	–	$4 \cdot 10^{-3}$	224
$C_{16}C_{10}DMABr$	$1.2 \cdot 10^{-5}$	$6 \cdot 10^{-4}$	$2 \cdot 10^{-3}$	256

Table 4 Values of interfacial tension of a 10 mM surfactant solution

surfactant	γ [mN/m]
$C_{14}C_8DMABr$	0.71
$C_{14}C_9DMABr$	0.17
$C_{14}C_{10}DMABr$	0.052
$C_{16}C_8DMABr$	0.45
$C_{16}C_9DMABr$	0.056
$C_{16}C_{10}DMABr$	0.033

and from interfacial tension measurements are the same, while for $C_{16}C_yDMABr$ the CMC from the interfacial tension measurement is always higher than that from the surface tension measurements. One reason for this might be that the surfactants are soluble in decane so that the actual concentration in the water phase is lower than the overall concentration.

The calculated values for the head group area per molecule are for all surfactants higher at the water/oil interface than at the air/water interface. The reason for this can be that the *n*-decane molecules intercalate between the two long chains of the surfactant so that the two chains spread out and the A_0-values rise.

A great influence of the alkyl chain length on the interfacial tension can be found at a surfactant concentration of 10 mM (Table 4). The dependence is illustrated in Fig. 7.

The values of interfacial tension are strongly influenced by the length of the second chain. When x is changed from 14 to 16, the interfacial tension values decrease by a factor of 2, and when the same number of methylene groups is added to the C_y-chain, the interfacial tension is lowered by a factor of 14.

The results of our work show that for a low CMC value, it is more effective to have one long chain, while for low interfacial tension it is better to have a double-chain surfactant with two more or less similar chains.

References

1. Langevin D (1986) Mol Cryst Liq Cryst 138:259
2. Ruckenstein E, Chi JC (1975) J Chem Soc Faraday 2, 71:1690
3. Reiss H (1975) J Colloid Interf Sci 53:61
4. Overbeek JThG (1978) Faraday Discuss Chem Soc 65:7
5. Aveyard R et al (1986) J Chem Soc Faraday Trans 82:125
6. Aveyard R et al (1986) J Chem Soc Faraday Trans 82:1755
7. Aveyard R et al (1988) J Chem Soc Faraday Trans 84:675
8. Verhoeckx GJ, de Bruyn PL, Overbeek JThG (1987) J Colloid Interf Sci 119:409
9. Doe PH, Wade WH (1977) J Colloid Interf Sci 59:525
10. Aveyard R, Lawless TD (1986) J Chem Soc Faraday Trans 82:295
11. Hoffmann H (1990) Progr Colloid Interf Sci 83:16
12. Harkins W, Jordan H (1930) J Am Chem Soc 52:1751

Progr Colloid Polym Sci (1996) 101:135–140

I. Grosse
H.-J. Jacobasch

Lubrication of synthetic fibers with surfactants – a complex surface-chemical process at the phase boundary solid/liquid

Dr. I. Grosse (✉) · H.-J. Jacobasch
Institute of Polymer Research
P.O. Box 12 04 11
01005 Dresden, FRG

Abstract The lubrication of fibers by surfactants is a surface-chemical process including several simultaneous events like wetting, spreading, and adsorption, which have to take place within the very short time of a few milliseconds.

Some theoretical aspects are given and the difficulties of their transfer to the real system of polymeric fibers with their bent, irregular, and inhomogeneous surfaces are discussed. Results are given for the adsorption of ionic surfactants: In solutions containing more than $1 \cdot 10^{-3}$ mol/l of surfactant, multilayers are created on the surface. Presence of nonionics increases the adsorption of cationic surfactant.

The investigation of wetting speed furnished only small differences at the same molar concentration of different surfactants. Wetting speed increases significantly with increasing concentration. It does not depend on surface tension of the solution.

As a special case of wetting, we considered heterocoalecence: spreading of oil droplets in surfactant solutions on solid surfaces as a model for lubrication by oil-in-water emulsions. A transition concentration c_T of surfactant has been proved depending on the nature of surfactant and the solid. This concentration is the limit between spreading and nospreading of the oil droplets indicating the suitability as detergent or as a part of lubricating oil-in-water emulsion.

Key words Polymer surface – fiber – surfactant – lubrication – wetting – adsorption

Introduction

"Lubrication" is a necessary process immediately after spinning of synthetic fibers. Freshly spun fibers cannot be processed without a layer of lubricant due to their static electrical charge and unsatisfactory slip-stick behaviour.

List of Abbreviations: PAN = Polyacrylonitrile; PA = Polyamide; PE = Polyethylene; PETP = Polyethylene terephthalate; VIF = Viscose fiber; NaDDS = Sodium dodecyl sulphate; DDPCl = Dodecyl pyridiniumchloride; HDPCl, CPCl = Hexadecyl (Cetyl) pyridiniumchloride; DDBDMABr = Dodecyl benzyldimethyl ammoniumbromide; HDBDMABr = Hexadecyl benzyldimethyl ammoniumbromide; FA-EO = Fatty alcohol ethoxylate.

Lubricants are composed from mixtures of surfactants and oils and are used either in diluted state as solution or emulsion or as concentrates. The conditions during lubrication and aftertreatment, e.g., time of contact, concentration of the lubricant or temperature are of great importance for the lubricating effect with respect to height and uniformity. The contact time depends on the used technology of lubrication. It decreases by increasing speed of the thread and decreasing contact length. In high speed spinning processes with velocities of more than 6000 m/min it is only few milliseconds (Fig. 1).

During this time several processes simultaneously have to take place at the interface between fiber and liquid, which are supported by capillary sucking of the

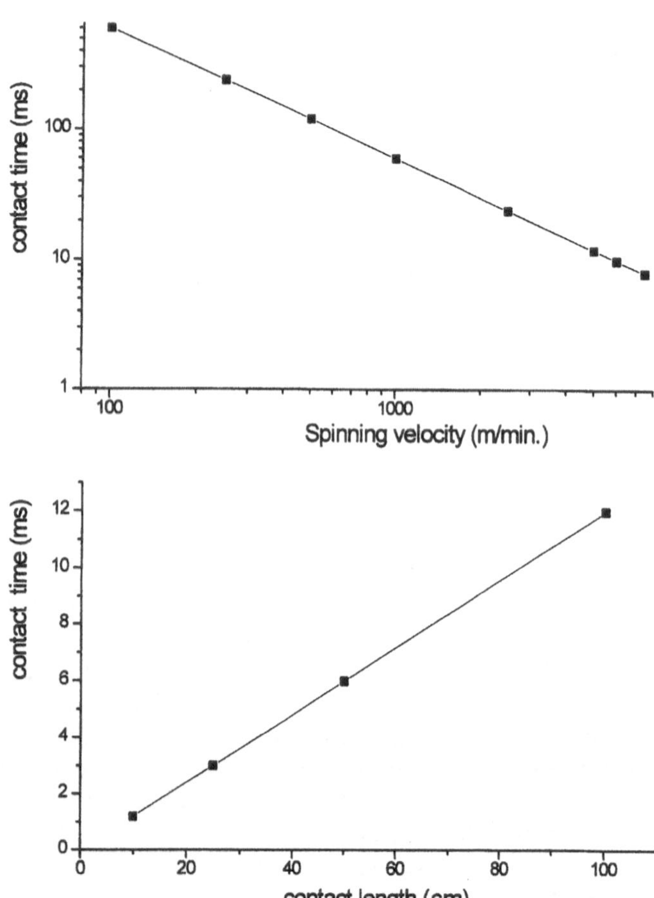

Fig. 1 Dependence of the contact time during lubrication from contact length and spinning velocity

multifilament yarns:

- wetting, spreading, adhesion, and adsorption.

Further processes occur in the limited volume of the liquid, which the fiber took away from the lubrication bath:

- adjustment of adsorption equilibrium of the surfactant molecules at the interface,
- competition adsorption between different components in solution,
- formation of micelle-like surface layers.

In thermal treatments after the lubrication an evaporation of water and other low-boiling components takes place. This is one reason for changes in the surface film. It also will be influenced by molecular and supermolecular changes in bulk and surface structures of the polymer, like crystallinity and porosity of the bulk. Thus some processes are possible like rearrangement of the molecules in the adsorption layer, diffusion or migration, dehydratation. This leads either to complete spreading on the surface or to formation of discrete "islands".

Theoretical basics

All wetting and spreading events at the phase boundary are governed by the laws of thermodynamics and kinetics. [1–3]. The well known mathematical equations (e.g. by Young, Laplace) [1–4] are valid only for smooth, energetically homogeneous surfaces and pure liquids. The real system of lubricant – in most cases surface-active substances dissolved in water – on fibers as a special polymer in an one-dimensionally extended form, cannot be recognized theoretically yet because of the deviation from ideal interface solid/liquid (Table 1). Models to include the inhomogeneities into the calculations are not relevant for conditions in practice and do not take into consideration the presence of surface-active agents in solution (Table 2): Wenzel's equation is a special case of Young's equation considering roughness R (Eq. (1), Cassie's equation considers different energetic states (Eq. (2))) [4].

Table 1 Specific features of the interface solid/liquid on fibers in lubrication liquids

Solids (polymers):
- strongly bent surface
- varying cross-sections (circular, trilobal, kidney-shaped, irregular)
- roughness caused by conditions in production (shape of cross-section, grooves and other irregularities) or by formation of crystallites (e.g. oligomers) on the polymer surface
- energetic differences (crystalline and amorphous parts, disperse and polar portions of surface strength)
- existence of adsorbed layers (water, gases) before.

Liquids (multicomponent systems):
- dissolved electrolytes and non-electrolytes (salts, organic additives)
- dissolved surface active agents enriching at the phase boundaries.

Table 2 Effect of inhomogenities on contact angle

$R(\gamma_{SV} - \gamma_{SL}) = \gamma_{LV} \cdot \cos\theta_w$ (Wenzel's equation) Eq. (1)

R = real surface/geometric surface

$\cos\theta_C = a_1 \cdot \cos\theta_1 + a_2 \cdot \cos\theta_2$ (Cassie's equation) Eq. (2)

$(a_1 + a_2 = 1)$
(a_1 and a_2: energetically different states, not measurable).

$\gamma_{sg} - \pi_e = \gamma_{sl} + \gamma_{lg} \cdot \cos\theta$ Eq. (3)

$\gamma_{sg} - \pi_e$ = interface tension of the solid with an adsorbed surfactant layer
π_e = "surface pressure" of the adsorbed substance on the solid surface: [1, 5, 6]

The effect of surfactants on wetting is attempted to be taken into account by the term π_e, the so-called "surface pressure" (Eq. (3)) [1, 5, 6].

Surfactants adsorb at the interface solid/liquid and create there mono or multilayers. The concentration of the adsorptive on the surface ("surface excess" Γ) is given by Gibb's equation (Eq. (4)) and correlates with the "surface pressure" π_e. The investigation of adsorption (adsorbed amount, heat of adsorption) gives statements about the affinity of surfactants to the solid surface.

$$\Gamma = -\frac{1}{RT}\cdot\frac{d\gamma}{d\ln c}\,. \tag{4}$$

In some systems (e.g., porous fibers, fibers with low crystallinity) also diffusion of parts of the lubricant into an outer fiber layer is possible. Consequently, in this case a time-dependent sorption will be observed as well as an increase in the adsorbed amount with temperature.

Results

The reported results are only a small contribution to the explanation of the processes at the real phase boundary fiber – surfactant solution.

To characterize the behavior of diluted surfactant solutions on polymer surfaces as models for lubricants the following methods are used:

– adsorption at fibers: estimation of the difference between origin and rest concentration, zetapotential measurements,

– wetting and spreading on foils: contact angle measurements, heterocoalescence between oil droplets and polymer surfaces.

Adsorption

The mass adsorbed on one area unit gives information about the surfactants' arrangement on the surface. Figures 2 and 3 show the adsorption isotherms of surfactants under different conditions on cellulosic (viscose) and polyacrylonitrile (PAN) fibers and the calculated place available per one molecule.

The results point to formation of multilayers at high concentrations if we assume the value of ca. 25–30 Å^2 for the space of one surfactant molecule in a so-called "molecule brush" [5, 7]. However, this result and the slight increase with temperature of adsorption at PAN-fibers can also be explained by diffusion.

Real lubrication systems mostly consist of surfactant mixtures: technical mixtures of different homologous substance or mixtures selectively produced for synergistic effects. Hence, the investigation of the adsorption behavior of single components in a mixture is very important for the explanation of the effects. So far it is difficult to estimate quantitatively and separately nonionic and cationic surfactants in mixtures. Figure 4 presents results of the adsorption of cationic surfactants from mixtures containing a nonionic ethoxylated fatty alcohol on PAN-fibers. Growing moieties of the nonionic lead to an increasing cationics' adsorption. So we could prove the results of previous zetapotential measurements (Fig. 5) [2, 6]: The addition of nonionic to very diluted solutions of cationic surfactant leads to a change of the slightly negative charge into the positive range. Consequently, significant synergistic effects of surfactant mixtures can even be concluded to occur at small differences not measurable by conventional methods.

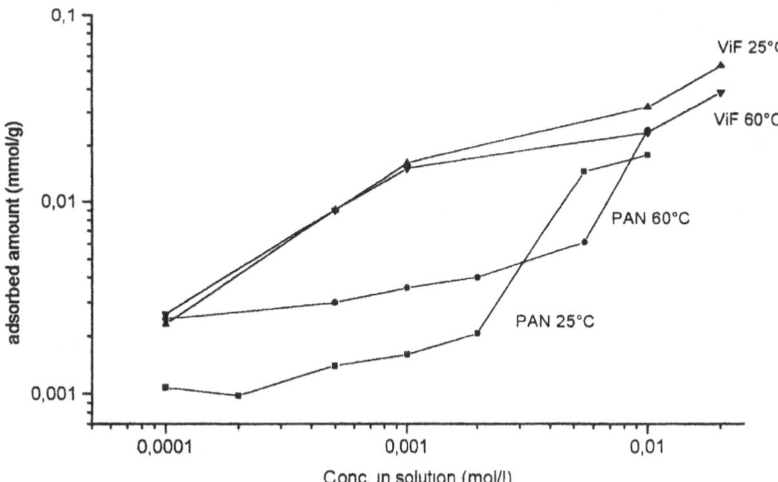

Fig. 2 Adsorption isotherms of Dodecyl-pyridinium-chloride on polyacrylonitrile and viscose fibers

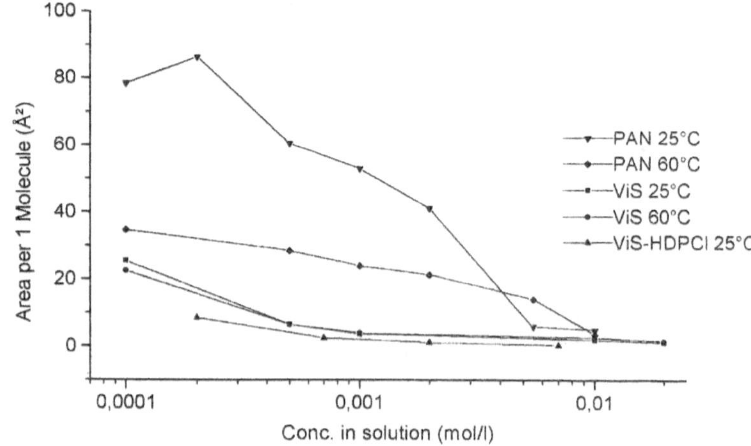

Fig. 3 Area available per 1 molecule of surfactant (DDPCl) on PAN and viscose fibers

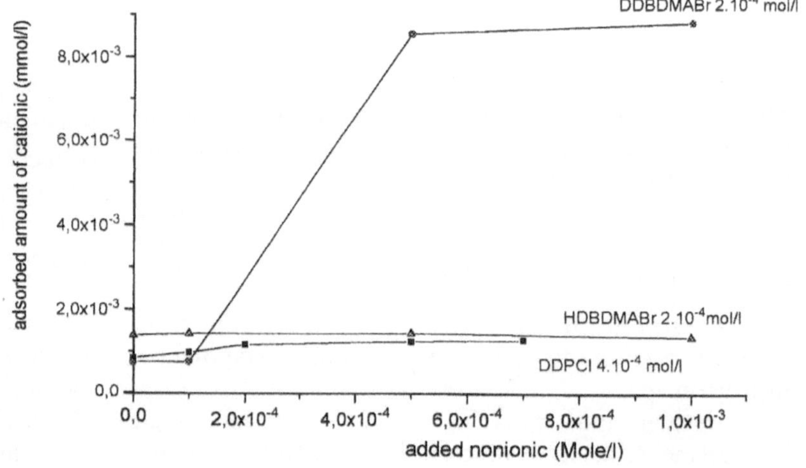

Fig. 4 Adsorption of cationic surfactants on PAN fibers in presence of increasing amounts on nonionic (fatty alcohol-EO)

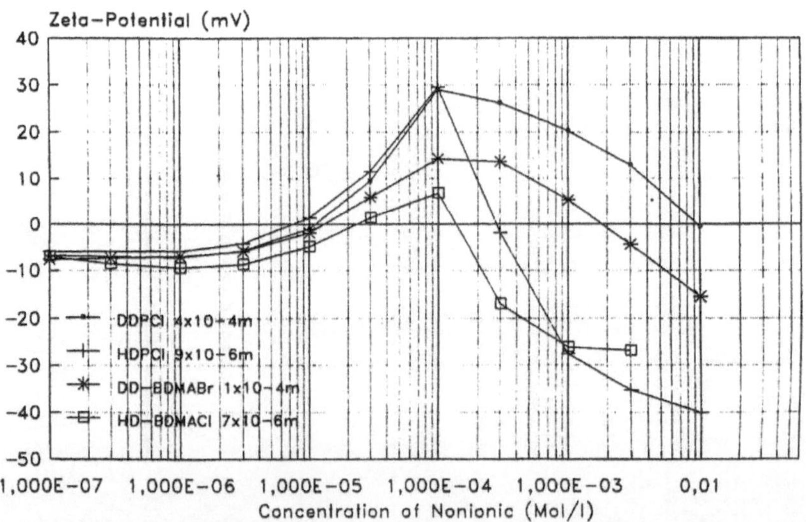

Fig. 5 Influence of nonionic surfactant (fatty alcohol-EO) on the zetapotential of PAN fibers in diluted solution of cationic surfactants

Wetting, spreading

For characterizing the wetting properties of fibers by liquids there are known different methods with advantages and disadvantages. In principle they also are applicable to surfactant solutions:

- attraction force between single fiber and solution (modified Wilhelmy plate-method) [8],
- wicking in fiber bundles (Washburn's method) [8],
- direct measuring of contant angle on single fibers (Grindstaff's method) [9],
- direct measuring of contant angle of a sessile drop only on polymer foils. This method is not suitable for fibers.

The last-named method has been used to estimate the wetting speed of surfactant solutions on untreated surfaces and also for rewetting of polyethylene terephthalate (PETP) surfaces treated with surfactants and dried before. There were found only poor differences in the kinetics of spreading of the tested surfactants presented as slope of the straight line $\Theta = f(t)$ in spite of great differences in surface tensions of these solutions. But the wetting significantly accelerates with increasing concentration of surfactant (Table 3). Complete spreading ($\Theta = 0$) is reached in a short time if there is a low value of contact angle at time $t = 0$.

An investigated special case of wetting was the spreading of oil droplets on polymer surfaces in surfactant solutions, the so-called heterocoalescence, as a model of the conditions during lubricating of fiber by a water-in-oil emulsion [10]. In some publications the opposite effect as a mechanism for oily soil release in detergency (rolling-up effect) [11–13] had been described.

Our investigations (measuring of the time between first contact polymer surface oil and spreading) have shown that there are significant differences in wetting depending

Table 3 Wetting and rewetting parameters of surfactant solutions on PET foil

			Wetting			Rewetting by water		
Surfactant	Concentration (Mol/l)	γ_{sl} Surface Tension (mN/m)	Θ Contact angle for $t = 0$ (°)	Slope ($tg\ \alpha$)	Time (s) for complete spreading ($\Theta = 0$)	Θ Contact angle for $t = 0$ (°)	Slope ($tg\ \alpha$)	Time (s) for complete spreading ($\Theta = 0$)
DDPCl	$1 \cdot 10^{-4}$	31.3	68.3	– 0.097	710			
HDPCl	$1 \cdot 10^{-3}$	35.1	43.1	– 0.113	370	54.2	– 0.073	740
HDPCl	$1 \cdot 10^{-4}$	27.7	66.5	– 0.118	600	65.6	– 0.049	1340
Ethoxylated fatty alcohol	$1 \cdot 10^{-3}$	29.6	45.1	– 0.086	250	29.1	– 0.046	630
Ethoxylated fatty alcohol	$1 \cdot 10^{-4}$	29.2	66.8	– 0.101	430	61.7	– 0.063	980
Ethoxylated fatty amine	$1 \cdot 10^{-4}$	41.6	62.1	– 0.111	560			
Ethoxylated alkyl phenylether	$1 \cdot 10^{-4}$	31.4	42.4	– 0.113	380			
NaDDS	$1 \cdot 10^{-4}$	39.1	66.0	– 0.101	654			

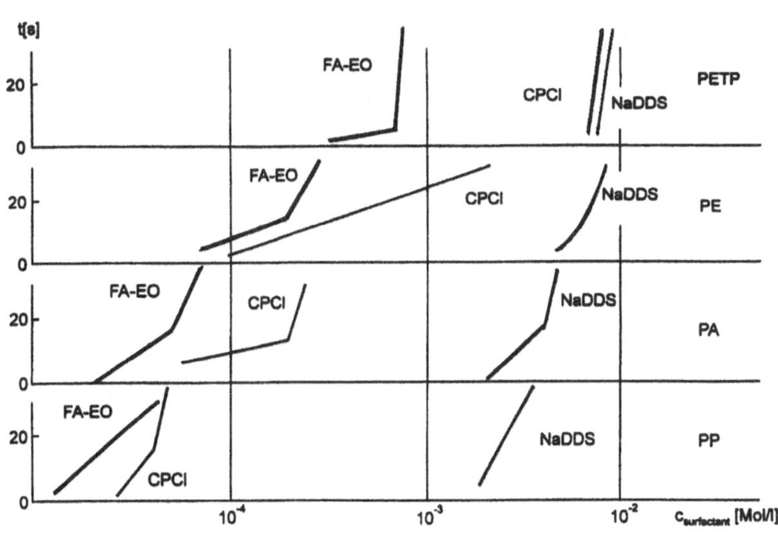

Fig. 6 Spreading time of oil droplets on polymer foils in surfactant solutions

on the species of polymer and the nature and concentration of surfactant (Fig. 6).

In Fig. 6 is represented the transition concentration c_T between spreading and nospreading. In surfactant solutions with concentrations higher than c_T the oil droplets are not able to spread on the polymer surface. Such solutions have a good detergency effect by removing oily soil. For an effective lubrication the surfactant concentration has to be smaller than transition concentration c_T. Nonionics in many cases were not suited for lubricant emulsions because of their low transition concentration. Among the cationic surfactants, we found species with very different effects, and it is possible to select the most suitable one. An increase in temperature was accompanied by a slight growth of the transition concentration.

The reasons for the heterocoalescence have not been fully cleared up yet. We could not find any simple correlations to other parameters of the surfactant solution like CMC or surface tension at the CMC. Also, the interface tension oil-solution does not have a simple connection with heterocoalescence.

Summary

This paper was meant to show the difficulties in creation of theoretical deductions and in their transfer to practical applications in the system synthetic fiber-lubricant. First results and derivations of the investigations of the different events at the solid–liquid interface in presence of surfactants like wetting, spreading, and adsorption point to the significance of further work in this field.

Acknowledgment The authors thank the "Sächsisches Staatsministerium für Wissenschaft und Kunst" for financial sponsorship of this work.

References

1. Miller CA, Neogi P (1985) Interfacial Phenomena. Surfactant Sci Ser Vol 17, Marcel Dekker Inc, pp 71
2. Jacobasch H-J (1984) Oberflächenchemie faserbildender Polymerer. Akademie-Verlag, Berlin
3. Schwabe K (1973) Physikalische Chemie. Akademie-Verlag
4. Neumann AW, Applied Surface Thermodynamics: Surface Tensions and Contact Angles (in press)
5. Ingram BT, Ottewil RH, In: Rubingh DN, Holland PM, Cationic Surfactants. Surfactant Sci Series Vol 37:87–140
6. Grosse I, Jacobasch H-J (1994) Tenside Surf Det 31, 6: 377–383
7. Sobisch T (1992) Tenside Surf Det 29: 199–202
8. Grundke K et al (1995) J Adhes Sci Technol 9, 3:327–350
9. Jacobasch H-J, Grosse I (1983) Colloids & Surfaces 7, 147–152
10. Grosse I, Müller H-J, being prepared
11. Stache H, Kosswig K (1981) Tenside Surf Det 18, 1:1–6
12. Stache H (1983) Textilveredlung 18, 1:3–12
13. Saito M, Otani M (1985) Textile Res J 55, 3:157–164

Progr Colloid Polym Sci (1996) 101:141–144

D.O. Hummel
T. Fröhlich

Computer-aided similarity and identity search with a *FTIR* library of surfactants and surfactant mixtures

Prof. Dr. D. Hummel (✉)
Im Lindenhof 15
53773 Hennef, FRG

T. Fröhlich
LabControl Inc.
Max-Planck-Str.179
50858 Köln, FRG

Abstract A digitized library of about 1100 standardized *FTIR* spectra of surfactants and related materials was evaluated with Lowry–Huppler algorithms and a combination algorithm specially developed for the detection of the components in mixtures. "Spectacle" by LabControl was used as the data management system. The *LH* algorithms 1 and 2 (sum of the absorbance differences and sum of the squared differences) were, in similarity searches, superior to 3 and 4 (derivatives). Both spectral identity of an analytical sample with components of the library and that belonging to a certain surfactant family can be established by a hit index.

An algorithm combining similarity and peak search was tested with intramolecular and intermolecular binary and multicomponent systems. In most cases the resulting list of the first 20 hits presented, in the highest ranks, the most similar spectra of mixtures represented in the library. Spectra of the components of the analytical mixture followed in lower ranks, usually down to rank 10. Tentative theoretical explanations for the results are given.

Key words Surfactants mixtures – *FTIR* spectrometry – computer-aided search – similarity identity

Definitions of spectral and structural similarity and identity

Identity, in a strict sense, means uniqueness; a tree is identical with itself. Two eggs are *equal*, two apples are *similar*. A horse and a cat are rather dissimilar though both are mammals. The problem is the quantization of similarity. An infrared spectrum is the two-dimensional presentation of the function $A(\tilde{\nu})$, all other parameters are kept constant. Most of the search algorithms use either peak data $(A, \tilde{\nu})$, i.e., peak tables, or completely digitized, standardized spectra. For the latter ones, the Lowry–Huppler algorithms [1] using the sum of absolute absorbance differences or the sum of squared absorbance differences are superior to the ones using the first derivatives of standard spectra [2]:

$$M_{ab} = \sum_{i} |s_i - r_i| \quad (1)$$

$$M_{sq} = \sum_{i} |s_i - r_i|^2 \quad (2)$$

$M \triangleq$ match value (measure of spectral dissimilarity); $i \triangleq$ absorbance for each resolution element; s/r = analytical/reference spectrum.

Decreasing values of M signify increasing spectral similarity. Since spectra of the same substance are never identical we have to define spectral identity as a high degree of similarity.

With the *LH*-algorithm 1, spectra having *M*-values $\leq 10^3$ can be considered to be *identical* [2].

There is a mutual relation between spectrum and structure only for oligoatomic structures. Due to the limited range of vibrational coupling in polyatomic molecules (multiple bonds, heavy atoms etc. interrupt coupling) much less than $3N-6$ ($N \triangleq$ number of atoms in the molecule) fundamental vibrations can be observed; certain structural information is lost. We may, however, consider the vibrational spectrum of a polyatomic molecule as a superimposition of the partial spectra of the partial structures [3,4]. This situation can be described as follows (according to clerc, modified):

spectrum = F (structure)
structure $\neq F^{-1}$ (spectrum)
partial spectrum $i = G_i$ (partial structure i) (3)
partial structure $i = G_i^{-1}$ (partial spectrum)
total structure $\neq \Sigma$ partial structures

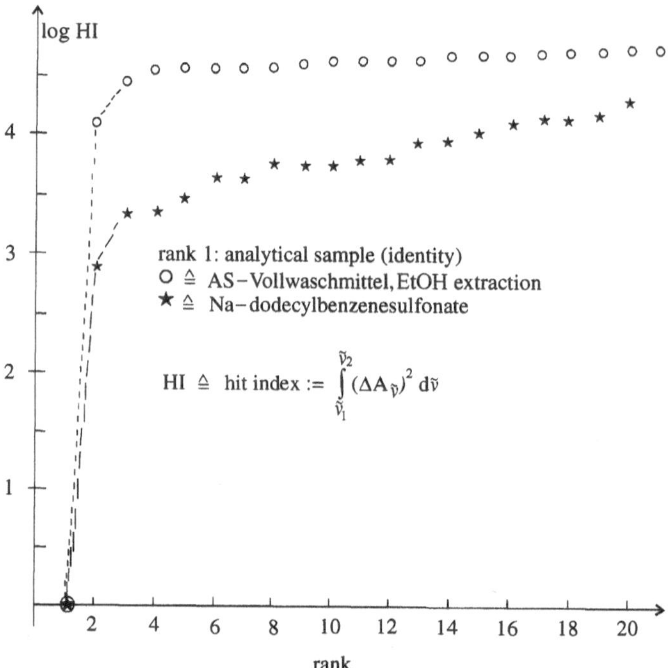

Fig. 1

Identification of chemically defined surfactants

The experiments were made with a library of about 1100 *FTIR* spectra of surfactants and related substances [6] together with the data management system SPECTACLE [7]. The spectral collection usually contained several examples of different origin and purity of each defined surfactant as well as numerous members of certain families (carboxylates, sulfonates etc.). Using the *LH* algorithm 1 (sum of absorbance differences) all of the runs were unambiguous. With a sodium fatty acid methyltaurate (Hostapon T, Hoechst)

$$R{-}CO{-}NH{-}CH_2{-}CH_2{-}SO_3^- Na^+$$

and within the first 10 ranks, two more like taurates occupied the ranks 2 and 3. 5 Na alkanesulfonates were found on 4 ... 9, 1 isethionate (O instead of NH) was on 10. No amide appeared within the first 20 ranks. With an isethionate (Hostapon KA) as analytical material three more isethionates were found within the first 10 ranks. Numerous other searches yielded similar results.

Structural families

In their molecule, all surfactants have a hydrophilic and a lipophilic part. We may say that they belong to a family if they coincide in one structural group. Thus, carboxylates, sulfonates, sulfates, phosphates, quaternary ammonium salts, ethylene oxide adducts, etc. are families. More rarely, the lipophilic part is used to characterize a family: fatty alkyl, lanoline, abietic, naphthalene, etc. derivatives. According to the concept of partial spectra and partial structures the spectra of members of one of these families should have one partial spectrum in common. This is rather easy to find in case of strongly polar and relatively small units (carboxylates, sulfonates etc.; empirical interpretation rules). In the case of larger units characterizing a family it may be helpful to determine a probability histogram with a library containing a large number of compounds belonging to the questionable family. Peaks in such a histogram usually designate the partial spectrum of the partial structure searched for.

It is interesting to see that hit indices determined by computer searches can also reveal the presence of spectral families [8]. The results of Fig. 1 were obtained with the *LH* algorithm 2. The two curves represent the dependence of $\log HI$ ($HI \equiv M_{sq}$ in Eq. (2)) from the rank for a complex mixture and for a material belonging to a family. In the upper curve the ethanol extract of a common washing powder was compared with the library containing, among others, a dozen of extracts from other washing powders. One of those was quite similar to the analytical sample ($\log HI \approx 4$), 6 of them kept a constant dissimilarity of 5. Finally, the curve reaches the dissimilarity mark of $\log HI = 5.5$ characterizing the spectra of all dissimilar organics. The lower curve shows the results with a Na dodecylbenzene sulfonate. There was one "identical" spectrum with $\log HI < 10^3$. The rest of the family follows with $\log HI$ values between 3.4 (still very similar) and 4. Finally the curve rises gradually to $\log HI$ values which do not belong to "family members".

Combination algorithm for the search of mixtures

Following the concept of partial spectra and partial structures (the first section) it should be possible to enhance the information obtained by computer search of mixtures by the use of an algorithm which combines similarity (full spectra) and peak search. Fröhlich [9, 7] has developed such an algorithm; it produces, in a simplified manner, a hit index

$$M_{\text{comb}} = \Sigma(S^2 + P^2)$$

with $S \triangleq$ rank by similarity search (full spectrum); $P \triangleq$ rank by peak search (peak table).

Checking this algorithm with surfactant mixtures, we found that the highest ranks were still occupied by the spectra having the closest similarity to the test material. Other than with the two simple LH algorithms 1 and 2 the next ranks were occupied by components of the mixture.

Alkylisethionates, $R{-}CO{-}O{-}CH_2{-}CH_2{-}SO_3^-Na^+$, may be considered to be an intramolecular mixture of aliphatic ester and aliphatic sulfonate. A search of a sodium fatty acid isethionate (Hostapon KA, Hoechst) with the combination algorithm puts three other isethionates on the ranks 2, 5 and 7. Five fatty acid esters were found on the ranks 3, 4, 6, 9, and 17, three alkanesulfonates on 8, 10, and 12.

Spectra of block copolymers of poly(oxypropylene) and poly(oxyethylene) with low content of POE can hardly be distinguished from the spectrum of pure POP [6]. The library used contains numerous PO–EO adducts and derivatives of these. A search with a PO–EO adduct containing 20% EO sequences (Genapol PF 20, Hoechst) puts 17 PO–EO adducts on the list; rank 19 was an EO graft copolymer and rank 20 was an oligo-propyleneoxide.

A rather extreme case was the residue of a $CHCl_3$ extraction of Persil (Henkel). The spectrum (Fig. 2, middle)

Fig. 2

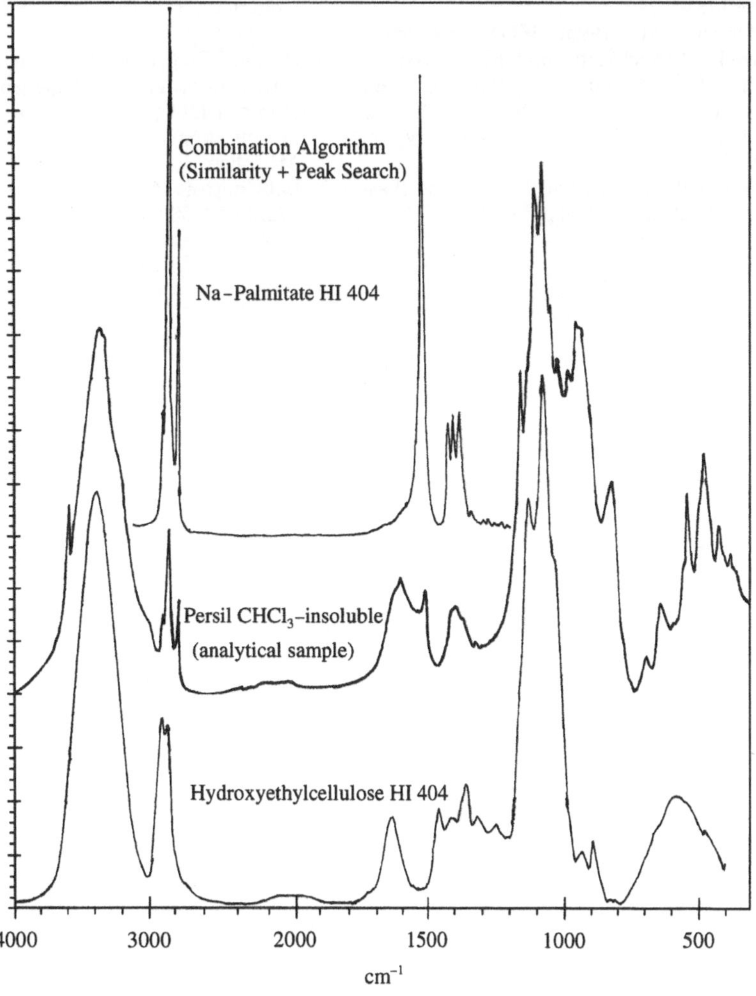

Table 1 Search results with the FTIR spectrum of a complex surfactant mixture. Analytical sample: Persil washing powder, $CHCl_3$-insoluble similarity search

HIT	Dissimilarity	Chemical composition
2	16900	hydroxyethyl cellulose
4	20300	ABS + alkylphenol-EO adduct
12	24100	*i*-hexylsulfosuccinate
16	25800	tributylphenol-EO adduct
18	26000	alkylbenzenesulfonate (Marlon A)
Similarity and peak search (combination algorithm)		
2	404	as above
3	404	Na palmitate
4	409	curd soap
9	425	Na laurylsulfate
11	436	laurylalcohol-EO adduct
19	500	alkylbenzenesulfonate (Marlon A)

shows some similarity with Na_2SO_4 and a cellulose derivative. No spectroscopist would dare to make a safe interpretation of this spectrum. Table 1 shows the results of a similarity search (above) and a combination search. In both cases hydroxyethyl cellulose was the nearest hit. Two soaps had similar values of M_{comb}; the similarity search did not put soaps on the list. No Na_2SO_4 was offered as possibility. In order to be sure, we made a search with Na_2SO_4; the Persil residue was not on the list. Undoubtedly, the results obtained with multicomponent mixtures need chemical verification. The main advantage of this approach is, however, to know *where to verify*.

Acknowledgment We gratefully acknowledge the support of our research by Stiftung Industrieforschung, Cologne. We thank Lab-Control GmbH, Cologne, for the data manager SPECTACLE.

References

1. Lowry SR, Huppler DA (1981) Anal Chem 53:889–893, ibid. (1983) 55:1288–1291
2. Amberg M (1995) FTIR-spektrometrische Ähnlichkeits- und Identitätssuche am Beispiel einer digitalisierten Bibliothek organischer Pigmente; Thesis, Math.-Nat. Fakultät, University of Cologne
3. Hummel DO, Atlas of Polymer and Plastics Analysis, 2nd ed., Vol 2b, p 93 ff
4. Roeges NPG (1994) A Guide to the Complete Interpretation of Infrared Spectra of Complex Structures; John Wiley & Sons, Chichester
5. Hummel DO (1996) Analyse der Tenside, Carl Hanser Verlag, München
6. Hummel DO (1996) FTIR Atlas of Surfactant Analysis, Carl Hanser Verlag, München
7. LabControl GmbH, Max-Planck-Str. 17a, D 50858 Cologne
8. Hummel DO (1994) GIT Fachz Lab 5:439–447
9. Fröhlich T, Anwendung kombinierter Korrelationskoeffizienten bei der computergestützten ir-spektroskopischen Identitäts- und Ähnlichkeitssuche von Multikomponentensystemen; Thesis, Math.-Nat. Fakultät, University of Cologne 1992

Progr Colloid Polym Sci (1996) 101:145–148
INTERFACES

K. Gloe
H. Stephan
T. Krüger
A. Möckel
N. Woller
G. Subklew
M.J. Schwuger
R. Neumann
E. Weber

Solvent extraction of toxic heavy metals with 8-hydroxyquinoline extractants from effluents

Prof. Dr. K. Gloe (✉) · H. Stephan
T. Krüger
Institut für Anorganische Chemie
Technische Universität Dresden
01062 Dresden, FRG

A. Möckel · N. Woller · G. Subklew
M.J. Schwuger
Institut für Angewandte
Physikalische Chemie
Forschungszentrum Jülich GmbH
52425 Jülich, FRG

R. Neumann · E. Weber
Institut für Organische Chemie
Technische Universität Bergakademie
Freiberg
09596 Freiberg, FRG

Abstract Interfacial activity and extraction data of position isomers of alkyl-8-hydroxyquinoline derivatives have been determined. The studies were performed to find possibilities for simultaneous separation of Cu^{2+}, Ni^{2+}, Zn^{2+}, Cd^{2+} and Hg^{2+} from percolating water model systems. 2-Nonyl-8-hydroxyquinoline shows high extractabilities for all metal ions. The addition of decanol to the organic phase gives a significant increase of extraction caused by alcohol solvation of the complex extracted.

Key words 8-hydroxyquinoline extractants – interfacial activity – percolating water – solvent extraction – toxic heavy metals

Introduction

The solvent extraction is a well-established unit process for separation and concentration of various metal ions [1, 2]. More recently, solvent extraction has become increasingly important in connection with environmental relevant separation and purification processes [3]. The reason for that lies in the advantageous approaches relating to both technological and economical viewpoints.

Relating to the choice of extraction systems for the separation of toxic heavy metals from effluents containing only low metal concentrations a great variety of requirements must be met in order to comply with the legal regulations. Thus a great number of physical and chemical properties of the extractant chosen have to adapted to the specific demands. Especially the complexing agent should have an extremely low solubility in the aqueous phase. On the other hand, the metal complex extracted must be highly soluble in the organic phase. That implies a structure for the complexing agent which is typical for surfactants. Along with other properties of the system, the interfacial properties influence the extraction rate and the characteristics of the dispersion [4]. The interfacial activity of the extractants must be well-balanced, because highly interfacial active agents produce dispersions that are slow to coalesce. Furthermore, extraction efficiency, separation selectivity and kinetic properties of extractants are of special interest.

Lipophilic 8-hydroxyquinolines seem to be suitable for the simultaneous separation of toxic heavy metals from neutral aqueous solution having a medium chloride content. This problem is relevant to the decontamination of metal pollutants from percolating water.

The aim of this work is to study the interfacial activity of chemical pure 8-hydroquinoline extractants 2–5, and to characterise the extraction efficiency towards Cu^{2+}, Ni^{2+}, Zn^{2+}, Cd^{2+} and Hg^{2+} from percolating water model

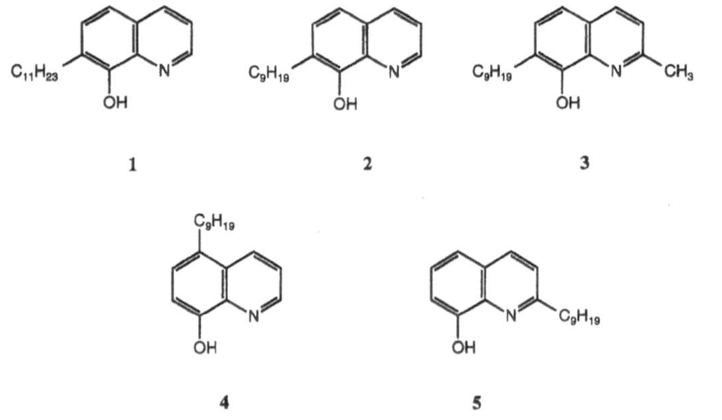

Fig. 1 Investigated compounds

systems. The results obtained with the chemical pure substances are compared with those of the commercially available complexing agent Kelex 100 1 [5, 6].

Experimental

Kelex 100 1 was provided by Witco GmbH. Compounds 2–4 were synthesized via Friedel-Crafts-acylation of 8-hydroxyquinoline and -quinaldine following Huang–Minlon reduction. 2-Nonyl-8-hydroxyquinoline 5 was obtained by direct introduction of the nonyl group with alkyllithium [7]. Other chemicals used were of the highest purity commercially available.

Interfacial tensions of reagent solutions in *n*-decane or *n*-decane/decanol-1 (v/v = 9/1) were measured at 20 °C using the Du Noüy ring method (tensiometer K 10; Krüss/Hamburg).

The extraction studies were performed at 25 ± 1 °C in 2 ml microcentrifuge tubes by means of mechanical shaking. The phase ratio $V_{(org)}:V_{(w)}$ was 1:1 (0.5 ml each). The shaking time was 30 min. In this time the equilibrium is achieved for Zn^{2+}, Cd^{2+} and Hg^{2+}, whereas Cu^{2+} and Ni^{2+} require about 2 h under the conditions chosen [8]. All samples were centrifuged after extraction. The determination of metal concentration in both phases was carried out radiometrically (isotopes from Medgenix Diagnostics) using γ-radiation measurement of ^{64}Cu, ^{65}Zn, ^{203}Hg in a NaI(T1) scintillation counter (Cobra II; Canberra–Packard), and β-radiation of ^{63}Ni, ^{115m}Cd in a liquid scintillation counter (Tricarb 2500, Canberra–Packard). Model percolating water solutions were prepared using 0.05 M tris(hydroxymethyl)aminomethane (TRIS)/HCl buffer containing 0.1 M NaCl in order to guarantee a constant species distribution of metal ions, and to avoid hydrolysis [9].

Results and discussion

Solvent extraction studies with the compounds 1–5 were performed in the system metal chloride-NaCl-buffer-water/extractant-diluent (decane or decane–decanol). Decane was selected as the diluent because paraffinic hydrocarbons have environmental compatible properties. The solubility of the chemical pure extractants 2–5 in decane is sufficiently in view of practical application – 0.08 M (4), ≥ 0.1 M (2, 3, 5), but in contrast to this the solubility of the metal complexes extracted is limited. Therefore decanol must be added as modifier to decane giving the necessary solubility in the organic phase.

Owing to the importance of interfacial properties in metal extraction, the interfacial tension was measured for the chemical pure compounds 2–5 in the system extractant-decane/water (Fig. 2). For the extractants investigated typical Gibbs isotherms were obtained, from which the surface excess, the interfacial molecular area and the critical micelle concentration can be estimated (Table 1) [10]. The surface excess and the interfacial molecular area are in the same order of magnitude as for hydroxyoximes which are used for copper extraction [11]. Generally, the interface is saturated at an extractant concentration of approximately 10^{-2} M in *n*-decane. The data listed in Table 1 proves the position of the alkyl group in the molecule has only a small influence on CMC, Γ and A_N. But, as shown in Fig. 2, the drop of interfacial tension with increasing extractant concentration depends significantly on the position of the alkyl group. The highest drop of γ-values was observed for the 2-alkyl derivative 5. It is interesting that

Fig. 2 Interfacial tension as a function of concentration for 8-hydroxyquinoline extractants 2–5 [ligand] = $1\cdot10^{-5}\ldots5\cdot10^{-1}$ M; Solid lines = system water//*n*-decane, dashed line = system water//*n*-decane-decanol-1 (v/v = 9/1)

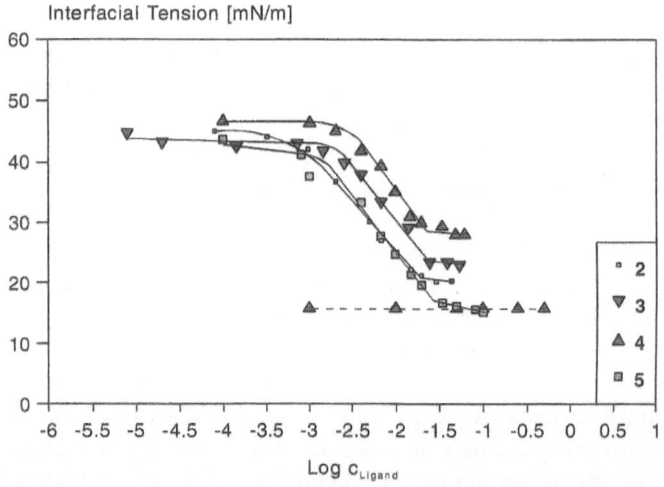

Table 1 Critical micelle concentration (CMC), interfacial tension (γ_{CMC}), surface excess (Γ) and interfacial molecular area (A_N) in the system water-decane

Extractant	CMC ($mol \cdot l^{-1}$)	γ_{CMC} ($mN \cdot m^{-1}$)	Γ ($mol \cdot m^{-2} 10^6$)	A_N (nm^2)
2	0.0157	21.6	2.9	0.57
3	0.0255	23.5	3.1	0.53
4	0.0255	27.9	3.6	0.45
5	0.0255	17.3	3.3	0.49

the interfacial tension values are independent on the concentration of the extractant in the presence of decanol. This result is obviously caused by strong hydrogen bonding between decanol and the extractant molecules leading to a pronounced arrangement of alcohol molecules at the interface.

The extraction properties of the hydroxyquinolines investigated differ in a characteristic manner on their structure. As shown in Fig. 3, the metal ions Cu^{2+}, Ni^{2+} and Zn^{2+} are almost quantitatively extracted from model solutions with Kelex 100 (1) dissolved in *n*-decane. In contrast to this, the extraction of Cd^{2+} and Hg^{2+} is rather poor. As expected, the extraction efficiency obtained for Kelex 100 is comparable with the chemical pure extractant 2. In comparison to 1 and 2 the extraction of Ni^{2+} is decreased for compound 3 and 5. The reason for this behavior is probably the lowering of the stability constant for nickel chelates if there is introduced an alkyl group in position 2 of 8-hydroxyquinolines [12]. Likewise, the quinaldine 3 is not capable to extract Cd^{2+} and Hg^{2+} from model percolating water. In contrast to the extraction behavior of 1–3 the separation of Cd^{2+} is significantly increased in the case of 4 and 5. However, compound 4 gives precipitates with all metal ions during the extraction experiments and consequently it is not suitable for practical application [13]. The reason for the favorable complexing properties of 5 for Cd^{2+} and Hg^{2+} is obviously connected with differences of the basicity of the nitrogen donor atom. PM3 calculations with MOPAC6 [7] have shown that the partial charge at the nitrogen atom of 5 is clearly increased comparing with 2 and 3, whereas the charge density at the oxygen atom of 2, 3 and 5 is only slightly different. This finding is in a good agreement with the acidity constants determined for different methyl-8-hydroxyquinolines [14]. Actually, compound 5 dissolved in decane-decanol shows the highest extractabilities for Cd^{2+} and Hg^{2+} comparing to the other derivatives (cf. Fig. 4). Generally, the addition of decanol-1 improves both the extraction efficiency and the loading of the organic phase with all metal ions. This is a result of incorporation of alcohol molecules into the complex extracted.

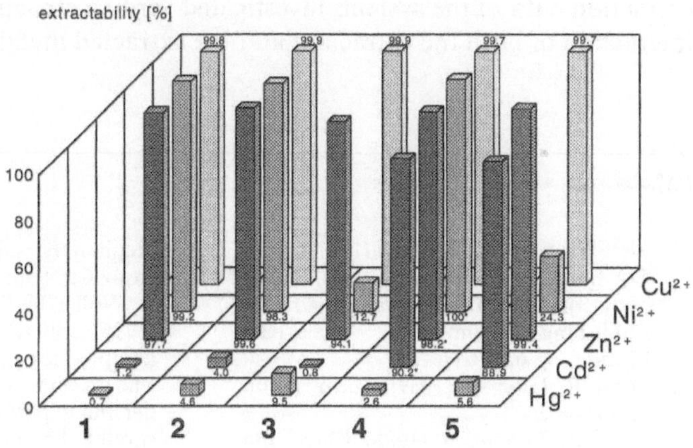

Fig. 3 Extractability of metal ions with 8-hydroxyquinoline extractants 1–5 (diluent: *n*-decane) $[MCl_2] = 1 \cdot 10^{-4}$ M; $[NaCl] = 1 \cdot 10^{-1}$ M; pH = 7.6 (TRIS/HCl-buffer); [extractant] = $1 \cdot 10^{-2}$ M in *n*-decane

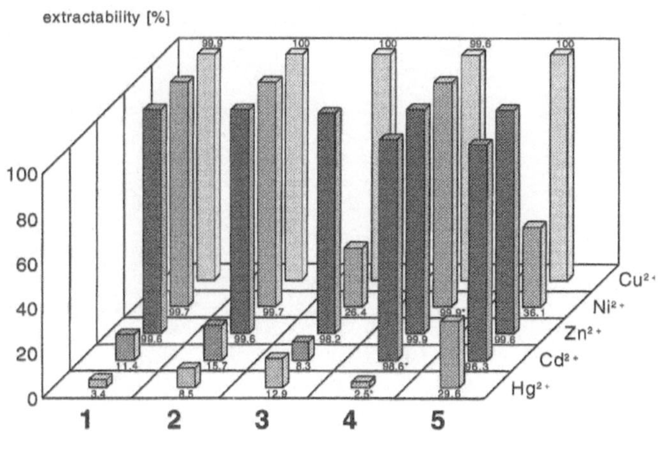

Fig. 4 Extractability of metal ions with 8-hydroxyquinoline extractants 1–5 (diluent: *n*-decane/decanol-1) $[MCl_2] = 1 \cdot 10^{-4}$ M; $[NaCl] = 1 \cdot 10^{-1}$ M; pH = 7.6 (TRIS/HCl-buffer); [extractant] = $1 \cdot 10^{-2}$ M in *n*-decane/decanol-1 (v/v = 9/1)

Conclusions

The results demonstrate the real possibilities of a simultaneous extraction of the toxic heavy metal ions Cu^{2+}, Ni^{2+}, Zn^{2+}, Cd^{2+} and Hg^{2+} from model percolating water systems using 2-nonyl-8-hydroxyquinoline dissolved in an *n*-decane/decanol-1 (v/v = 9/1) mixture. The advantage of this isomer is related to their stronger complex formation with Cd^{2+} and Hg^{2+} and is based on the higher basicity of the pyridine nitrogen donor atom. The interfacial and extraction data of the systems investigated verify a strong interaction of both the extractant and the extracted metal complex with decanol leading to a sharp increase of extractabilities.

References

1. Flett DS, Cox M, Melling J (1991) In: Lo TC, Baird MHI, Hanson C (eds) Handbook of Solvent Extraction. Krieger Publishing Company, Malarbar-Florida, pp 629–647
2. Mühl P, Gloe K (1989) Chem Techn 41:457–462
3. Marr R, Draxler J (1988) Chem Ing Techn 60:348–359
4. Cox M, Flett DS (1979) Proc Int Solv Extr Conf, ISEC 1977, CIM Spec Vol 21:63–72
5. Ashbrook AW (1975) Coord Chem Rev 16:285–307
6. Gareil P, de Beler S, Bauer D (1989) Hydrometallurgy 22:239–248
7. Stephan H, Gloe K, Krüger T, Chartroux C, Neumann R, Weber E, Möckel A, Woller N, Subklew G, Schwuger MJ (1996) Solv Extr Res Devel (Japan) 3: accepted for publication
8. The formation of hydrolyzed species under the experimental conditions chosen should be responsible for the slow achievement of the extraction equilibrium in the case of copper and nickel; Boumezioud M, Tondre C, Lagrange P (1988) Polyhedron 7:513–521
9. Metal ions of percolating water are complexed by polyhydroxy carboxylic acids possessing additional nitrogen donor atoms (humic and fulvic acids); Sigg L, Stumm W (1993) Aquatische Chemie Teubner-Verlag, Stuttgart, p 201
10. Lange H (1975) Tenside Detergents 12:27–34
11. Szymanowski J (1993) Hydroxyoximes and Copper Hydrometallurgy. CRC Press, Boca Raton-Ann Arbor-London-Tokyo, pp 145–168
12. Friedrich A, Schilde U, Uhlemann E (1986) Z Anorg Allg Chem 534: 199–205
13. Uhlemann E, Weber W, Fischer C, Raab M (1984) Anal Chim Acta 156:201–206
14. Friedrich A, Bukowsky H, Uhlemann E, Gloe K, Mühl P (1987) Anal Chim Acta 193:373–379

Progr Colloid Polym Sci (1996) 101:149–156

H. Dautzenberg
G. Arnold
B. Tiersch
B. Lukanoff
U. Eckert

Polyelectrolyte complex formation at the interface of solutions

H. Dautzenberg (✉) · G. Arnold
B. Lukanoff · U. Eckert
Research Group "Polyelectrolyte Complexes", University of Potsdam
Kantstraße 55
14513 Teltow, FRG

B. Tiersch
Research Group "Environment Saving Use of Plastics"
University of Potsdam
Kantsraße 55
14513 Teltow, FRG

Abstract The reaction between oppositely charged polyelectrolytes at the interface between their aqueous solutions, being the basis for the formation of flat membranes or microcapsules, has been studied with sodium cellulose sulphate and poly(diallyldimethylammonium chloride) as reaction components. By supplemental monitoring of the kinetics of membrane formation at thin ring layers additional information was obtained supporting the proposed mechanism of the process. In the very beginning at the boundary a semipermeable primary membrane is formed, which controls further membrane growth by its diffusion resistance. Osmotic pressure differences resulting from difference counterion concentrations inside and outside the capsule additionally remarkably affect the process. Capsule properties depend directly and indirectly on the polymer-chemical features of the used polyelectrolytes and vary in a wide range. For each specific combination the capsule properties can be additionally controlled by the main process variables as reaction time, polymer concentrations and presence of low molecular electrolytes.

Key words Polyelectrolyte complex formation – microcapsules – poly(diallyldimethylammonium chloride) – sodium cellulose sulphate – model studies

Introduction

The reaction between oppositely charged polyelectrolytes in aqueous solution often results in the formation of solid structures. If droplets of an aqueous polyanion solution are introduced into an aqueous solution containing a polycation, three situations can be met. Firstly, the droplets very quickly disperse, the solution becomes turbid, and a precipitate of small irregular particles is formed when the composition of the solution comes close to a stoichiometric relation of the total charges of both the polyelectrolytes [1]. Secondly, with solutions of sufficiently high concentrations the droplets are preserved [2] due to an immediate formation of a solid-like network structure at their surface. Thus microcapsules consisting of a liquid core surrounded by a semipermeable capsule wall are formed. Thirdly, because of the interfacial tension or the relation between the density of the solutions, sometimes the droplets are not able to cross the surface of the polycation solution. Instead they spread over a limited area of the surface and a membrane is formed at this region leaving the polyanion solution floating on top.

In the same way as microcapsules, flat membranes can be built if a thin layer of the polyanion solution is immersed into or covered with an excess of the polycation solution. The polyelectrolytes react at the interface thus forming a membrane stretching over the area of contact.

Polyelectrolyte complex (PEC) flat membranes have found increasing interest for separating liquids by pervaporation [3]. The preparation of microcapsules by PEC formation proved to be a powerful tool for encapsulating of sensitive biological subjects. It was possible to avoid the use of organic liquids, toxic reagents and drastic reaction conditions, that means, to maintain physiological conditions allowing even the encapsulation of sensitive viable cells without noticeable damage. Since the late 1970s, several methods and combinations of polyelectrolytes have been proposed for the encapsulation of biological matter [4–8]. The encapsulation offered still further advantages in comparison with other methods like entrapment into gel networks by polymerization or ionotropic gelation, as, for instance, the possibility to create capsule walls with definite cut-offs.

Sodium cellulose sulphate (NaCS) and poly(diallyldimethylammonium chloride) (PDADMAC) proved to be a very promising combination [9]. It has been applied to the encapsulation of very different kinds of biological matter without causing any damage to it. Moreover, capsules with excellent mechanical properties were obtained [10–15].

For application, however, it is not sufficient to avoid the loss of activity, to prevent the release of encapsulated biological material or to have good mechanical properties. To yield optimum results the immobilized systems have to be given a streamlined property profile, and this has to be done under restrictions arising from the matter to be immobilized.

To obtain the necessary information for optimizing the immobilization procedure and for controlling the capsule properties, we studied the process of capsule formation in detail. We investigated the dependence of practically relevant capsule properties on the polymerchemical characteristics of the reacting polyelectrolytes as well as on several process variables. To overcome difficulties in the interpretation of the experimental results caused by interfering osmotic effects, we additionally monitored the membrane growth at the interface between the reacting solutions in model studies with a newly developed method suppressing osmotic effects to a minimum.

In this contribution some experimental results are presented and the conclusions, which can be drawn so far, are summarized.

Materials and methods

As anionic reaction component different homogeneously [16] as well as heterogeneously [17–19] prepared NaCS samples were used with degree of substitution (DS) values ranging from 0.3 to 0.65 and viscosities (aqueous solution, 1 wt-% = $\eta_{1\%}$) from 15 mPas to more than 500 mPas. As can be seen from Figure 1, the samples showed a rather narrow molecular weight distribution (MWD). As cationic reaction component served different PDADMAC samples, which according to Figure 2 represent four types of PDADMAC: PDADMAC of low molecular weight (M_w) with narrow and broad MWD (MIK 7, MK 10 and B40), PDADMAC of higher molecular weight (T32) and PDADMAC of medium molecular weight and broad MWD (MGV).

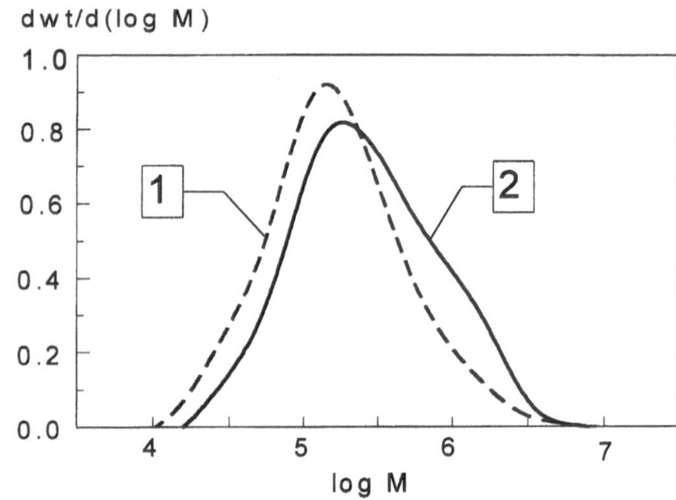

Fig. 1 Molecular weight distribution of sodium cellulose sulphate samples (0.5 m $NaNO_3$, HEMA BIO linear, 0.8 ml/min) 1-homogeneously prepared high viscous NaCS; 2-heterogeneously prepared lower viscous NaCS

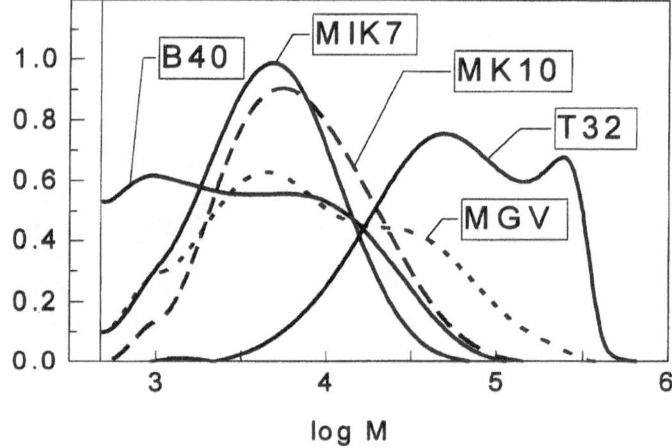

Fig. 2 Molecular weight distribution of poly(diallyldimethylammonium chloride) samples (0.5 m $NaNO_3$, HEMA BIO 100, 0.8 ml/min)

Two sets of experiments were carried out. To form capsules the aqueous NaCS solution was pressed through a capillary with a constant flow. The droplets of about 15 mg, falling due to their own weight, were introduced into a stirred aqueous PDADMAC solution containing, if necessary, an additional amount of sodium chloride as low molecular electrolyte. After different residence times in the precipitation bath capsules were taken out for investigation. The degree of NaCS conversion was determined either by weighing or spectrophotometrically (binding of methylene blue) [20]. Generally, the size and the mechanical strength of the capsules were measured [21]. Microphotographs were taken to reveal the shape of the capsules. The morphology of the capsules wall was investigated by transmission electron microscopy of ultrathin sections and by scanning electron microscopy. Information about the cut-off of the capsule wall was obtained polarographically after equilibrating with polyethylene glycols of different M_w. For additional model studies, each time a droplet of a NaCS solution (2 to 3 mg) was pressed between two glass plates to a spot of about 0.1 mm in height and 10 mm in diameter and surrounded by a tenfold amount of PDADMAC solution. In that way the borderline between the two solutions was held in a fixed position. The broadening of the reaction zone was microscopically monitored by using a TV camera and a crosshair system.

Results

Process of capsule formation

The process of capsule formation starts with a spontaneous creation of a solid-like precipitate at the droplet surface. This semipermeable primary membrane is likely the result of a phase separation process succeeding the reaction between the oppositely charged polyelectrolytes in the border region of the contacting aqueous solutions. After this initial step the reaction becomes diffusion-controlled. The PEC formation proceeds only to that extent the reaction components can cross the primary membrane. As we found, capsule wall growth exclusively takes place in the NaCS containing solution, that means inside the capsule. It is prevented entirely if the cut-off of the primary membrane is too low in comparison with the molecular weight of the penetrating polycation. Figure 3a shows the typical microgranular structure of a primary membrane formed under conditions where the capsule wall growth stops (PDADMAC of higher molecular weight, absence of low molecular electrolytes). As can be seen from Figure 3b,

Fig. 3 Electron micrographs of typical capsule walls. a-TEM of an ultrathin section (conditions of capsule preparation: 2 wt-% homogeneously prepared NaCS, 2 wt-% PDADMAC type T32, 10 min reaction time, water as reaction medium); b-the same as a, but prepared in physiological NaCl solution; c-SEM of a fracture surface after different reaction time (conditions of capsule preparation: 4 wt-% heterogeneously prepared NaCS, 2 wt-% PDADMAC B40, 15 min or 180 min reaction time)

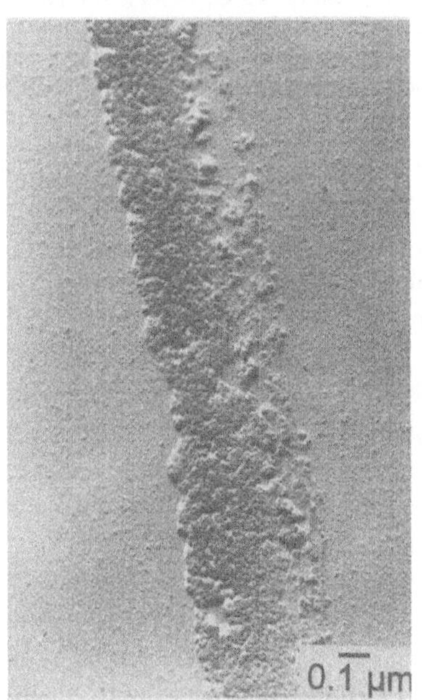

a)

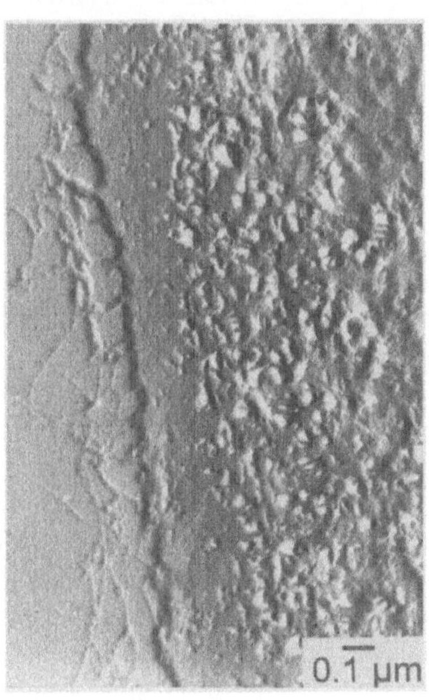

b)

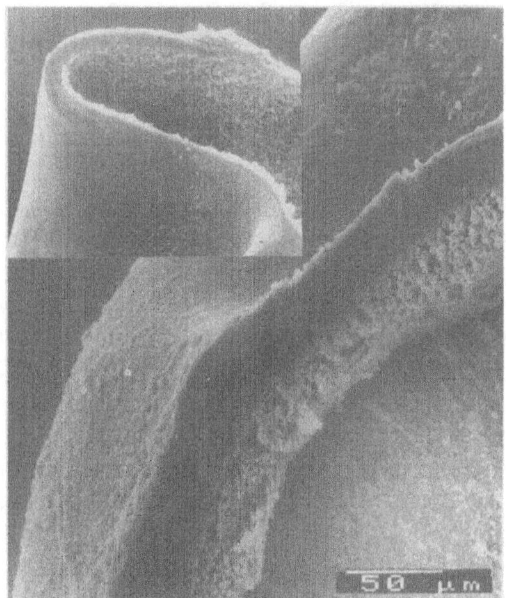

c)

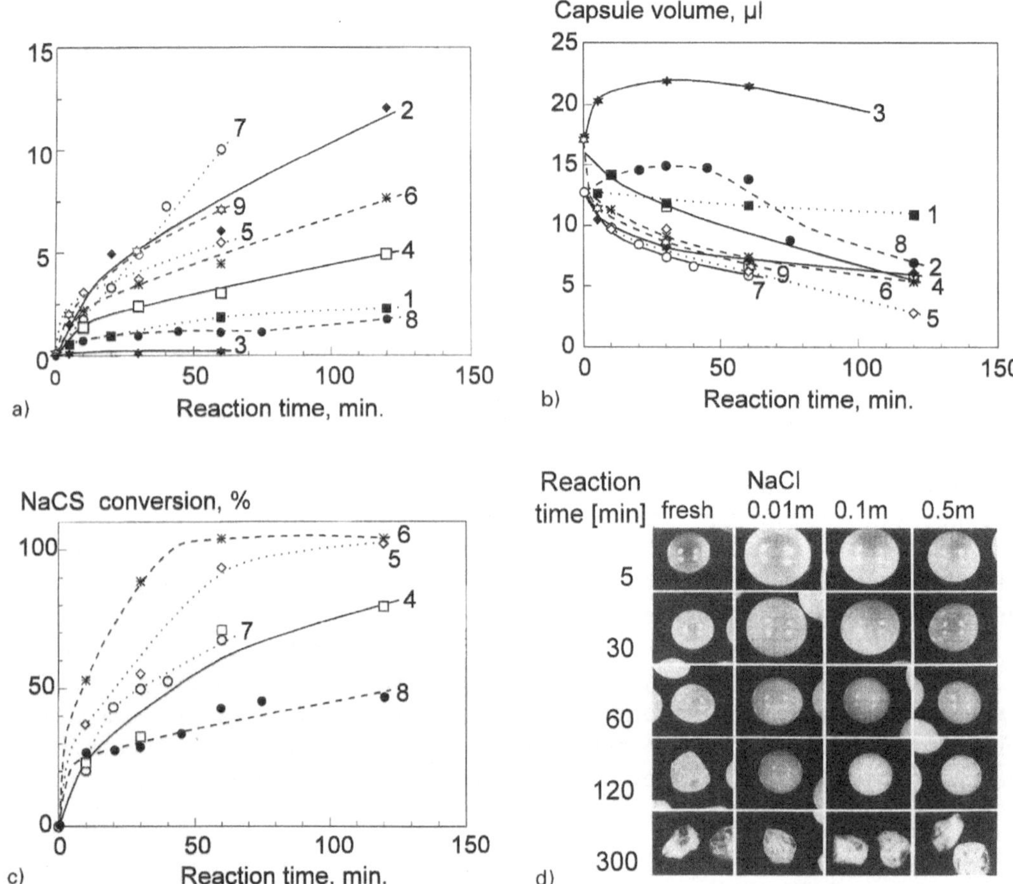

Fig. 4 Capsule properties in dependence on the reaction time. a)-Mechanical strength, b)-Capsule volume, c)-Degree of NaCS conversion conditions of capsule preparation: 1-■-2 wt-% NaCS R73, $\eta_{1\%} = 430$ mPas/-2 wt-% PDADMAC T32; water; 20 °C; 2-◆-2 wt-% NaCS R73, $\eta_{1\%} = 430$ mPas/2 wt-% PDADMAC T32; physiol. NaCl sol.; 20 °C; 3-★-4 wt-% NaCS Mi, $\eta_{1\%} = 35$ mPas/2 wt-% PDADMAC T32; water; 20 °C; 4-□-4 wt-% NaCS STE, $\eta_{1\%} = 35$ mPas/2 wt-% PDADMAC B40; water; 20 °C; 5-◇-4 wt-% NaCS STE, $\eta_{1\%} = 35$ mPas/2 wt-% PDADMAC B40; physiol. NaCl; 20 °C; 6-*-4 wt-% NaCS STE, $\eta_{1\%} = 35$ mPas/2 wt-% PDADMAC B40; water, 40°C; 7-○-4 wt-% NaCS 1162, $\eta_{1\%} = 50$ mPas/2 wt-% PDADMAC B40; water; 20°C; 8-●-5 wt-% NaCS 1129, $\eta_{1\%} = 16$ mPas/2 wt-% PDADMAC B40; water, 20°C; 9-☆-2 wt-% NaCS Mi, $\eta_{1\%} = 35$ mPas/6 wt-% PDADMAC T32; water, 20 °C; d-size and shape of capsules after recovering from the precipitation bath or after 24 h storage in NaCl solution of different concentrations (conditions of capsule preparation similar to curve 3 mentioned above)

presence of salt during the process of capsule formation facilitates the penetration of the polycation. If the molecular mass of the polycation is low enough, capsule wall growth continues until all the NaCS inside the capsule is converted into PEC. Capsule walls obtained in this way generally have a pronounced asymmetric structure with a compact membrane outside and a loose packaging inside (Fig. 3c).

The practically relevant properties of the capsules, like their mechanical strength or the transparency, cut-off and diffusion resistance of the capsule wall, depend on the thickness and the morphology of the capsule wall. The capsule wall thickness is determined by the rate of NaCS conversion. Besides, it is limited by the amount of NaCS available per mm^2 surface of the final capsule and therefore depends on the droplet volume at the beginning, on the initial NaCS concentration therein and on the final size of the capsule.

Figure 4 presents some data on the process of capsule formation that reveal the general trends in the dependence of the capsule properties on the polymerchemical features of the polyelectrolytes as well as on the conditions of capsule preparation. Without discussing the correlations in detail we can conclude:

High viscous NaCS samples (homogeneously prepared) can be combined with PDADMAC samples of higher molecular weight and yield capsules of high mechanical strengths at comparatively low concentrations of the reacting polyelectrolytes (Fig. 4a, curve 1). Presence of salt (0.1 to 0.15 m sodium chloride) facilitates the diffusion of the polycation. The reaction goes on faster and after the same reaction time the mechanical strength is higher (curve 2). Low viscous NaCS samples (heterogeneously prepared) have to be used at higher concentration to get capsules at all. With PDADMAC samples of high M_W at the common concentration of about 2 wt-% the reaction often comes to a standstill very soon, thus leaving the capsules extremely fragile but fully transparent (curve 3). If PDADMAC of low M_W is applied, even at low concentrations of 2 wt-% capsule wall growth generally continues until all NaCS is consumed, as long as the reaction is not stopped deliberately (curve 4). This, however, is only true for PDADMAC samples containing portions of very low M_W. If samples of low M_W and narrow MWD are used, like sample MIK 7, capsule wall growth also often stops soon, so that only capsules of low stability result. Presence of salt again accelerates the process (curve 5) and leads to higher capsule stability. The same effect is caused by increased temperature (curve 6). To achieve optimum mechanical strength, the NaCS concentration must be adjusted at its optimum, which is higher the lower the molecular weight of the sample (curves 7 and 8). Any stability level that can be achieved is lower the smaller the chain length of the polyanion. The PDADMAC concentration has a very strong impact. Its increase can force the reaction to take place to a sufficient extent also between NaCS samples of low molecular weight and PDADMAC samples of high molecular weight (curve 9).

The dependency of the mechanical strength or any other capsule parameter on a certain process variable is different for each specific combination of the reacting polyelectrolytes. There are no generally valid quantitative relations. That makes an evaluation of the experimental data extremely difficult. One reason for this is the circumstance that the situation in the border area of the reacting solutions not only changes according to the progressing conversion of the polyelectrolytes, but also because of osmotic effects. As Figures 4b and d demonstrate, capsules swell or shrink during the process of their formation in agreement with the progressing degree of NaCS conversion. For that an osmotic pressure difference between the solution inside and outside the capsules is responsible. This difference is caused by the counterions of the separated polyelectrolytes. Due to the establishing DONNAN equilibrium it reduces, as can also be shown by rough calculations [22], with increasing concentrations of low molecular electrolytes. Under swelling conditions water streaming into the capsule enlarges the capsule surface and

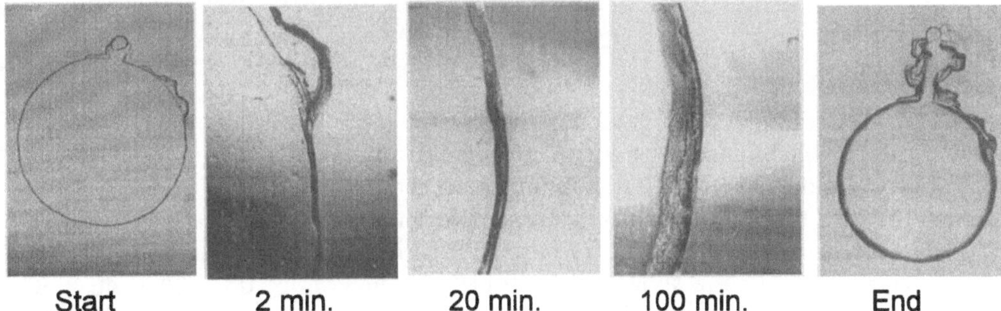

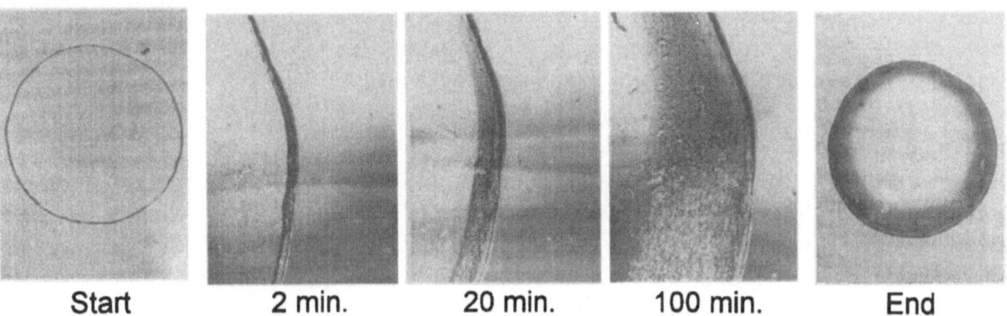

Fig. 5 Membrane growth due to polyelectrolyte complex formation between NaCS and PDADMAC at the border of the two reacting solutions (Magnification start and end 1:2; membranes after different reaction time 1:16)

reduces the NaCS concentration near the capsule wall. As a result the capsule wall growth is decelerated and the material has to stretch over a wider area. Under shrinking conditions the wall material is compressed and simultaneously shifted to the interior of the capsule. The way the PDADMAC has to travel to reach regions of high NaCS concentration is shortened.

Capsule swelling happens as long as the counterion concentration inside the capsule ($\pi_{\text{inside}} = c_{\text{inside}} * R * T$) is higher than the counterion concentration outside the capsule ($\pi_{\text{outside}} = c_{\text{outside}} * R * T$). It is particularly pronounced, if the NaCS conversion cannot continue. Otherwise, the osmotic pressure inside the capsule continuously drops and finally capsules start shrinking. Capsule shrinkage takes place from the beginning, if the reaction starts at $\pi_{\text{outside}} > \pi_{\text{inside}}$. After transferring into media free of polycation, the capsules differently swell according to the residual amount of NaCS left inside and the resulting pressure difference (Fig. 4d). Capsules formed only under shrinking conditions reach the highest values of mechanical strength. Capsules, which only swell, stay fragile. As we can see from Figure 4c the swelling and shrinking behaviour of the capsules and the increase of their mechanical strength is closely but not strongly connected with the NaCS conversion. A sufficient degree of NaCS conversion is a prerequisite for an acceptable capsule stability. The NaCS conversion is faster the lower the NaCS concentration, and higher according to the PDADMAC concentration, and can be accelerated by presence of sodium chloride or by increased temperature.

Model studies

In Figure 5 two examples are given for the shift of the reaction front at the boundary between the aqueous solutions of NaCS and PDADMAC. It confirms the observation that the PEC formation exclusively takes place within the NaCS solution. The thin layer of 0.1 mm in height reacts very similarly to the surface of a droplet. Again, a primary membrane is very quickly formed and it separates the two solutions. But osmotic effects are small because pressure balance can be achieved due to the fragility of the thin PEC ring enclosing the NaCS solution. In case of an excess pressure inside, like in series A, the primary membrane breaks at a weak point and incoming water drifts out (see start and end). If the pressure difference is small, the ring stays free of defects (start and end in series B). As not shown here, an excess pressure outside creates dents in the ring layer. Generally, however, the whole ring is kept in a fixed position. As expected, the ring layer grows faster at higher PDADMAC concentration (series B). The microphotographs of the ring layer also

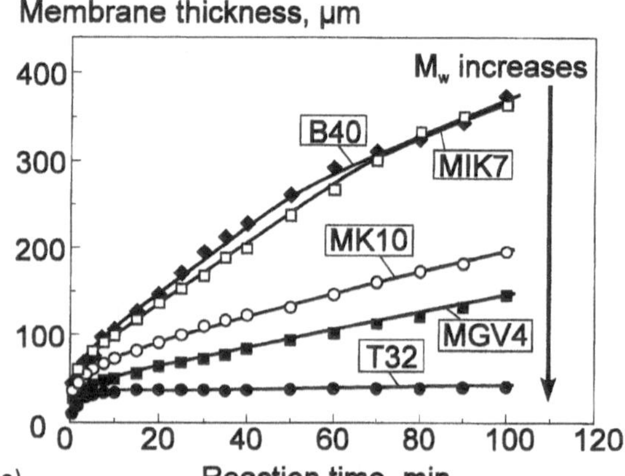

a)

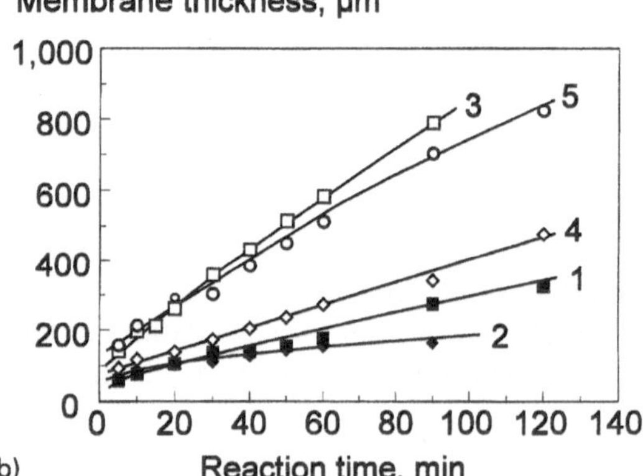

b)

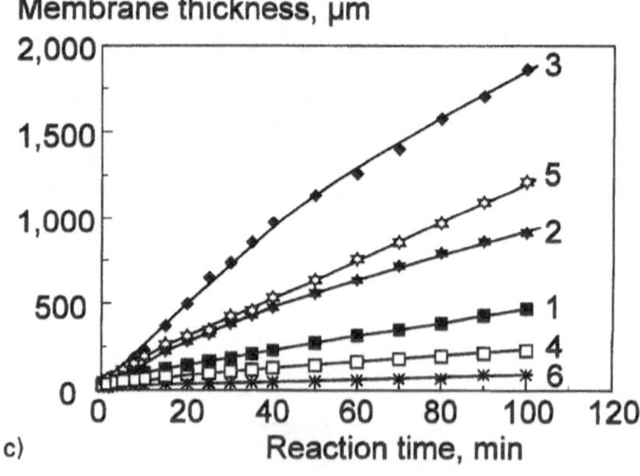

c)

Fig. 6 Kinetics of the membrane growth at the border of the reacting NaCS and PDADMAC solutions. a. Variation of the PDADMAC type (see Fig. 2): 4 wt- % NaCS Mi26, 2 wt-% PDADMAC. b. Variation of the NaCS type and concentration, 2 wt-% PDADMAC; 1-■-2 wt-% NaCS 1181 ($\eta_{1\%} = 19.5$ mPas; DS = 0.63); 2-◆-4 wt-% NaCS 1181; 3-□-2 wt-% NaCS 1172 ($\eta_{1\%} = 36.5$ mPas; DS = 0.35); 4-◇-4 wt-% NaCS 1172; 5-○-2 wt-% NaCS KW58 ($\eta_{1\%} = 430$ mPas; DS = 0.35). c. Variation of the polymer concentrations; NaCS Mi26 ($\eta_{1\%} = 36$ mPas; DS = 0.50); PDADMAC B40 (one example with PDADMAC T32 is included for comparison). 1-■-2 wt-% NaCS, 2 wt-% PDADMAC; 2-★-2 wt-% NaCS, 4 wt-% PDADMAC; 3-◆-2 wt-% NaCS, 6 wt-% PDADMAC; 4-□-4 wt-% NaCS, 2 wt-% PDADMAC; 5-☆-4 wt-% NaCS, 6 wt-% PDADMAC; 6-*-2 wt-% NaCS, 6 wt-% T32

reveal structural differences. The slower growing ring at 2% PDADMAC obviously is somewhat more dense than the fast growing one at 4 wt-% PDADMAC. This is only valid for the secondary wall. Direct conclusions on the structural of the primary membrane that controls the whole process by its diffusion resistance are not possible from these images.

From Figure 6 it becomes clear that the character of the PDADMAC is extremely important for the course of the reaction. The higher the Mw of the PDADMAC sample, the slower the membrane grows in thickness (Fig. 6a). According to the experiments of capsule formation, membrane growth stops very soon if PDADMAC of high Mw at comparatively low concentration (2 wt-%) is applied. However, the two samples of low Mw (broad and narrow MWD) do not behave as differently as they do regarding capsule formation.

Of particular interest are conclusions which can be drawn about the impact of the NaCS quality on the course of the reaction. They explain the behavior of the NaCS samples in the process of capsule formation. Figure 6b gives some evidence that the diffusion process id decelerated, if the DS of the NaCS sample is too high (curve 1 and 2 for DS = 0,65 in comparison to curve 3 and 4 for DS = 0,35). At low NaCS concentration the membrane growth is mostly faster than at higher concentration. This is more pronounced with the samples of lower DS. The viscosity does not directly affect the reaction (curves 3 and 5). This demonstrates on the one hand that the diffusion characteristic of the primary membrane is determining and on the other hand that differences found for the capsule formation are at least partly osmotically caused.

Figure 6c informs on the concentration dependence of the reaction. Roughly, we can say that the reaction is accelerated by increasing PDADMAC concentration (curves 1; 2; 3 or 4; 5) and, as we already mentioned, is decelerated by increasing NaCS concentration (curves 1; 4 or 3; 5). Further, we see that the diffusion limitation is a decisive factor. With a high molecular PDADMAC sample even at 2 wt-% NaCS and 6 wt-% PDADMAC the reaction rate remains on a very low level (curve 6). This is in contrast to the behavior of the same combination of samples in the capsule formation, where the wall growth was clearly sped up by increased PDADMAC concentration. Possibly, the reduction of the capsule volume due to osmotic draining of water plays an important part.

Conclusions

The formation of microcapsules based on the reaction between oppositely charged polyelectrolytes at the interface of their aqueous solutions is a sophisticated process. It is controlled by the diffusion resistance of the primary membrane formed immediately after the solutions have come into contact and additionally affected by osmotic effects. Depending on the relation of the counterion concentrations inside and outside the capsule, in the second step of the process, capsule wall growth that is restricted to the area inside the capsule takes place either under swelling ($\pi_{\text{inside}} > \pi_{\text{outside}}$) or under shrinkage ($\pi_{\text{inside}} < \pi_{\text{outside}}$) of the capsules accompanied by a change of the reacting surface area and concentration changes additional to those caused by the consumption of the reacting polyelectrolytes. According to this mechanism the course of the process as well as the properties of the obtained capsules strongly depend on the polymer-chemical characteristics of the used polyelectrolytes. Most important parameters are the M_w and the DS of the sodium cellulose sulphate being the anionic reaction component inside the capsule and on the M_w and MWD of the PDADMAC being the cationic counterpart in the precipitation bath. Each combination of samples shows its specific dependence for the capsule properties, like mechanical strength, transparency, morphology or cut-off of the capsule wall, on the process variables as reaction time or temperature, polymer concentrations and presence of low molecular electrolytes (sodium chloride). The mechanical strength as one of the most important features for practical purposes generally increases with progressing NaCS conversion during storage of the capsules in the precipitation bath. However, only fragile capsules are obtained if the reaction stops early due to the relation between the cut-off of the primary membrane and the M_w and MWD of the PDADMAC preventing the necessary diffusion. Otherwise the capsule properties can be deliberately controlled by the process variables.

The model studies on the kinetics of the reaction at the interface between the polyelectrolyte solutions confirm the proposed mechanism of capsule formation. Due to the possibility to separate the diffusion-controlled part of the reaction from the osmotically controlled one clearer correlations are obtained. It can be decided whether or to what extent a specific feature of the used polyelectrolyte or a process variable directly or via the resulting osmotic conditions only indirectly affects the process of capsule formation. Thus it can be supposed that the DS of the NaCS also has a direct impact on the process of the capsule wall growth, while this seems not to be true for the viscosity of the solution.

Acknowledgment We gratefully acknowledge the funding of parts of this work by the "Deutsche Forschungsgemeinschaft".

References

1. Philipp B, Kötz J, Dautzenberg H, Dawydoff W, Linow KJ (1991) In: Mitchell Jr J Applied Polymer Analysis and Characterization Part III Carl Hanser Verlag Munich Vienna, New York Barcelona, pp 281–310
2. GB 21 35 954A (1983)
3. Schwarz HH, Paul D, Apostel R (1993) Filtrieren & Separieren 6:309–311
4. Lim F, Sun AM (1980) Science 210:908–910
5. Goosen MFA, O'Shea CM, Gharapetian H et al (1985) Biotech Bioeng 27:146–150
6. DE 36 15043A1 (1986)
7. Chang TMS (1987) Biomat Art Cells, Art Org 15:1–20
8. Gharapetian H, Maleki M, O'Shea GM et al. (1987) Biotech Bioeng 30:775–779
9. Dautzenberg H, Loth F, Fechner K et al (1985) Makromol Chem Suppl 9: 203–210
10. Pommerening K, Ristau O, Rein H et al (1983) Biomed Biochim Acta 42: 813–823
11. Braun K, Besch W, Jahr H et al. (1985) Biomed Biochem Acta 44:143–147
12. Torner H, Kauffold P, Götze M et al (1986) Arch exper Vet med Leipzig 40:541–554
13. Braun K, Kauert C, Weber A et al (1988) Z exp Chir Transplant künstl Organe 21:58–64
14. Merten OW, Dautzenberg H, Palfi GE (1991) Cytotechn 7:121–130
15. Stange J, Mitzner S, Dautzenberg H et al (1993) Biomat Art Cells & Immob Biotech 21:343–352
16. DD WP 15 25 65 (1978)
17. DE 40 21 049A1 (1990)
18. Lukanoff B, Dautzenberg H (1994) Das Papier 6:287–298
19. Dautzenberg H, Lukanoff B, Neubauer K (1995) In: Cellulose and cellose derivatives: Physico-chemical aspects and industrial applictions: Published by Woodhead Publishing Ltd Abington Hall: Abington, Cambridge CB1 6AH England, pp 435–445
20. Dautzenberg H, Holzapfel G, Lukanoff B (1993) Biomat, Art Cells & Immob Biotech 21:399–405
21. Dautzenberg H, Krause M, Lukanoff B (1994) Internat Workshop Bioencapsulation IV, pp 9–16
22. Dautzenberg H, Lukanoff B, Eckert U et al, Polyelektrolytes Potsdam '95 "Berichte der Bunsen-Gesellschaft" (submitted)

Progr Colloid Polym Sci (1996) 101:157–165

K. Stana-Kleinschek
V. Ribitsch

Surface properties of textile cellulose as a function of processing steps

K. Stana-Kleinschek
Technical University of Maribor
Department of Textile Chemistry
Maribor, Slovenia

Dr. V. Ribitsch (✉)
Institut für Physikalische Chemie
Universität Graz
Heinrichstraße 28
8010 Graz, Austria

Abstract Natural cellulose fibers comprise several non-cellulose compounds and cationic trash which causes problems during different adsorption processes such as dyeing, printing, final fiber finishing and coating. Therefore the pretreatment (classical NaOH, demineralization, oxidative bleaching) is the most important step in cellulose textile finishing.

Alternative ways to describe the success of different processes in fiber pretreatment which result in distinct surface charge and hydrophilicity are the determination of electrokinetic properties and the water uptake of fibers and textile materials.

The zetapotential was determined by streaming potential measurements as a function of the pH and the surfactant concentration in the liquid phase. The water uptake was observed measuring the changes of the ultrasound intensity caused by the water penetration into dry cellulose fabric.

The degradation and removal of hydrophobic non-cellulose compounds which cover the primary hydroxyl and carboxyl groups of the cellulose polymer is clearly shown by an increase of the negative ZP. This observation correlates well with the penetration measurements showing improved hydrophilicity for example after NaOH treatment or extraction. The progress of the fiber processing (cleaning, pretreatment) is reflected by the surface charge as well as the hydrophilicity of the fiber.

Key words Cellulose fibers – zetapotential – hydrophilicity – ultra sound velocity – pretreatment

Introduction

Textile cellulose purification processes

Due to the trends during the last few years to wear ecologically friendly materials, natural cellulose fibers are the most popular textile fibers. The cellulose cotton fibers have several essential advantages (costumer friendly) in comparison to synthetic textile polymers (PES, PA, PAC), but they also show a disadvantage, their technological finishing process is much more complex.

Because of their chemical composition of approximately 90% cellulose and 10% different non cellulose compounds as hemicellulose, pectins, proteins, waxes, pigments and mineral salts, many different steps are necessary during the pretreatment–purification process [1].

Almost all non cellulose compounds, which creates problems – poor absorbency, poor rewettability – are located in the primary shell of the fiber, the fiber surface. It is therefore this part of the fiber which is modified during the textile pretreatment processes. [1, 2].

The textile pretreatment processes used in the cotton finishing industry, cleaning the row fabrics by means of

Table 1 Average cotton composition

Content	% in the cross-section of the fiber	% in the primary shell
Cellulose	88 to 96	52
Pectins	0.7 to 1.2	12
Waxes	0.3 to 1.0	7
Proteins	1.1 to 1.9	12
Ashes	0.7 to 1.6	3
Other comp	0.5 to 1.0	14

Table 2 Pretreatment operations of cotton fibers [6]

Processing stage	Purpose of processing	Kind of process
Alkaline treatment using NaOH	– removal of non cellulose compounds – swelling of the cellulose – disintegration of seed husks – cleaning and pre-brightening – removal of disturbing metals	extraction swelling
Acid treatment using different mixtures of acid sequestering agents and surfactants	– removal of alkaline earth and heavy metals – cleaning and prebrightening – improvement of wetting properties	demineralization

extraction, deminerlization, oxidation and swelling processes, comprise at least two operations:

- alkaline or new ecologically friendly acid treatment
- chemical bleaching.

The typical purification stages of cotton pretreatment are represented in Table 2.

All chemical purification and oxidation processes are performed in aqueous environment, the main steps are degradation of non cellulose compounds and oxidation of natural pigments. During these processes active groups on the cellulose polymer surface are recovered, causing swelling of the polymer and modifications of the electrokinetic properties and adsorption abilities of the fibers.

Electrokinetic properties of the fiber-forming cellulose polymers

Electrokinetic properties are generated by the electrochemical double layer created at the interface between charged solid materials such as fibers or fabrics and aqueous solutions containing various electrolyte ions, acid complex forming agents, surfactants, oxidants or dyes. The net charge at the surface of a material in contact with a polar medium is governed by three processes:

- association/dissociation of surface chemical groups,
- adsorption of ionic species,
- dissolution of ions from the material into solution.

The distribution of the electrical charges at the interface is different from that in the bulk phase. The simplest model describing the situation is that of Stern by which the charges at the surface are compensated by ions of opposite charge (counter-ions) in solution forming two different layers. The layer closest to the surface (the Stern layer), is considered immobile while the outer layer (the diffuse layer), allows diffusion of ions through thermal motion. Electrokinetic behavior depends on the potential at the slipping plane (shear plane) between the charged surface and the electrolyte solution. This potential is called the electrokinetic or zetapotential [ZP], it is frequently assumed to be identical with the Stern potential [3]. The ZP is generally measured as a function of the pH showing a plateau in the basic region where all the acid groups are dissociated or OH^- ions or other anions are adsorbed. Reducing the pH causes association of acid groups and/or desorption of anions and adsorption of cations. The ZP of fiber-forming polymers such as cotton shows negative values because of acidic dissociable groups.

It was recognized that the pK values of the dissociable groups can be calculated from the ZP-pH function (using the plateau value of the ZP and the pH at which the ZP is half of the plateau value) [4]. In this way calculated pK values of cotton fibers indicate that the dissociation of acid groups is mainly responsible for the negative ZP.

Specific ion adsorption, another event possibly determining the surface charge, is recognizable in the course of zeta potential – electrolyte concentration function and depends on the surface properties of the polymer, as well as on the type and kind of electrolyte and adsorption behavior of the solvent.

Measurement of the ZP of materials of irregular size and shape such as textile fibers, hair, polymer films, paper pulp and coarse dispersions of mining materials, can be performed most satisfactorily using the streaming potential method. A review of the theoretical background, and methodical details of ZP measurements on fibers was given by Jacobasch [5].

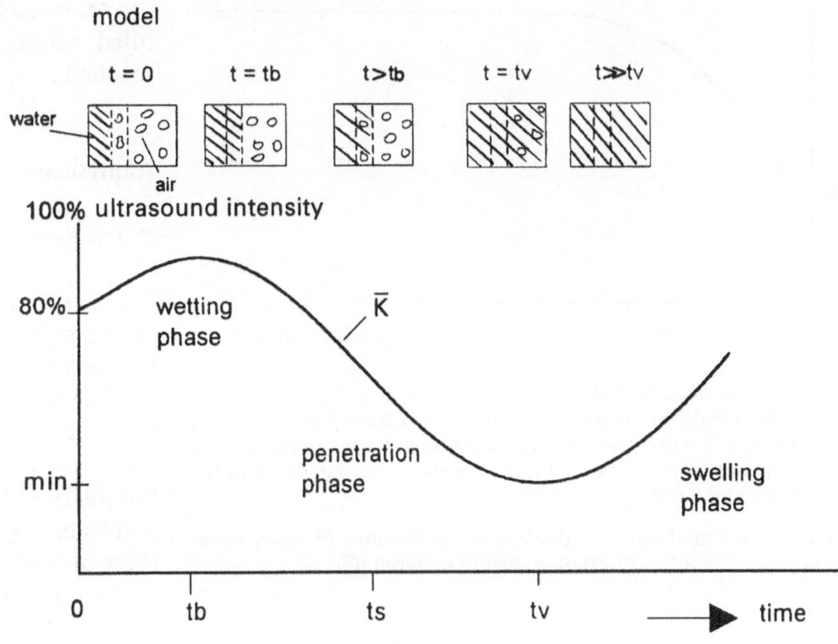

Fig. 1 Determination of the dynamic of water penetration process into dry cellulose material-paper-by ultrasound intensity [7].
Ultrasound transmission:
- in water → very good
- in air → poor
- in water air mixture → very poor

Water uptake properties of fibers – reason for different adsorption ability

The mentioned purification processes cause the degradation of the hydrophobic primary shell of cellulose polymers. These phenomena not only change the electrokinetic properties, but also the water uptake – hydrophilicity – of cellulose fibers. The resulting increased water uptake of fibers is indispensable in environmentally friendly finishing processes – whose performance is determined by adsorption processes applied during dyeing or printing [6].

The water penetration into paper which is also a cellulose product of different molecular structure as cellulose textile fibers can be described by three phases:

- the wetting phase: the solvent contacts the surface but is not yet soaked into the material
- the penetration phase: the penetration of water into the fiber mesh is in progress
- the swelling phase: the penetration process is finished; the fiber swelling, starts.

This phenomena can be observed measuring the intensity of an ultrasound signal transmitting through the material as a function of time. Because of different ultrasound expansion speed in air (dry material) compared to water soaked material, the ultrasound signal is changed during the water adsorption – uptake process.

The model of ultrasound intensity describing the dynamic of the water uptake and penetration process (into paper) is represented in Fig. 1.

Phase 1: The water uptake of cellulose materials is determined by the slow replacement of air present at the fiber surface and in the pores between the fibers by penetrating water. At the beginning of the penetration phase an air – water mixture is obtained. The reduction of ultrasound transmission is typical for the first step of the water penetration process. The intensity of ultrasound signal correlates with the surface roughness and the amount of air located at the surface. The duration of this phenomena is described by the time t_b and determined by the surface hydrophilicity of material.

Phase 2: The second part of the curve is correlated with the penetration speed. The highest speed of water penetration is obtained at the time t_s, the slope $\bar{K}$ of the penetration curve describes the speed of penetration.

Phase 3: The value t_v represents and describes the point at which no further liquid can penetrate into the material.

Compared to the structure of paper fibers the size and molecular structure of textile cellulose fibers is much more compact and complicated. Plane textile fabrics are combined in different ways (different types of thread binding in textile goods), which is not the case in paper sheets. Because of these differences in micro and macro structure – a fabric of cellulose textile fibers contains much more air

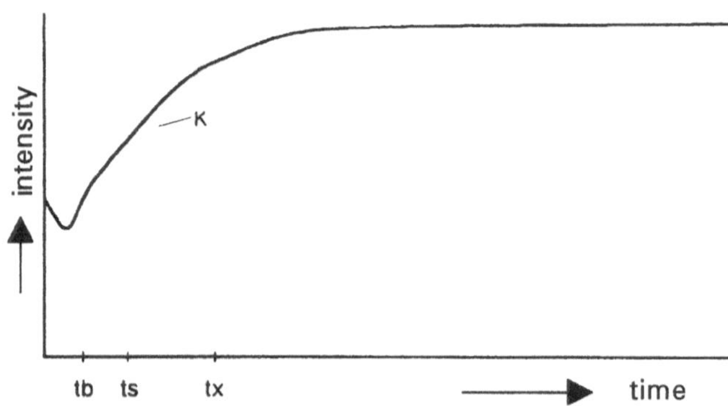

t_b – removal of surface air
t_s – time at which the highest speed of water penetration is observed
K – slope of the penetration curve describing the penetration speed
t_x – time necessary for the solution to arrive at the back side of the textile material

Fig. 2 Ultrasound intensity describing the dynamic of water penetration process into dry cellulose textile material [7]

than a plane sheet of paper and is therefore more hydrophobic – the course of the ultrasound signal describing the water uptake has to be interpreted in another way.

A model describing the dynamic of the water penetration process into textile dry materials measured by ultrasound velocity is represented in Fig. 2.

Material and methods

The material used for analysis was 100% pure cotton.

Preparation methods

Removal of non cellulose compounds using textile pretreatment processees

– *Alkaline treatment*
(20 g/l NaOH; pH = 11,5; t = 90 min; T = 95 °C)

– *Demineralisation (acid treatment)-using sequestering agents (i.e., polyphosphonic acids) in acid environment*
(2 ml/l sequestering agent; pH = 2,7; t = 30 min; T = 98 °C)

Removal of individual non cellulose compounds applying specific methods

– *Petrol–ether extraction*
(Soxhlet extraction; min. 6 circles/h)

– *HCl treatment*
(0.1 M HCl; 20 h; 20 °C)

After each treatment the fibers were washed in distilled water until a conductivity less than 3 μs/m was reached.

Analytical methods

Determination of the zetapotential

The streaming potential method was used as it has been shown to be the most appropriate electrokinetic technique to study the ZP of fiber systems [9, 10].

The streaming potential of every sample was measured in 0.001 N KCl, the surface conductivity was not taken into account. The ZP was calculated according to the Smoluchowski equation and the method of Fairbrother and Mastin [8]. The here mentioned ZP values are always these obtained at the constant part of the ZP – pH function in the alkaline region at pH = 9.

The ZP was investigated as a function of the ionic strength (Fig. 3), pH (Figs. 3 and 4) and as a function of surfactant concentration (Fig. 5). The surfactant (N–Cetylpyridiniumchlorid–N–CPC) concentration was increased stepwise until the fiber was oppositely charged.

The pK values were determined according to the following equation [5]:

$$pK = pH_{1/2ZP_p} + 0.4343 \cdot \frac{F \cdot ZP_p}{2RT}$$

$pH_{1/2ZP}$ corresponding pH value where ZP potential = 1/2 ZP of plateau
ZP_p ZP of plateau
F Faraday constant
R gas constant
T temperature

Instrument: Electrokinetic Analyzer EKA, P. Paar KG, fiber and flat plate cell.

Determination of water uptake – adsorption ability

The water uptake-adsorption ability-of different purified textile cellulose fibers was determined measuring the time dependence of an ultrasound transmission signal [7].

Instrument: Dynamic penetration measuring system DPM 20, EMCO – Elektronische Meß und Steuerungstechnik GmbH, Leipzig.

Results and Discussion

The ZP as a function of ionic strength and pH of different purified cellulose fibers shows the known picture – it decreases with increasing ionic strength due to the compression of the electrical double layer (Fig. 3).

The exception is raw cellulose material (untreated) containing the intact primary shell of fibers. Because of different non cellulose compounds (wax, pectins), which are present on the fiber's surface, probably different processes takes place at the same time like association/dissociation of surface groups, adsorption and especially the dissolution of ions from the fiber surface. This is the reason for the different ZP–ionic strength and pH function.

It is known that hydrophobic non cellulose substances interfere with different adsorption processes (e.g., dyeing with direct dyes) nevertheless, the influence of different pretreatment processes (degradation of non cellulose compounds) on the electrokinetic properties – ZP of cellulose textile fibers was never investigated. The influence of cleaning processes on the ZP is represented in Fig. 4.

Textile processes	*Methods removing specific compounds only*
– *Alkaline treatment* removing all non cell. compounds except waxes	– *Petrolion–ether extraction* removing waxes
– *Demineralisation* removing cationic trash	– *HCl treatment* supposed to remove all non cell compounds

The natural cellulose fibers are negatively charged (ZP = – 11 mV) due to the presence of carbonyl and hydroxyl-groups. In the case of raw (untreated) material these groups are covered by non-cellulose compounds present in the primary shell of the natural fiber. The textile finishing processes are applied in order to degrade, remove or complex these compounds and this causes an increase of the negative ZP depending on the degree of purification. This is clearly represented by the above shown results, the increase of the negative ZP can be explained by the improved accessibility of anionic groups.

The classical NaOH treatment degrades and removes practically all non-cellulose compounds except waxes which remain to about 50% on the fiber. This process results in a 40% increase of the negative surface charge expressed by a ZP of –14.5 mV. 2% NaOH used during pretreatment does not modify cellulose from structure I to II, but it causes swelling of the surface layers. This increases the size of the active surface, but the amount of

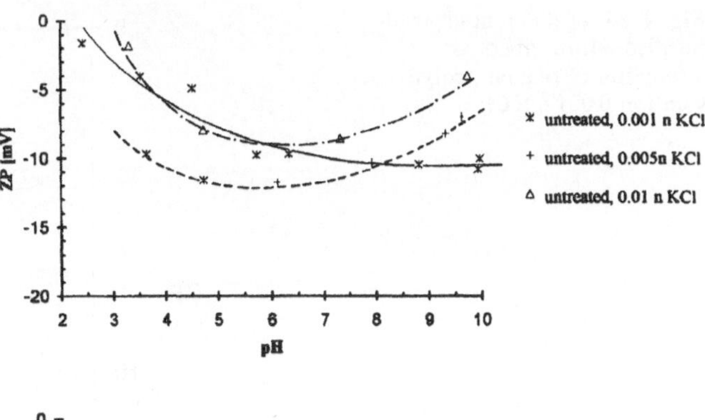

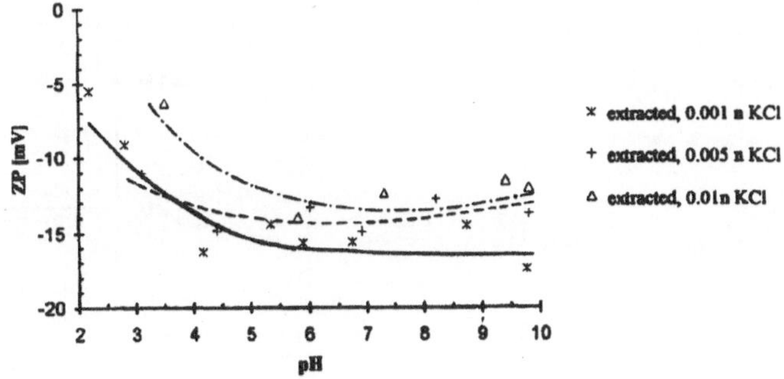

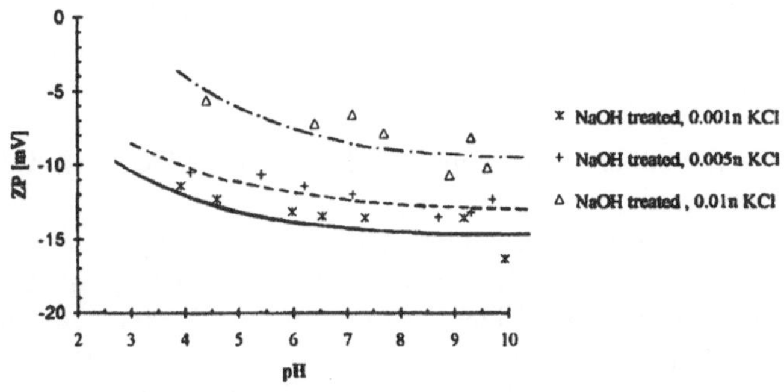

Fig. 3 ZP as a function of ionic strength and pH of different pretreated cellulose fibers

dissociable groups is not changed. The swelling itself causes a reduction of the ZP, because of the shift of the shear plane into the liquid phase.

Petrol ether extraction is used to remove the waxes entirely. This treatment causes the highest ZP (–17.4 mV) of all processes applied to cellulose fibers. If the NaOH treated fibers undergo an additional petrol ether treatment, the same ZP–pH function is obtained as in the case of the petrol ether treated samples. It can be concluded from these results that the waxes are the one component mainly influencing the surface charge of cellulose fibers.

A very similar result (–16.5 mV) is obtained by so-called acid treatment using complex forming agents (i.e., poly-phosphonic acid). All cations on the surface

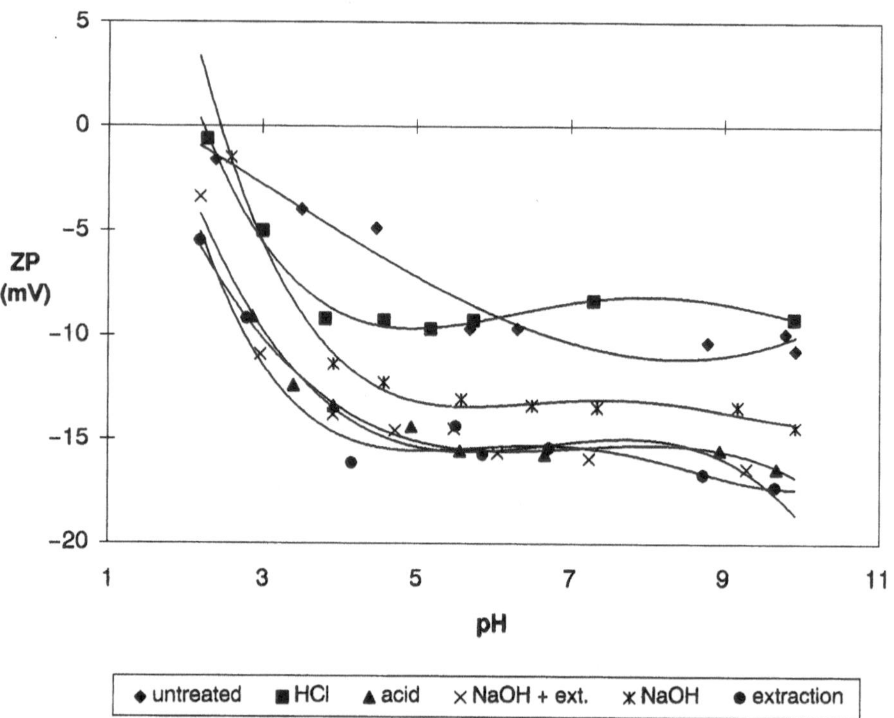

Fig. 4 ZP of different cleaned-purified cotton fibers as a function of pH; electrolyte solution 0.001 *n* KCl

of the fibers are complexed and dissolved by this treatment.

The non textile typical HCl treatment which should remove the non-cellulose compounds causes a completely different situation. The ZP–pH function is changed and the ZP plateau values (-9, 5 mV) are even lower than that of the raw material (-11 mV). This observation can be explained by the formation of hydrocelluloses in acid medium (pH = 2.5). During the primary step of the hydrolyzation H_3O^+ cations and/or –COH– groups are present [11] and the dissociation equilibrium is shifted in such a way that the number of negative surface groups is decreased, causing a reduction of the negative ZP.

The ZP–ionic strength functions (Fig. 3) as well as the ZP–pH functions (Fig. 4) indicate that the electrokinetic surface properties are dominated by the non-cellulose compounds, especially the waxes, even if their mass is less then 5–10 % of that of the total fiber. The pH–ZP function is not changed qualitatively by the treatment, the ZP values of the plateau region are a function of the degree of removal of non cellulose compounds and can therefore be used to describe the progress of these process steps. The isoelectric points are shifted towards lower pH values during the cleaning process, due to the increased accessibility of dissociable surface groups (IEP shifted from pH 2–3 to pH 1.5).

The surfactant adsorption process shows a similar picture (Fig. 5). The removal of waxes and cationic trash by both extraction and acid (using complex forming agents) treatments increases the number of charged groups and causes therefore a higher amount of N – CPC necessary to obtain charge reversal concentration (CRC).

The NaOH treatment offers better accessibility due to fiber swelling and enlargement of the primary intrafibrillar spaces and therefore faster adsorption of cationic surfactant.

The same adsorption ability is shown by materials which were first NaOH cleaned and afterwards extracted, which is not indicated by the pH–ZP function. The degradation of hydrophobic waxes and fatty acids has greater influence on electrokinetic properties of materials then the enlargement of primary intramicellar spaces, caused by NaOH treatment.

The non typical HCl cleaning process probably causes a chemical modification of cellulose (–CHO– groups were formatted), in this case the amount of N–CPC necessary to obtain the charge reversal is the highest one. Because of chemical modification of cellulose, pH–ZP and surfactant concentration – ZP functions of HCl cleaned material do not correlate. Different effects are observed in this case, on the one hand the modification of the dissociable groups on the surface by the ZP–pH function (pK values are small) and on the other hand the modification of adsorption sites.

Progr Colloid Polym Sci (1996) 101:157–165

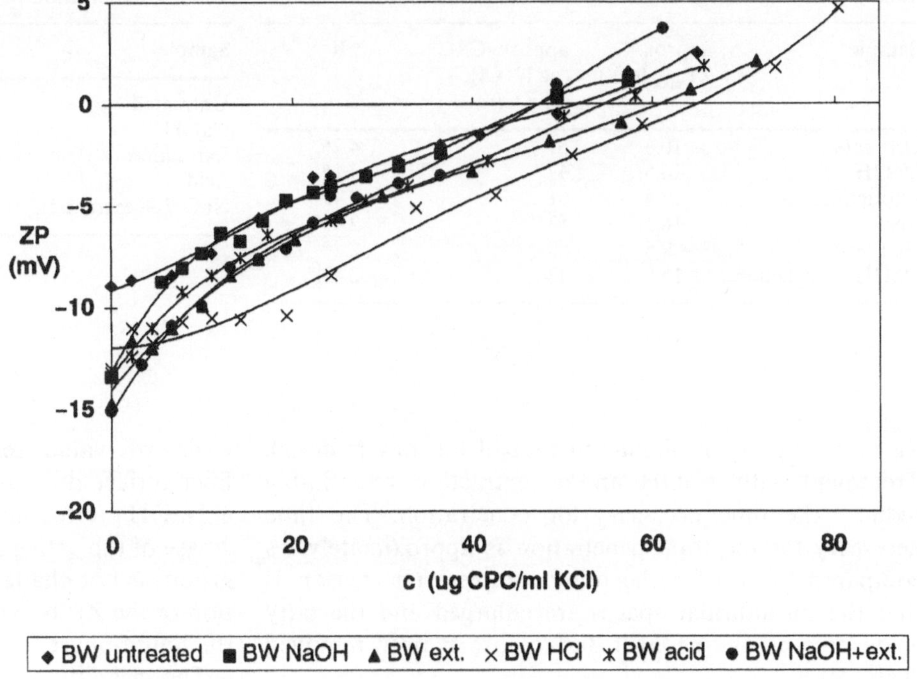

Fig. 5 ZP of different cleaned-pretreated cotton fibers as a function of surfactant concentration; pH = 8.7; 0.001 *n* KCl

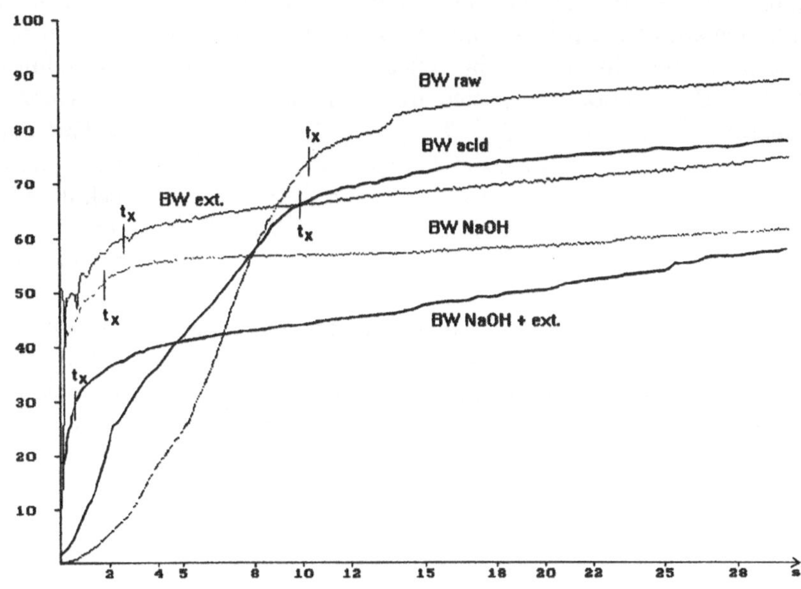

Fig. 6 The ultrasound signal describing the water uptake of differently cleaned cellulose textile materials

Because of their structure textile fabrics contain more air (between fibers and in the pores) as paper and are therefore more hydrophobic; this is shown by the dynamic of water penetration measured by ultrasound intensity (Fig. 6) as a function of penetration time.

The clear difference in the slopes described by the factor $\bar{K}$ indicate the differences in penetration speed. The raw samples and the acid cleaned samples are different compared to the samples treated by other methods. After acid treatment, even if the cations are removed,

Table 3 Elektrokinetic properties of different purified cellulose fibers

Sample	approx. ZP_{max} (mV)	approx. CRC (μg N-CPC/ ml KCl)	pK
Untreated	− 10.8	51	4.25
NaOH	− 14.5	47	3.15
Extraction	− 17.4	61	2.49
Acid	− 16.5	53	2.63
HCl	− 9.5	65	2.80
NaOH + extraction	− 16.5	47	2.49

Table 4 Penetration parameters determined by ultrasound intensity

Sample	$\bar{K}$	t_x (s)
Untreated	1.7	10.4
NaOH	8.9	1.6
Extraction	8.0	2.5
Acid	1.9	9.8
NaOH + extraction	13.4	0.7

the water uptake is similar to that of the raw material. Treatment with NaOH and/or extraction remarkably reduces the time necessary for penetration. The time necessary for the total penetration is approximately 2 s compared to 10 s for the other samples. The reason is that the intrafibrillar spaces are enlarged and the fatty acids present on the fiber surface are mainly removed (Table 4).

These results are on the one hand the proof that the removal of fatty acids has a pronounced influence on the adsorption abilities of cellulose fibers. On the other hand, it is the proof that the assumption drawn from the ZP–pH and ZP-surfactant concentration functions namely – the enlargement of the primary intrafibrillar spaces in cellulose polymers – has greater influence on adsorption abilities as degradation of specific hydrophobic substance (waxes).

Conclusion

- The ZP_{max} in the ZP–pH function increases due to increased accessibility of dissociable groups by removal of wax and cationic components by extraction and acid cleaning processes.
- The pK values reflect the chemical modification of the fiber surface due to HCl treatment.
- NaOH pretreatment–purification process – causes an increase of the active surface, but the amount of dissociable groups is not changed. The swelling itself causes a reduction of the ZP because of the shift of the shear plane into the liquid phase.
- The cleaning process shifts the IEP to higher acidity showing the dominance of dissociation processes in the ZP–pH function.
- The enlargement of the primary intrafibrillar spaces in cellulose polymer has greater influence on surfactant adsorption ability as the degradation of specific hydrophobic substances (waxes)–increased accessibility of active groups on cellulose polymer surface determinate by ZP as well as by water penetration results.
- The correlation between ZP–pH and ZP-surfactant conc. functions is not the same by chemical modified cellulose (HCl cleaning–causing formation of new groups). This is proven by the shift in pK values and the differences in surfactant adsorption.
- The zetapotential and water penetration measurements are appropriate tools to observe the kind of surface modification, the progress of the chemical process, the adsorption ability and the interaction of the fibers with components of the liquid phase.

References

1. Rath H (1972) In: Lehrbuch der Textilchemie. Springer-Verlag, Berlin, S50–55
2. Rösch G (1988) In: The practice of cotton fabric preparation, Reprint from Textil praxis international 44:4–8
3. Schurz J, Jorde Ch, Ribitsch V, Jacobasch HJ, Körber H, Hanke R (1985) Das Elektrokinetisches Meßsystem EKM. Symposium Elektrokinetische Erscheinungen Dresden 85:221–226
4. Kitahara A, Watanable A (1984) In: Electrical phenomena at interfaces, Marcel Dekker Inc, New York, 299–303
5. Jacobasch HJ, Grosse I (1987) Textiltechnik 37:266–2679
6. Espinosa-Jimenez, Gonzales Caballero F (1991) Cellulose Chem Technol 65–77
7. Gruner G (1993) Deutsche Papierwirtschaft 24:4–9
8. Fairbrother F, Mastin H (1924) J Chem Soc 75:2318–2323

Progr Colloid Polym Sci (1996) 101:157–165

9. Ribitsch V, Jorde Ch, Schurz J, Jacobasch HJ (1988) Progr Colloid & Polymer Sci 77:49–54

10. Ribitsch V, Jacobasch HJ, Boerner M (1991) In: Advances in Measurement and Control of Colloidal Processes. Butterworth–Heinemann, 345–365

11. Honeman J, Recent Advances in the Chemistry of Cellulose and Starch, London, Heywood a Comp LCD, 1959, 75–91

Progr Colloid Polym Sci (1996) 101:166–169

INTERFACES

R. Zimehl
B. Ruffmann
A. Vózár
I. Dékány

Characterization of polymer particles: I. Competitive liquid sorption and calorimetry

Dr. R. Zimehl (✉) · B. Ruffmann
Institute of Inorganic Chemistry
University of Kiel
Olshausenstraße 40-60
24098 Kiel, FRG

A. Vózár · I. Dékány
Department of Colloid Chemistry
Attila József University
Aradi Vértanuk tere 1
6720 Szeged, Hungary

Abstract Three classes of polymer particles are investigated by selective liquid sorption from 1-propanol-water mixtures. Class 1 are gel polymerized polar (hydrophilic) dextran spheres (Sephadex), class 2 are macroreticular polar (hydrophilic) ion exchangers (Chelite) and class 3 are emulsion polymerized latex particles. The enthalpy of displacement isotherms accompanying the sorption processes are determined by different calorimetric techniques (flow microcalorimetry, titration calorimetry, immersion calorimetry). The enthalpy isotherms reveal the differences in polarity of the polymer network.

Key words Polymers – sorption – microcalorimetry – networks – latices

Introduction

Sorption of binary liquid mixtures by natural or synthetic polymer networks is often accompanied by specific interactions of the adsorptive with internal surfaces, which makes the sorption very selective. Dékány and co-worker [1, 2] investigated in detail the structure of organic layers positioned between the basal planes of hydrophobic layer silicates and the interplanar swelling and liquid sorption. In the present study we investigate the sorption of water from 1-propanol-water mixtures on polymer networks of different polarity and swelling ability. The aim of our investigations is to examine the influence of the polymer particle structure and the polarity of the network on the adsorption excess isotherms and to illustrate the relation between the enthalpy of displacement (derived from microcalorimetric measurements) and the material content in the adsorption volume.

Theoretical

Experimental data of adsorption from binary solution are often expressed as composite isotherms, in which the excess amount adsorbed of one component is plotted as a function of its equilibrium concentration. The excess amount adsorbed of component 2, $n_2^{\sigma(n)}$, is usually calculated from its concentration change in bulk solution, Δx_2 and can be related to the composition of the adsorbed liquid Z by several procedures [2, 3]. Sorption from binary liquid mixtures is accompanied by a heat effect which can be most directly obtained by flow microcalorimetry [4]. The enthalpy of displacement, $\Delta_{12}H$, is given by

$$\Delta_{12}H = \Delta_{12}H_b - \Delta_{mix}H = (\Delta n_1^s h_1^s) + (\Delta n_2^s h_2^s) + \Delta_{12}H^{se}, \quad (1)$$

where the overall heat effect, $\Delta_{12}H_b$, is corrected for the mixing heat in the bulk phase, $\Delta_{mix}H$, and it can also be written in terms of the molar enthalpies of the components in the adsorption volume, h_1^s and h_2^s [2, 4]. The function $\Delta_{12}H^{se}$ denotes an excess enthalpy for the sorbed state. Everett [5] has suggested that the heat of immersion of a solid by a binary liquid mixture is given most simply by:

$$\Delta_w H = x_1^s \Delta H_1^0 + x_2^s \Delta H_2^0 , \quad (2)$$

where ΔH_1^0 and ΔH_2^0 are the heat of immersion of the adsorbent in the pure liquids 1 and 2, respectively. Both equations can be readily combined to give

$$\Delta_{12}H = \Delta_w H - \Delta_w H_1^0 = x_2^s(\Delta_w H_2^0 - \Delta_w H_1^0) + \Delta_{12}H^{se} \quad (3)$$

The function $\Delta_{12}H = f(x_2^s)$ is determined by the composition of the sorbed phase (x_2^s), the difference in the immersion enthalpies of the pure components ($\Delta_w H_2^0 - \Delta_w H_1^0$) and the excess enthalpy of the sorbed phase, $\Delta_{12}H^{se}$.

Experimental

Materials

The adsorption of 1-propanol (index 1) and water (index 2) was studied with different polymer particles:

Class 1 Nonionic sorbent Sephadex G-25 and LH-20 (Fluka, Germany), cross-linked dextran gels; particle size 0.05–0.1 mm

Class 2 Chelite-S and Chelite-C (Serva, Germany), copolymers of styrene and divinylbenzene, highly crosslinked macroporous material with hydrophilic mercapto groups (Chelite-S) and aminodiacetic acid groups (Chelite-C); particle size 0.05–0.1 mm

Class 3 Latex particles based on styrene copolymerized with different amounts of hydrophilic comonomers [6]. Latex SPP is a copolymer of polystyrene/co-poly N-(3-sulfopropyl)-N-methacrylamidopropyl-N, N-dimethylammonium-betaine, latex KSS is a copolymer made from styrene and styrene sulfonate.

Before adsorption measurements, polymer samples were washed with alcohol and water, dried with acetone and stored in a vacuum desiccator at 353 K for some days.

Methods

Adsorption measurements were performed in well-sealed test tubes at room temperature. Amounts of 5–10 cm^{-3} of the binary liquid mixture were added to 0.2–0.5 g of vacuum dried (353 K) polymer samples. The change of concentration of water in the bulk liquid (Δx_2) was measured by a Zeiss liquid interferometer after equilibrium periods of 48 h.

Calorimetric determinations of integral displacement enthalpies were made by using a flow microcalorimeter (LKB 2107) at 298 ± 0.01 K on prewetted polymer samples (50–200 mg).

The enthalpy of immersion was recorded with a 2225 Precision solution Calorimeter used in a 2277 Thermal Activity Monitor thermostat at 298 ± 0.01 K. The vacuum dried polymer sample (≈200 mg) was introduced in the 2225 PSC which has been previously filled with 200 ml of the given liquid.

In some cases enthalpy changes were conducted by monitoring the heat flux during titration of latex dispersions or suspended polymer particles in a titration unit of an isothermal 2277 Thermal Activity Monitor.

The standard deviation of the calorimetric measurement by the above methods varied between ±9% and ±12%, depending on the polymer sample, the composition of the liquid mixture and the applied method.

Results and discussion

First the selective sorption of 1-propanol-water mixtures on the polymer particles was studied. The excess isotherm $n_2^{\sigma(n)} = f(x_2)$ for the hydrophilic particles (Chelite-C, Sephadex G-25) and 1-propanol(1)-water(2) mixtures is U-shaped, and water is preferentially sorbed over the entire range of concentrations. For samples Chelite-S and Sephadex LH-20, $n_2^{\sigma(n)}$ is slightly negative at $x_2 \geq 0.9$, and the excess isotherm is S-shaped with an azeotropic point at $x_2 \approx 0.89$ (Figs. 1 and 2).

This change in the shape of the excess isotherms is very pronounced for the latex sediments. The excess isotherm for the less hydrophilic KSS latex particles is S-shaped with an extended linear portion (Fig. 3) and an azeotropic point at $x_2 \approx 0.6$. Again the more hydrophilic SPP latex has a U-shaped isotherm and water is preferentially sorbed over the entire range of concentrations.

Generally at higher water mole fractions, 1-propanol is partitioned in preference towards the less polar polystyrene particles and the hydrophobically modified dextran beads. The isotherms in Figs. 1 to 3 clearly indicate that the uptake of water differs significantly with the nature of the polymer networks. The shape of the isotherms is a function of the particle structure and the polarity of the polymer framework.

The displacement of 1-propanol by water is further illustrated by the enthalpy changes $-\Delta_{12}H = f(x_2)$ for

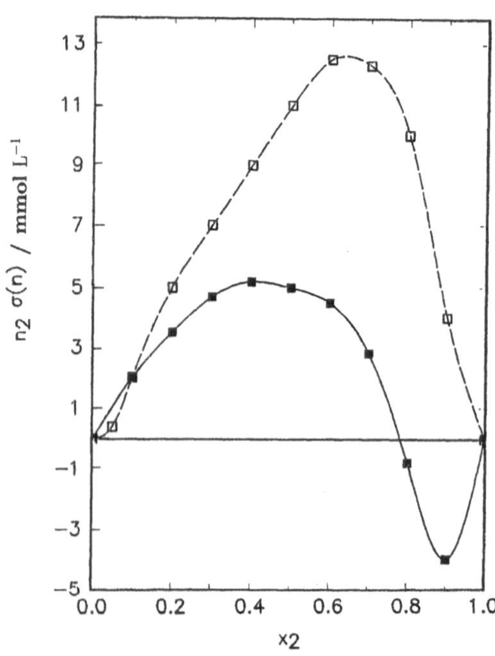

Fig. 1 Excess isotherm on nonionic dextran beads in 1-propanol(1)-water(2) mixtures ····⊡···· Sephadex G-25, –■– Sephadex LH-20

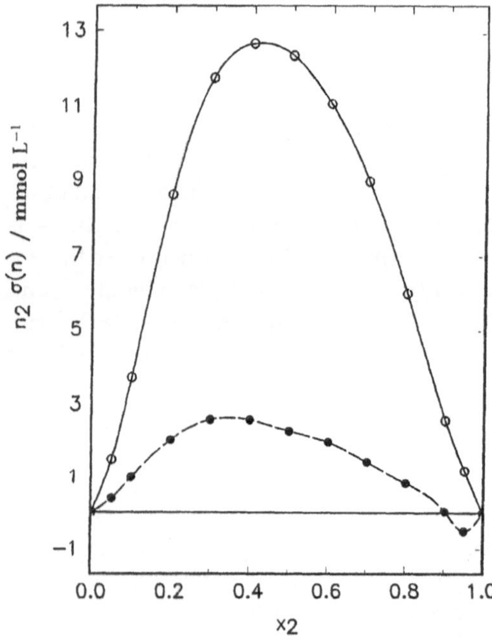

Fig. 2 Excess isotherm on macroporous polystyrene particles in 1-propanol(1)-water(2) mixtures –○– Chelite-C, –●– Chelite-S

two classes of polymer particles in Fig. 4. For the polystyrene particles (Chelite-C, Chelite-S) the enthalpy of displacement increases monotonously up to $x_2 \approx 0.9$. As the excess isoterm of sample S is S-shaped, the curve decreases slightly at mole fractions of water $x_2 \geq 0.9$. For the two

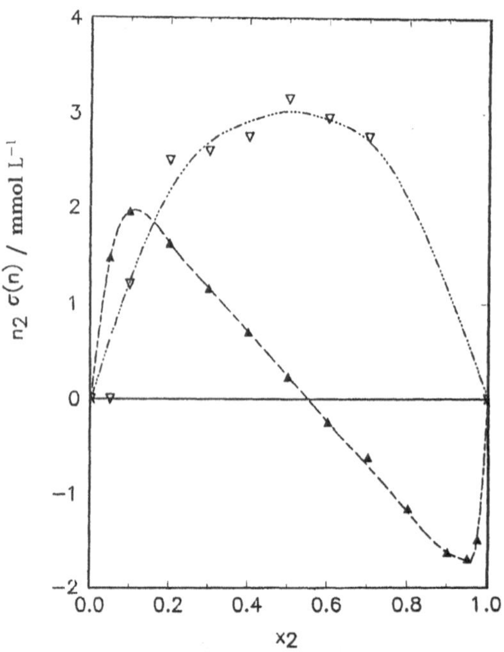

Fig. 3 Excess isotherm on latex sediments in 1-propanol(1)-water(2) mixtures. ··▽·· polystyrene/co-poly N-(3-sulfopropyl)-N-methacrylamidopropyl-N, N-dimethylammonium-betaine latex particles –▲– polystyrene/co-polystyrene sulfonate latex particles

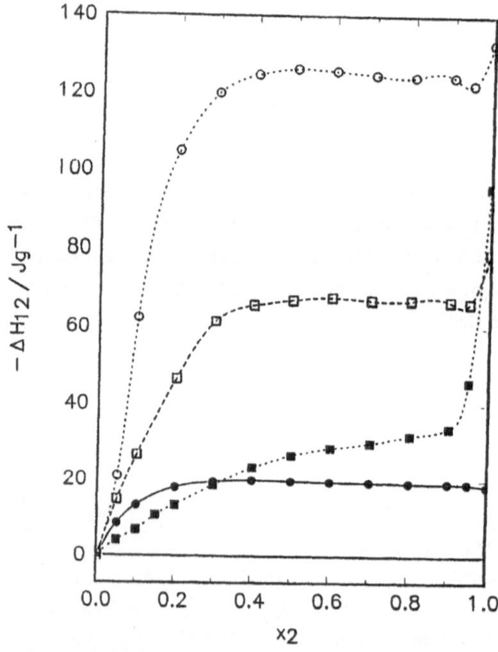

Fig. 4 Enthalpy of displacement isotherms determined by flow microcalorimetry on different polymer particles. Notation as in Figs. 1 and 2

similar crosslinked dextran beads $-\Delta_{12}H$ increases monotonously in the first steps of the displacement process but has a noticeable upward section at higher mole fractions of water.

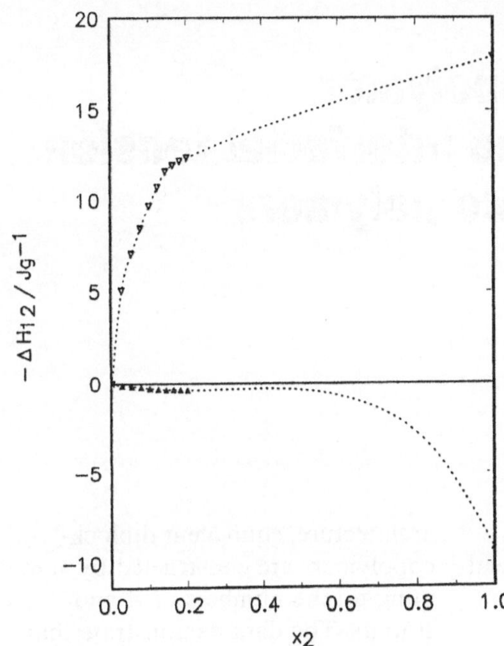

Fig. 5 Enthalpy of displacement isotherms determined by titration and immersion calorimetry acccording to Eqs. (2) and (3). Notation as in Fig. 3

As the mean size of the latex particles is essentially too small for accurate flow calorimetry, we determined the enthalpy of displacement isotherm $-\Delta_{12}H = f(x_2)$ for the two latex samples by a combination of titration and immersion calorimetry [2, 7]. Again the difference in polarity and internal particle structure of the two latex sediments is well reflected by the displacement enthalpy (Fig. 5).

Conclusions

Hydrophilic macroporous resin Chelite-C and much more the dextran beads represent a rather classical type of polymer network with a certain degree of permanent porosity. In the water-rich region these particles are highly swollen xerogels and consists of two components: water acts as the dispersion medium and the charged polymer network is the dispersed component. The macromolecular network of the resin is contracted during drying and loses a part of its porosity. When the dried resins are contacted with 1-propanol-water mixtures, swelling and selective sorption of water occurs. The less hydrophilic network (Chelite-S) represent a true macroreticular resin (aerogel). After drying the particles retain their porosity. In 1-propanol-water mixtures the pore volume is filled with bulk liquid. Water is preferentially sorbed over a wide range of concentrations but there is also a very modest enrichment of water for $x_2 \geq x_2^a$.

The general shape of $-\Delta_{12}H$ suggests that the preferential sorption of water on the different networks is accompanied by a significant enthalpy decrease. Unfortunately, the cumulative displacement enthalpy of 1-propanol by water can be measured with reliable accuracy only up to $x_2 \approx 0.9$. At higher water concentration the mixing heat effect, $\Delta_{mix}H$, is too high in comparison with the displacement for flow calorimetry measurement.

Acknowledgment We are grateful to the Alexander von Humboldt Stiftung for financial support. Special thanks are due I. Abraham for ultracentrifugation and to T. Marosi and K. Kuhn for kind assistance.

References

1. Dékány I, Szántó F, Weiss A (1989) Colloids and Surfaces 41:107–121
2. Dékány I (1992) Pure & Appl Chem 64:1499–1509
3. Kipling JJ (1965) Adsorption from Solutions of Non-Electrolytes. Academic Press, London-New York, pp 32–62
4. Dékány I, Zsednai A, László K, Nagy LG (1987) Colloids and Surfaces 23:41–55
5. Everett DH (1964) Trans Faraday Soc 60:1803–1813
6. Zimehl R, Lagaly G, Ahrens J (1990) Colloid Polym Sci 268:924–933
7. Zimehl R, Ruffmann B, Lagaly G, to be published Colloid Polym Sci

Progr Colloid Polym Sci (1996) 101:170–171
© Steinkopff Verlag 1996

U. Jorzik
M. Wagner
B.A. Wolf

Effect of block copolymer architecture on the interfacial tension between immiscible polymers

U. Jorzik · M. Wagner
Prof. Dr. B.A. Wolf (✉)
Institut für Physikalische Chemie
Johannes Gutenberg-Universität Mainz
Jakob-Welder-Weg 13
55099 Mainz, FRG

Abstract The effect of block copolymer additives on the interfacial tension (σ) is studied for the systems polydimethylsiloxane/polyethyleneoxide (*A*/*B*) and polyethylmethylsiloxane/polypropyleneoxide by means of a sessile-drop and a pendant-drop-apparatus. Diblockcopolymers (*A-block-B*), triblockcopolymers (*A-block-B-block-A*) and "tooth-brush" like copolymers (*A* backbone, *B* brushes) served as additives.

Measurements of σ were performed as a function of concentration and temperature; $\sigma^{plateau}$, the maximum reduction of the interfacial tension, was calculated for each copolymer. To compare the efficiency of additives of different molecular architecture, equivalent diblockcopolymers are constructed by summing up the numbers of *A* and *B* units. The data demonstrate that $\sigma^{plateau}$ – keeping the number of *A* units constant – does not depend on the number of *B* units in case of the present copolymers. A plot of $\sigma^{plateau}$ versus the number of *A* units shows that the maximum reduction is independent of the particular architecture of the copolymer and can be estimated from the mere knowledge of the number of *A* units, the copolymer contains.

Key words Polymer blends – interfacial tension – block copolymers – copolymer architecture

The interfacial tension σ was measured using a sessile-drop (SED) and a pendant-drop (PED) apparatus. The experimental setup and the preparation of the samples have been described elsewhere [1, 2]. The following homopolymers were used, giving in brackets the abbreviation and the weight-average molecular weights in kg/mol: polydimethylsiloxane (PDMS 177), polyethylmethylsiloxane (PEMS 31), poly(ethylene oxide) (PEO 41), and poly(propylene oxide) (PPO 2). The copolymers (the indices indicating the numbers of monomeric units) were DMS_8–EO_{27}, DMS_x–EO_{37}–DMS_x ($x = 4, 16, 23, 32$), DMS_x–EO_{77}–DMS_x ($x = 4, 15, 32$) and so-called "toothbrush" copolymers consisting of a DMS-backbone and EO-brushes on five consecutive DMS-units at one end of the backbone. Their structure is $Si(CH_3)_3$–$(OSi(CH_3)_2)_x$–$[O$–$Si(CH_3)/(CH_2)_3$–$(OCH_2CH_2)_{15})H)]_5$–$Si(CH_3)_3$, where $x = 13$ or 20.

Interfacial tensions were measured as a function of temperature by means of the SED method for the systems PDMS 177/PEO 41 (70–130 °C) and PDMS 177/PEO 41/Additive (ADD) (70–130 °C) and by means of the PED method for the systems PDMS 177/PEO 41 (70–130 °C), PEMS 31/PPO 2 (50–120 °C) and PEMS 31/PPO 2/ADD (50–120 °C).

Within experimental error σ depends linearly on temperature. Reading the values for 100 °C (regression analysis), the evaluation is performed in terms of the composition of the droplet phase, expressed as the base mole fraction x_{Add} of the additive (number of monomeric units of the copolymer molecules divided by all monomeric units contained in this phase). The data points are fitted using the following Langmuir-analogue equation (1) and relation (2), reported by Tang/Huang [3]

$$\sigma = \sigma_0 - \frac{F_1 \cdot x_{Add}}{1 + F_2 \cdot x_{Add}} \quad (1)$$

$$\sigma = (\sigma_0 - \sigma_S) \cdot \exp(-F_3 \cdot x_{Add}) + \sigma_S \quad (2)$$

σ_0 and σ are the interfacial tensions without and with additive, respectively; F_1, F_2, F_3 and σ_S ($= \sigma_{saturation}$) constitute fit parameters.

Equations (1) and (2) allow the calculation of system specific parameters. The maximum reduction of interfacial tension ($\sigma^{plateau}$) and the characteristic additive concentration (x^{ch}_{add}) are related to F_1 and F_2 by

$$x^{ch}_{Add}(\text{Langmuir}) = (1 + F_2)^{-1} \quad (3)$$

$$\sigma^{plateau} = \frac{F_1}{1 + F_2} . \quad (4)$$

Analogously, σ_S is directly contained in Eq. (2) and x^{ch}_{Add} (Tang/Huang) is given by $(1/F_3)$. $\sigma^{plateau}$ and σ_S are of comparable magnitude, x^{ch}_{Add} (Langmuir) and x^{ch}_{Add} (Tang/Huang) differ markedly. Only $\sigma^{plateau}$ is used for the following discussion.

A comparison of the influences of the molecular architecture of the additives on the reduction of σ is performed by means of the systems PDMS 177/PEO 41/ADD. To this end the triblock and the toothbrush copolymers are replaced in an experiment in thought by equivalent diblockcopolymers consisting of the same number of DMS- and EO-units. These data together with the corresponding $\sigma^{plateau}$ are listed in Table 1.

Keeping the number m of DMS-units in the equivalent diblockcopolymers constant (for example at a value of 8), one compares $\sigma^{plateau}$ for a variable number of EO-units. It can be easily been that $\sigma^{plateau}$ is almost independent of the number of EO-units for the additives under consideration. This statement also holds true for $m = 30$ and 64. The conclusion is drawn that $\sigma^{plateau}$ does practically not depend on the length of the (ethyleneoxide)-block for the present systems. Figure 1 shows how $\sigma^{plateau}$ varies with the total number m of the DMS-units.

Table 1 Number of DMS- and EO-units of the different additives and the corresponding values for $\sigma^{plateau}$

Copolymer	DMS-units	EO-units	$\sigma^{plateau}$ [mN/m]
DMS_8-EO_{27}	8	27	6.7
$DMS_{14}(MS(-EO_{15}))_5$-DMS	20	75	2.6
$DMS_{21}(MS(-EO_{15}))_5$-DMS	28	75	3.6
DMS_4-EO_{37}-DMS_4	8	37	7.3
DMS_{16}-EO_{37}-DMS_{16}	32	37	1.2
DMS_{23}-EO_{37}-DMS_{23}	46	37	0.8
DMS_{32}-EO_{37}-DMS_{32}	64	37	0.8
DMS_4-EO_{77}-DMS_4	8	77	8.6
DMS_{16}-EO_{77}-DMS_{15}	30	77	1.0
DMS_{32}-EO_{77}-DMS_{32}	64	77	0.7

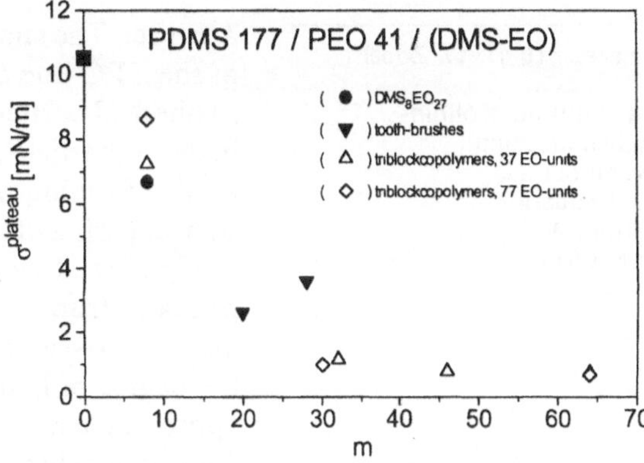

Fig. 1 $\sigma^{plateau}$ as a function of the number of DMS-units of the additives, (■) gives the interfacial tension of the binary system

The following conclusions can be drawn: i) Influences of the number of EO units have already died out at a value of 27, ii) the molecular architecture of the copolymers is not decisive for the maximum reduction of interfacial tension in the system PDMS 177/PEO 41/ADD, iii) consequently, it is possible to estimate the value of $\sigma^{plateau}$, i.e., the maximum reduction that can be achieved with a certain additive, from the mere knowledge of the number of DMS-units the copolymer contains.

References

1. Wagner M, Wolf BA (1993) Polymer 34(7):1460
2. Uzman M, Song B, Runke T, Cackovic H, Springer J (1991) Makromol Chem 192:1129
3. Tang T, Huang B (1991) Polymer 35(2):281

Progr Colloid Polym Sci (1996) 101:172–177
© Steinkopff Verlag 1996 INTERFACES

H. Stechemesser
G. Zobel
H. Partzscht

Dewetting of solid surfaces

Dr. H. Stechemesser (✉) · G. Zobel
H. Partzscht
Max-Planck-Institut für Kolloid- und Grenzflächenforschung
Arbeitsgruppe an der TU Bergakademie Freiberg
Chemnitzer Straße 40
09599 Freiberg, FRG

Abstract The spontaneous movement of the TPC-line across the surface of a spherical particle after rupture of the between particle surface and gas bubble forming thin liquid film has been experimentally studied by means of high-speed video technique at the pendant drop. The time dependence of the TPC-velocity correlating with the degree of hydrophobicity can be approximated by a simple exponential relationship. The analysis of the experimental data on the basis of the molecular-kinetic theory demonstrates that this theory, also for the spontaneous movement of the TPC-line across the particle surface, leads to physical reasonable conclusions as comparison with the molecular parameters of the forced movement shows.

With the presented measuring method was created the requirements for systematic studies of the dewetting kinetics on a bubble/particle system under influence of interfacial forces.

Key words Dewetting – three-phase contact movement – bubble/particle attachment – high-speed video technique

Introduction

For the selective attachment of one component of a heterogeneous mixture of solid particles on gas bubbles are identified as essential, three elementary processes [1]:

1) The approach of particle to bubble in the flow field, 2) the formation of thin liquid film between them, its rupture and the formation of a three-phase contact (TPC), and 3) the stabilization of bubble/particle aggregates under formation of a equilibrium contact area.

The interaction between bubble and particle take place in two extreme cases in dependence on direction of bubble movement in reference to the bubble surface: 1) sliding of the particle in tangential direction [1, 2], and 2) collision in normal direction [3, 4].

The temporal term introduced through the process of thinning of the liquid film, followed by the mentioned further processes, controls with it the coalescence of heterodispersed particles. Times are associated with these processes, whose co-operation forms the temporal condition for a successful coalescence [4]:

$$t_{\mathrm{Con}} \geq t_{\mathrm{I}} = t_{\mathrm{F}} + t_{\mathrm{TPC}} \,. \tag{1}$$

The contact time, t_{Con} is characterized by the time of interaction between bubble and particle as a result of collison and/or sliding. The induction time, t_{I}, is the sum of the time necessary for drainage of the liquid film forming between bubble and particle, t_{F}, and the time necessary for expansion of the three-phase contact forming after rupture of the film, t_{TPC}.

The difficulty in observing the film rupture and the movement of the TPC-line lies in the velocity of the process [5]. Both contact time and induction time are in the range of a few milliseconds up to 100 ms and the time for dewetting, i.e., the movement of the TPC-line is not longer than 10 ms.

The aim of this work was to find out the conditions of the experimental requirement to produce evidence of the TPC-velocity at the spontaneous dewetting at the model of the pendant drop and to produce evidence of the influence of the hydrophobic/hydrophilic property of the particle surface, i.e., the difference between the equilibrium and the receding contact angle as the driving force for the dewetting process.

Materials

The measurements of contact angle were accomplished on plates and the flotation as well as the measurements of the TPC-velocity were accomplished with glass spheres (called ballotini) of a diameter between 280 and 300 μm. All these materials consist of soda-lime-glass (VEB Glaswerke Ilmenau, Germany). The experiments were performed in pure Milli-Q water at pH 5 and temperature 22 °C. The cleaning of the material resulted in a sulphuric acid/perhydrol mixture (7:3) at 90° up to 120 °C during 30 min.

After the cleaning, repeated rinsing in Milli-Q water, the methylation was performed subsequently in a solution of $4 \cdot 10^{-2}$ mole/l trimethylchlorosilane (Merck, Darmstadt, Germany) in cyclohexane (Merck; analytical grade) at different periods of times. After twice-repeated rinsing in cyclohexane and short-term drying over an electro-burner, the implementation results directly.

Methods

Method of pendant drop

Inverse microscope ICM 405 (Carl Zeiss, Oberkochen) (1); Capillary tube (3); Through-illumination lamp (4); High-speed video system "SPEEDCAM" (Weinberger AG, Dietikon, Switzerland) consists of: high-speed camera (2); video monitor (5); camera control system (6); high-frequency computer (7).

The general arrangement of equipment is shown in Fig. 1. The main part of the equipment is the capillary glass tube (3). With it the so-called pendant drop method is realized. The inner diameter of the capillary tube is 8 mm. The capillary tube was cleaned with hot chromo-sulphuric acid, washed with bi-distilled water, and then filled with Milli-Q water. A weak convex air–liquid interface (meniscus) is produced at the bottom of the capillary tube. A particle falls within the capillary tube downwards, pushes against the meniscus and forms there a liquid film after some reflections. The liquid films thins itself to the critical rupture thickness, and the three-phase contact moves across the particle surface with high velocity (v_{TPC}).

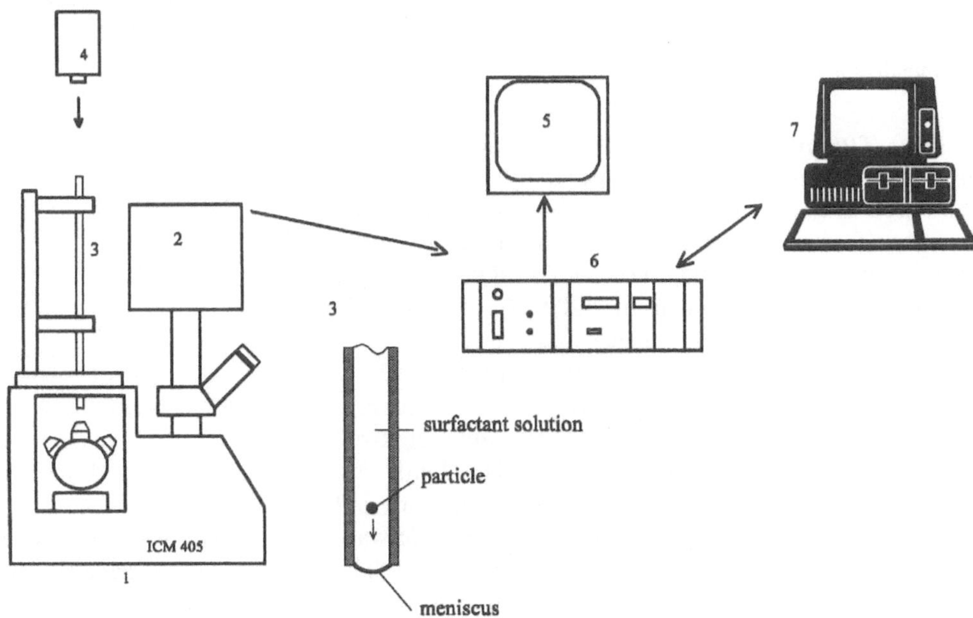

Fig. 1 Schematic diagram of the experimental set-up for the observation of the TPC-movement

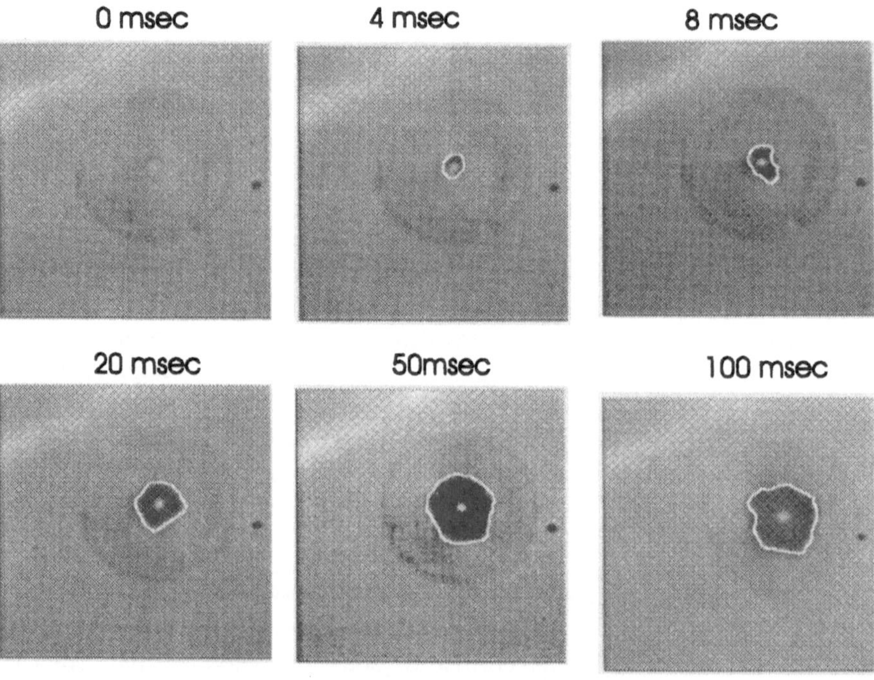

Fig. 2 Cutting of the picture sequence of the TPC-movement on the surface of a 10 min methylated particle (radius: 146 μm); white border – contour of the TPC; black area – breakthrough area of the particle through the meniscus ("naked" particle surface)

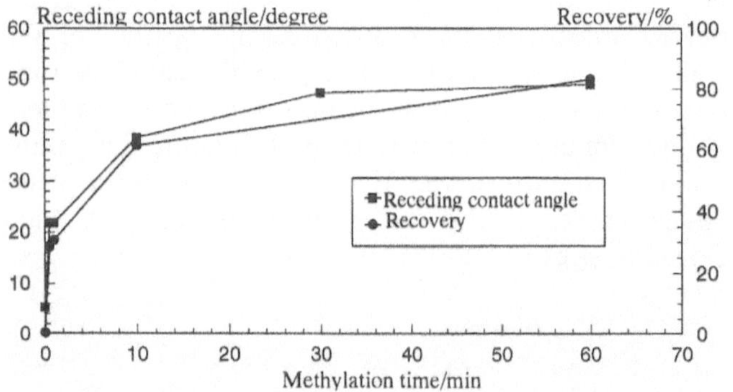

Fig. 3 Receding contact angle and flotation recovery in dependence of the methylation time

The TPC-movement is observed vertically to the liquid/gas interface and recorded by means of the high-speed video system. With two high-speed cameras, the image frequency can be changed from 265 to about 6000 pictures per second. The video images can be digitized at a resolution from 128×16 to 256×256 pixels, depending on the image frequency. The time resolution can be changed from 0.179 ms to 3.77 ms. The investigations of the TPC-movement were accomplished with an image frequency of 1025 s^{-1} and a digitized resolution of 128×128 pixels (Fig. 2). The lightening of the TPC-line (Fig. 2-white border) results with aid of picture treatment using the image analysis software OPTIMAS (BioScan, Inc., Dayton, Edmons, WA)). The TPC-velocity is calculated from the alteration of the TPC-radius on time. The TPC-radius itself is calculated in form of the circle equivalent radius from the breakthrough area (Fig. 2).

Contact angle measurements

The receding contact angle was measured by using the Contact-Angle Measuring System G 40 (KRÜSS GmbH, Hamburg). Freshly prepared glass plates (1 cm × 1 cm) were dried under a stream of nitrogen gas. After this the contact angles were measured by using the sessile drop technique with pure Milli-Q water.

Flotation experiments

Flotation tests were conducted in a modified Hallimond tube [6]. In each experiment 5 g freshly prepared ballotini were floated with nitrogen bubbles (diameter: 1 μm) for 1 min.

Experimental results and discussion

As shown in Fig. 3 both the receding contact angle as well as the flotation recovery correlate strongly with the degree of methylation, i.e., with the degree of hydrophobicity of the particle surface. With it the requirements are given to investigate the TPC-velocity in dependence on the degree of hydrophobicity.

The results of the TPC-velocity in dependence of the time (Fig. 4) show that the degree of methylation (hydrophobicity) after approximately 10 ms is insignificant.

The experimental data can be approximated by the following empirical relationship between the TPC-velocity on the time:

$$v_{TPC} = v_0 \cdot \exp(-t/\tau) - c \cdot t \tag{2}$$

where v_0 is the initial velocity, τ the decay time constant, and c represents any constant.

From the existing results (Table 1) it can be derived that the initial velocity of the TPC-expansion increases with the degree of hydrophobicity, but the decay time constant does not show a tendency assignment. Hence further investigations have to give clarification. The constant c in Eq. (1) is so small that this time part has no influence on the bubble/particle interaction and can be neglected.

At present, the theoretical models for description of the dynamic wetting and dewetting processes on systems with forced TPC-movement are classified into three groups:

- molecular-kinetic model by Blake and Haynes [7]
- hydrodynamic models by Voinov [8] and Cox [9]
- combined molecular-hydrodynamic model by Petrov [10].

The aim of this work is to examine whether a physically meaningful interpretation of the experimental results obtained by means of the spontaneous TPC-movement is possible by using the molecular-kinetic model.

This model describes the TPC-movement as a stress modified molecular rate process involving adsorption of the advancing phase and concurrent desorption of molecules of the receding phase. When a shear is applied to the TPC, the out-of-balance force initiates TPC-movement in the direction of the applied shear. If it is assumed that the shearing force is provided by the out-of-balance forces

Fig. 4 The TPC-velocity in dependence of the time at different degrees of hydrophobicity

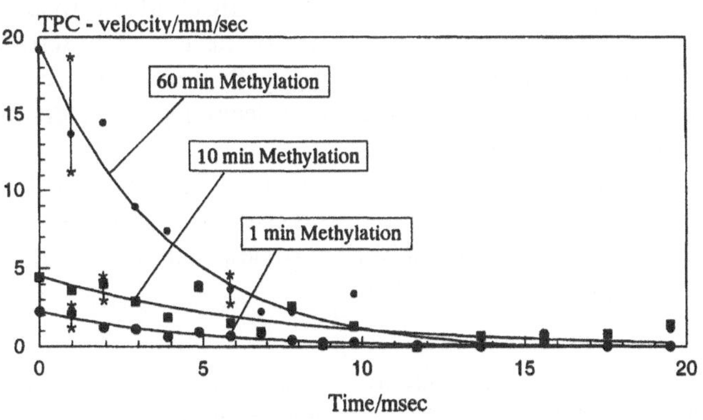

Table 1 Fitting values of the TPC-velocity according to Eq. (2). The values are the averages of 15 measurements

Methylation time min	Particle radius, R_p μm	Initial velocity, v_0 mm/s	Decay time constant, τ ms	c mm/s^2
1	147.9	2.23	4.08	0.00005
10	142.1	4.5	6.97	0.0007
60	141.3	19.53	3.74	0.019

acting on the TPC-line, and further, that there is no dissipation of this force other than in molecular displacement, then the following equation describes the relationship between contact angle and TPC-velocity, v_{TPC} [11]:

$$v_{TPC} = 2 \cdot \kappa_0 \cdot \lambda \cdot \sinh\{+/- \sigma_{lv} \cdot (\cos \Theta_s - \cos \Theta_d)/2 \cdot n \cdot k \cdot T\} , \tag{3}$$

where κ_0 is the adsorption/desorption rate of molecules on the TPC-line under quasi-state conditions, λ is the distance between adsorption sites, n is the surface density of these sites, Θ_s and Θ_d are the static and the dynamic contact angles, respectively, and σ_{lv} is the surface tension.

The negative sign corresponds to the receding movement. At receding movement and with the condition that $\sigma_{lv} \cdot (\cos \Theta_r - \cos \Theta_s)/2 \gg k \cdot T$ the following equation is valid:

$$\Delta \cos \Theta = \cos \Theta_r - \cos \Theta_s = (2 \cdot n \cdot k \cdot T/\sigma_{lv}) \cdot \ln(v_{TPC}/\kappa_0 \cdot \lambda) , \tag{4}$$

where Θ_r is the receding contact angle. The dynamic contact angle can be calculated according to the method given by Mingins–Scheludko [12] from the experimental data r_t and R_p:

$$\Theta_r = \arcsin(r_t/R_p) , \tag{5}$$

where r_t is the time dependent TPC-radius, and R_p the particle radius.

If it is further assumed that the distribution of adsorption sites is isotropic, then $\lambda = n^{-1/2}$ and Eq. (4) is reduced to a two-parameter, single, variable equation. A good fit of the obtained dynamic contact angle data to Eq. (4) is the precondition to get physically meaningful values of κ_0 and λ. In Fig. 5 are presented our experimental results.

In Table 2 the fitted parameters, the equilibrium contact angles after receding according to Eq. (5), and the activation free energy of dewetting, ΔG_w according to Eq. (6) [11] are summed up for the time range between 0 to 10 ms of the TPC-movement across the glass surface. The activation free energy, ΔG_w is given by

$$\Delta G_w = -N_0 \cdot k \cdot T \cdot \ln(h \cdot \kappa_0/k \cdot T) , \tag{6}$$

Table 2 Values of the parameters obtained by fitting on the molecular-kinetic model

Methylation time min	Θ_s degree	λ nm	κ_0 s^{-1}	n cm^{-2}	ΔG_w $kJ\,mole^{-1}$
1	4.9	10.3	6.10×10^3	9.4×10^{11}	39.5
10	19.5	2.17	1.84×10^5	2.1×10^{13}	30.8
60	37.8	1.33	4.72×10^5	5.7×10^{13}	28.1

where h and N_0 are the Planck constant and the Loschmidt number, respectively.

We can see that the estimated values of λ and κ_0 depending on methylation time, i.e., depending on the degree of hydrophobicity, are physically meaningful.

i) On the basis of investigations by Herzberg [13] the number of the active centers of quartz, which can interact with trimethylchlorosilane (TMCS) molecules, should be 2.6 sites/cm^2 in the form of free OH-groups. That corresponds to a surface density of $n = 2.6 \cdot 10^{14}$ sites/cm^2, i.e., a distance of active centers $\lambda = 0.62$ nm.

That is also in agreement with results of Sedev [14] for the forced movement of the TPC-line across a quartz surface totally methylated with dimethylchlorosilane, where the corresponding values of λ and κ_0 are 0.71 nm and $1.1 \cdot 10^6\,s^{-1}$ respectively. Because of the chemical composition of the used soda-lime-glass where, in comparison to the quartz surface, less active centers for the exchange with TMCS molecules are available, the λ-values must be larger, as shown in Table 2.

ii) There is also a satisfying correlation between the density of active sites on the glass surface (Table 2) and the driving force for the movement of the TPC-line (Fig. 5 and Table 1). With increasing degree of methylation the hydrophobicity is rising and for this the equilibrium contact angle after receding (Fig. 3), i.e., the driving force of the TPC movement increases. The distance between the active centers and with it the activation free energy of dewetting becomes smaller and, consequently, the initial velocity of the TPC-movement becomes higher.

It is likewise meaningful, that with increasing of the degree of methylation and with it the surface density of the active sites, n, the quasi-equilibrium rate constant, κ_0, must become larger.

In summary, it was shown phenomenologically that the application of the molecular-kinetic theory on the spontaneous movement of the TPC-line leads to physically reasonable connections between the driving force, distinguished by $\Delta \cos \Theta$, and the TPC-velocity.

Fig. 5 $\Delta \cos \Theta = \cos \Theta_r - \cos \Theta_s$ in dependence of the TPC-velocity at different degrees of hydrophobicity

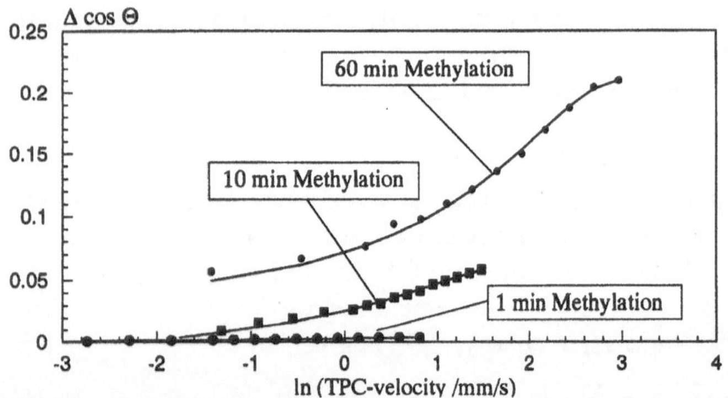

Conclusions

The movement of the TPC-line across the particle surface after rupture of the between particle surface and gas bubble forming thin liquid film can be investigated by means of the high-speed video technique at the pendant drop.

It was shown that for the movement of the TPC across the particle surface only the period of the first 10 ms after the beginning of the rupture of the thin liquid film is decisive for the process. The time dependence of the TPC-velocity can be approximated by a simple exponential relationship, in which the initial velocity correlates with the degree of hydrophobicity characterized by the receding contact angle. No correlation with the degree of hydrophobicity could be found for the decay constant.

The analysis of the experimental data on the basis of the molecular-kinetic theory demonstrates that this theory also for the spontaneous movement of the TPC-line across the particle surface leads to physically reasonable conclusions as comparison with the molecular parameters of the forced movement shows.

With the presented measuring method was created the requirements for systematic studies of the spontaneous dewetting kinetics on a bubble/particle system under influence of interfacial forces.

Acknowledgments The authors thank the Deutsche Forschungsgemeinschaft for financial support of this project.

References

1. Schulze HJ (1989) In: Laskowski JS (ed) Frothing in Flotation. Gordon and Breach Science Publishers, New York, pp 43–76
2. Dobby GS, Finch JA (1986) J Coll Interf Sci 109:493–498
3. Ye Y, Miller JD (1989) Int J Min Process 25:199–240
4. Stechemesser H (1989) Freiberger Forschungshefte A 790:81–100
5. Schulze HJ, Radoev B, Geidel Th, Stechemesser H, Töpfer E (1989) Int J Min Process 27:263–278
6. Schubert H (1986) Aufbereitung fester mineralischer Rohstoffe, VEB Deutscher Verlag der Grundstoffindustrie, Band II, Leipzig, pp 427
7. Blake TD, Haynes JM (1969) Coll Interf Sci 30:421–423
8. Voinov OW (1976) Mechanika Zidkosti i Gaza 5:76–84
9. Cox RG (1986) J Fluid Mechan 168:169–194
10. Petrov PG, Petrov JG (1992) Langmuir 8:1762–1767
11. Hayes RA, Ralston J (1994) Langmuir 10:340–342
12. Mingins T, Scheludko A (1979) J Chem Soc, Faraday Transact 1, 75:1–6
13. Herzberg WJ, Erwin WR (1970) J Coll Interf Sci 33:172–176
14. Sedev RV, Petrov JG (1992) Colloids & Surfaces 62:141–151

Progr Colloid Polym Sci (1996) 101:178–183

K.-U. Fulda
D. Piecha
B. Tieke
H. Yarmohammadipour

Monolayer characteristics of monodisperse core-shell latex particles prepared by soap-free emulsion copolymerization

K.-U. Fulda · D. Piecha
Prof. Dr. B. Tieke (✉)
H. Yarmohammadipour
Institut für Physikalische Chemie
der Universität zu Köln
Luxemburger Straße 116
50939 Köln, FRG

Abstract The monolayer characteristics of monodisperse core-shell latex particles at the air–water interface and on solid substrates were investigated. Particles were prepared by soap-free emulsion copolymerization of styrene and acrylic acid (1), styrene and 2-hydroxyethylmethacrylate (2), styrene and glycidylmethacrylate (3), and styrene and acryloxyethyl trimethylammonium chloride (4). They consist of a hydrophobic polystyrene core and a hydrophilic shell of the respective poly(meth)acrylate. Diameters lie in the range of 0.2 to 0.5 μm. Particle films at the air–water interface were studied by spreading ethanolic latex dispersions and measuring the π-A curves. For the same sort of particles, π-A curves vary strongly with the pH and the ion concentration of the subphase, and for different particles the π-A curves differ on the same subphase. For example, 1 forms real monolayers only on strongly acidified subphase, while 4 forms monolayers only on fairly concentrated salt solution. 3 with less hydrophilic shell than 2 tends to form 3-D aggregates rather than 2, which forms real monolayers. Particle mono- and multilayers on solid substrates were also studied. Monolayers were prepared using Langmuir–Blodgett (LB) deposition and electrostatic self-assembly (EA). Scanning electron micrographs of LB monolayers indicate a fairly regular, dense packing of the particles. LB deposition of a second layer leads to much less ordered films. EA of particles 1 onto cationic substrates proceeds via fractal growth of two-dimensional particle clusters at the interface. A substrate coverage of about 50% can be reached. Alternating EA of anionic particles 1 and cationic particles 4 leads to irregular 3-D particle clusters on the substrate but not to regular multilayers.

Key words Core-shell latex particles – monolayers – π-A isotherms – Langmuir–Blodgett films – electrostatic self-assembly

Introduction

Latex particles are used for a variety of applications such as synthetic rubber, coatings, paints, adhesives, additives for construction materials, flocculants, rheological modifiers and carriers in the field of diagnostics and drug delivery systems. Conventional latex particles consist of a hydrophobic polymer core and an amphiphilic shell of emulsifier and/or stabilizer molecules. They can be easily dispersed in aqueous media and – if special precautions concerning ion concentration of the subphase are taken – they also exhibit amphiphilic properties and form monolayers on an aqueous subphase. Previous studies on the spreading behavior were restricted to conventional latex particles [1–4] and models of latex particles [5]. π-A isotherms obtained are contradictory. This may partly be due to a cospreading of the amphiphilic emulsifier and/or stabilizer molecules at the air–water interface.

In order to obtain reliable information on particle organization at air–liquid and air-solid interfaces, it is therefore necessary to work with core-shell latex particles with the hydrophilic shell being covalently attached to the hydrophobic core. This has recently been shown studying salt-induced particle aggregation in two dimensions [6, 7], surface pressure induced ordering in Langmuir films and formation of Langmuir–Blodgett films [8]. Such particles are easily prepared by soap-free emulsion (co)polymerization [9–12]. Monodisperse spheres of 0.2 to 0.5 μm in diameter are usually obtained. Here we describe the interfacial arrangement of a variety of core-shell latex particles with different hydrophilic shell. Monolayers at the air–water interface were studied by measuring the π-A isotherms. Monlayers on solid substrates were prepared upon Langmuir–Blodgett (LB) deposition and electrostatic self-assembly (EA). Particle organization is characterized by scanning electron microscopy (SEM). Attempts to prepare multilayer structures are also reported.

Experimental section

Materials

Core-shell particles were prepared as previously described [9–12]. Details of preparation conditions and particle properties are compiled in Table 1.

Methods

Monolayers were spread from ethanolic dispersion (spectroscopic grade, concentration 10 mg mL^{-1}) onto a pure water subphase (Milli Q plus) using a Lauda film balance FW2 equipped with film lift. pH values were adjusted with hydrochloric acid or sodium hydroxide solution. π-A isotherms were measured at a compression rate of 1 cm min^{-1} corresponding to a change in trough area by 20 cm^2 min^{-1}. LB-layers were deposited on glass substrates. The surface pressure was 25 mN m^{-1} and the subphase temperature was 15 °C. Dipping rate was 2 mm min^{-1}. EA particle layers were prepared by electrostatic adsorption from aqueous dispersion (concentration 8.3 mg mL^{-1}) of pH 5.3 onto glass supports precoated with four layers of polystyrene sulfonic acid (PSS, Janssen, MW 70 000) and four layers of polyallylamine hydrochloride (PAH, Aldrich, MW 50 000) in alternating sequence using EA [13]. SEM studies were carried out using a Philips XL 40 microscope. Particle films have been studied on glass or aluminum supports after a thin gold layer was evaporated.

Results and Discussion

The particles

The spherical particles were prepared by soap-free emulsion copolymerization [9–12]. Styrene was used as hydrophobic monomer. This guaranteed the formation of tough and stable cores with a glass temperature of about 110 °C. The shell-forming hydrophilic monomers were widely varied. Cationic, anionic and non-ionic (meth)acrylates were

Table 1 Preparation conditions and characteristic properties of the particles

No.	Monomers	Molar ratio	Monomer concentration [mol L^{-3}]	Initiator/ weight%	Reaction time [h]	Particle diameter d [nm]	Polydispersity $[d_w/d_n]$***
1a	sty*/acrylic acid	10.8	1.18	KPS**/0.50	6	458 ± 12	1.002
1b	sty*/acrylic acid	8.8	0.98	KPS**/0.55	7	214 ± 12	1.004
2	sty*/2-hydroxyethyl methacrylate	8.8	0.75	KPS**/1.35	1	210 ± 24	1.019
3	sty*/2.3-epoxypropyl methacrylate	8.2	1.29	KPS**/0.80	2	218 ± 12	1.009
4	sty*/acryloxyethyl trimethyl-ammonium chloride	20.5	1.01	AIAP**/0.56	18	204 ± 8	1.004

* sty = styrene.
** KPS = potassium peroxodisulfate.
** AIAP = 2.2-azobis(2-amidinopropane) hydrochloride.

*** $d_w = \frac{\sum_i n_i d_i^4}{\sum_i n_i d_i^3}$; $d_n = \frac{\sum_i n_i d_i}{\sum_i n_i}$.

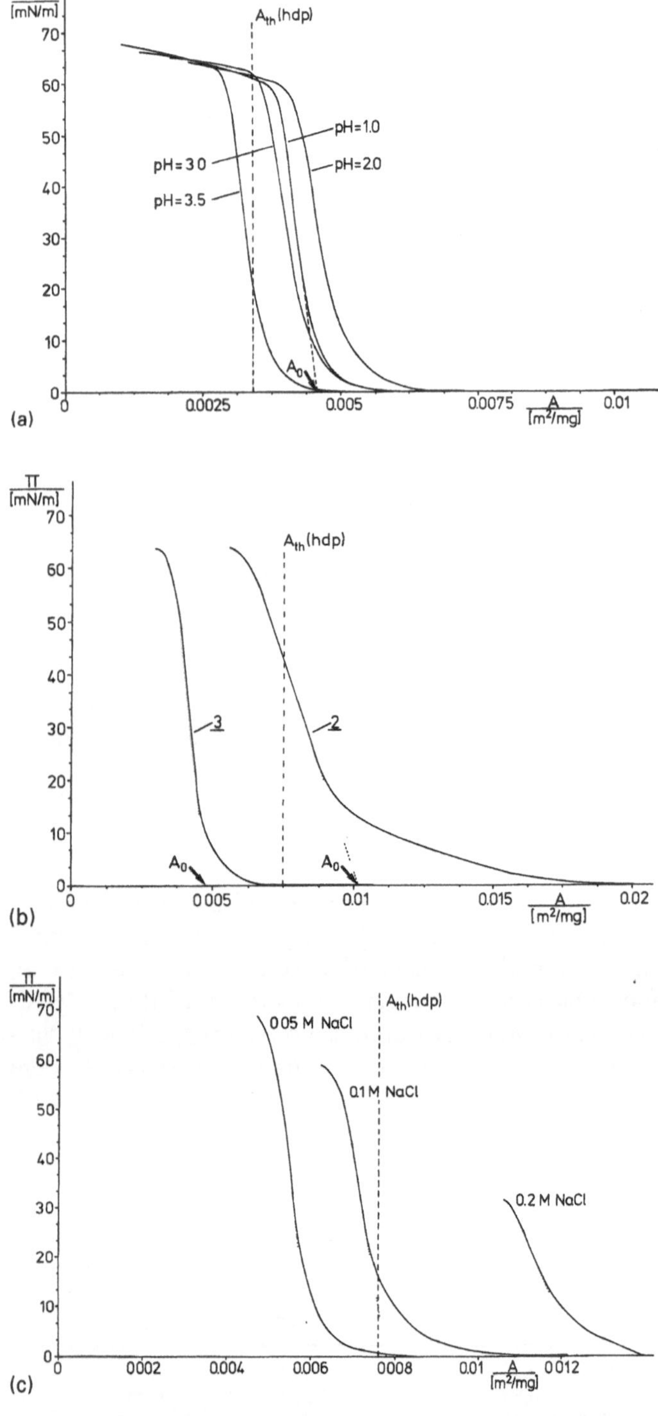

Fig. 1 π-A isotherms of (a) 1a on aqueous subphase of different pH values, $T = 15\,^{\circ}\mathrm{C}$, (b) 2 and 3 on aqueous subphase of pH 1.8, $T = 20\,^{\circ}\mathrm{C}$ and (c) 4 on aqueous subphase of different NaCl concentrations, pH 5.5, $T = 15\,^{\circ}\mathrm{C}$. A_{th} values are calculated using particle diameters from SEM pictures

used. This allowed to study interfacial properties on a wide scale. In Table 1 preparation conditions and characteristic data of the particles are compiled.

Monolayers at the air-water interface

Particles listed in Table 1 were spread from ethanolic dispersion onto a pure water subphase. In all cases particle films were obtained if special precautions concerning pH and ion concentration of the subphase were taken. Typical π-A curves are shown in Fig. 1. Qualitative shapes are almost the same as those previously reported for monolayers of polystyrene monomolecular particles [14]. On compression, the surface pressure is monotonously increased until collapse occurs at π-values of 60 to 65 $\mathrm{mN\,m^{-1}}$. The A_0 value is the limiting area of the particle film defined by extrapolating the maximum slope of the π-A curve to zero surface pressure. A_{th} is the theoretical area occupied by a particle monolayer with hexagonal dense packing (hdp). It can be easily calculated, if particle diameter and mass of the spread material are known. The ratio of A_{th}/A_0 indicates the ratio of the minimum area, which a particle monolayer can occupy at the interface, to the area the particle monolayer actually occupies. Consequently, real monolayers can only exist if the ratio is smaller than 1. π-A curves in Fig. 1a and c show that for a distinct sort of particles the ratio of A_{th}/A_0 varies strongly with the pH and the ion concentration of the subphase, and that for a distinct subphase the ratio also varies with the chemical nature of the particle shell (Fig. 1b). Monolayers of 1 are only obtained on strongly acidified subphase. At pH-values >3 ionization of the COOH groups becomes significant. As a consequence, particle solubility increases and they begin to disappear into the subphase as indicated by the decreasing A-value. The very different A_{th}/A_0 ratios observed for 2 and 3 can be attributed to the different hydrophilicity of the particle shells. 3 with epoxide units in the shell is considerably less hydrophilic than 2 with hydroxy groups and therefore tends to form 3-D particle aggregates rather than 2, which forms real monolayers. π-A curves of 4 show a strong dependence on the salt concentration of the subphase. At low salt concentration it can be assumed that the cationic particles are very soluble and simply disappear in the subphase. At very high salt concentrations, the particles are only little soluble and probably form two-dimensional aggregates with fractal dimension [7] displaying large free areas between the particles. Consequently, the ratios A_{th}/A_0 must be large at low salt concentration and small at high salt concentration, as it is in fact observed.

Monolayers on solid supports

Several methods are suitable to coat a solid support with latex particles. The simplest way is to drop a dispersion onto the substrate and to evaporate the aqueous phase.

The residual particles form highly ordered substrate coatings which, however, are not controlled in thickness. As demonstrated below, regular substrate coatings of defined thickness can be obtained more easily by Langmuir–Blodgett (LB) deposition of particle monolayers from the air–water interface or by electrostatic adsorption ("self-assembly") from an aqueous dispersion.

Langmuir–Blodgett layers

If the monolayer at the air–water interface is kept under a constant surface pressure of about 25 $mN\,m^{-1}$, the film area decreases only slightly until a constant value is obtained. The decrease may originate from a partial ordering of the particles at the interface. The compressed monolayers is now stable and can be easily transferred onto a glass support utilizing the LB technique (Fig. 2a). In this way, LB monolayers of all particles listed in Table 1 can be prepared. However, the spherical shape of the particles causes that the contact with the substrate surface is only small and consequently the adhesion is poor. Two methods can be applied to improve the adhesion. One method is to briefly anneal the monolayer slightly below the glass transition temperature of the core polymer. The particles soften and the contact area increases, while spherical shape and mutual particle arrangement are preserved. Another method is to build up the LB monolayers on the highly charged substrates also used for preparation of EA layers (see the experimental part). In Figs. 2b and c, scanning electron micrographs of particle LB-monolayers are shown. Particle arrangements are in good agreement with those concluded above from the π-A curves. Particles 1 with strongly hydrophilic shell form monolayers with an occupied area of about 75% (Fig. 2a). The monolayer consists of small particle domains, preferentially with hdp, and uncovered regions are in between. Particles with nonionic shell such as 3 tend to form multilayered aggregates in addition to monolayer structures (Fig. 2c).

(a)

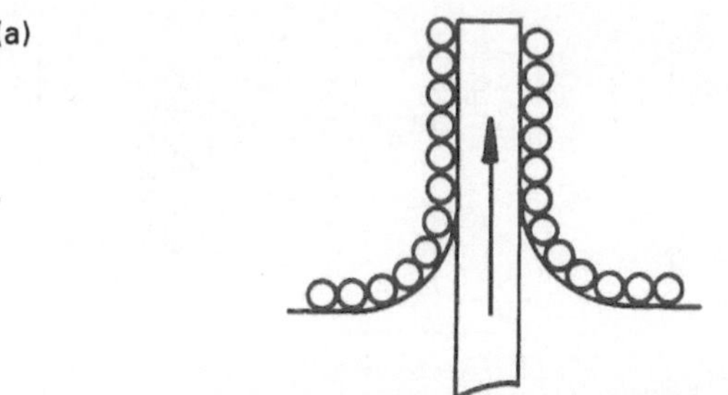

(b)

(c)

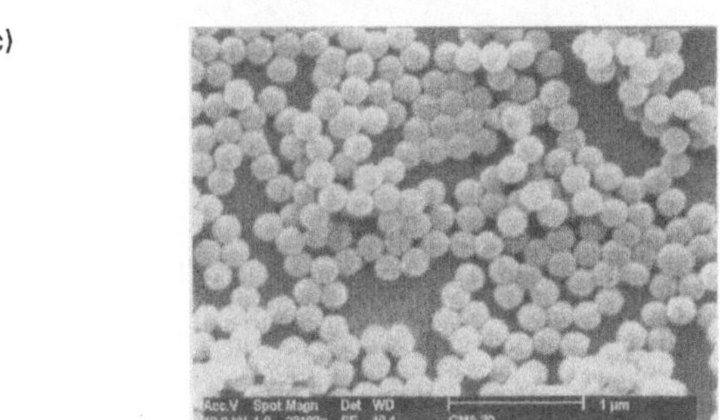

Fig. 2 Scheme of LB-monolayer preparation (a) and scanning electron micrographs of LB monolayers of 1a on glass (b) and 3 on alumina (c)

Electrostatically adsorbed layers

As a possible alternative to prepare particle monolayers on solid substrates, we also studied electrostatic adsorption (EA) of latex particles onto oppositely charged surfaces. The method is technically applied to modify fibers and paper, for example [15]. Dipping of highly charged glass substrates (see experimental part) into aqueous latex dispersions of opposite charge led to particle adsorption within a few minutes (Fig. 3a). However, the occupied area was strongly affected by the pH of the aqueous phase. For particles 1a, EA was optimum at a pH value of 5.3 and an occupied area of 50% could be reached. The SEM picture indicates the fractal dimension of the predominantly two-dimensional particle clusters 1b at the interface (Fig. 3b).

Multilayers on solid supports

Besides monolayer formation we also investigated the deposition of a second layer onto solid supports. LB and EA method were tried as schematically shown in Figs. 4a and 5a. LB transfer is feasible if precautions for a good adhesion of the already deposited layers are taken. As

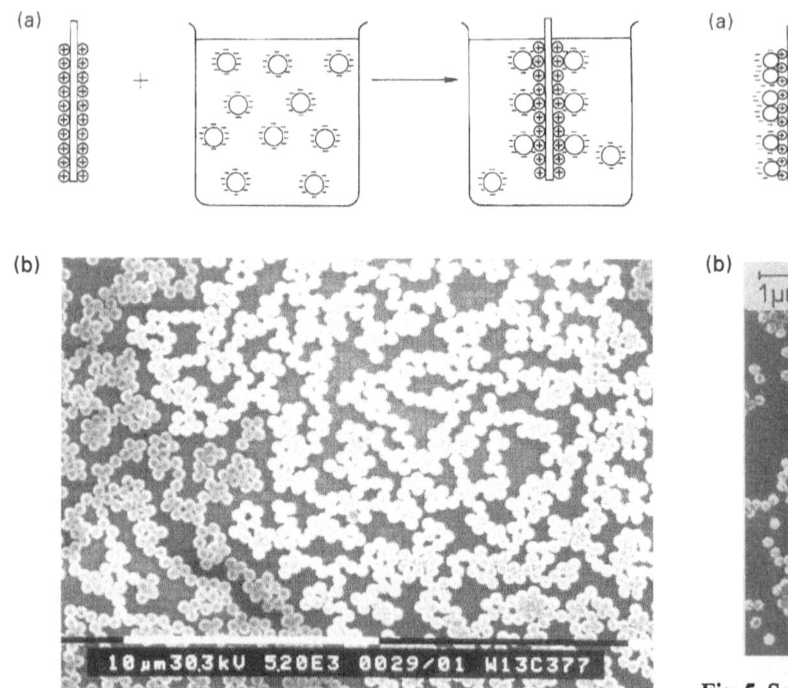

Fig. 3 Scheme of EA of a particle monolayer (a) and scanning electron micrograph of an EA-monolayer of particles 1a (b) adsorbed on a positively charged glass substrate (for precoating see experimental part)

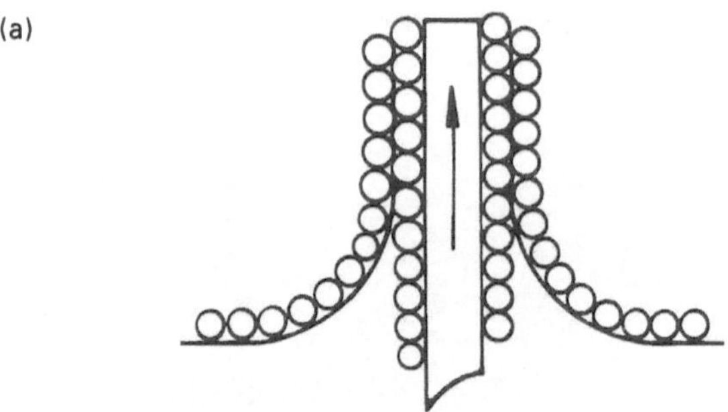

Fig. 4 Scheme of LB-deposition of a second particle layer (a) and scanning electron micrograph of a two-layer LB film of 1a (b)

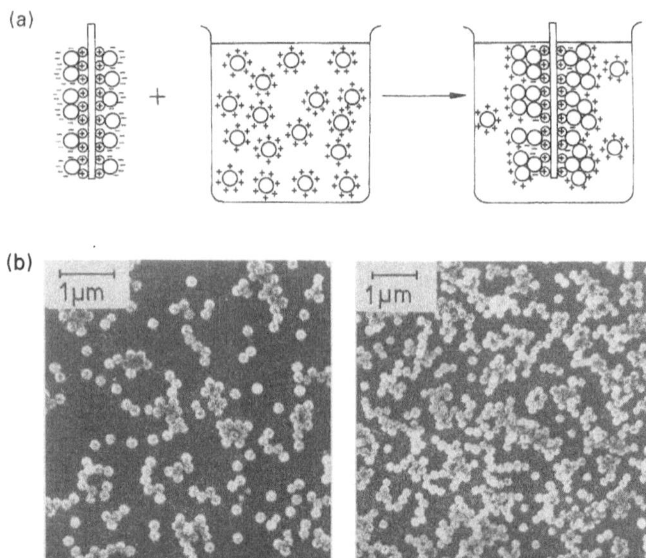

Fig. 5 Scheme of EA of a second particle layer (a) and scanning electron micrographs of an EA monolayer of particles 1b on a charged substrate (b) and the same sample after additional EA of oppositely charged particles 4 (for precoating of the substrate see the experimental part)

described above, a short annealing slightly below the glass transition temperature of the core polymer is sufficient. Figure 4b shows a LB film of particles 1a obtained after two-fold dipping of the substrate through the monolayer. Clearly the second layer is very disordered. One reason may be that particles from the second layer partly fill the holes in the first layer, which generally increases the height differences and disorder in the two-layer assembly. Deposition of a second layer was also tried by the EA method. A cationic substrate partially covered with anionic particles 1b adsorbed by the EA method (see Fig. 3b) was dipped into a latex dispersion of cationic particles 4. Figure 3c shows that particles 4 were coadsorbed and formed three-dimensional clusters with the already adsorbed particles 1b (Fig. 3c). The overall substrate coverage increased but the general state of order decreased. A regular two-layer arrangement could not be obtained.

Conclusions

Our studies show that core-shell latex particles prepared by soap-free emulsion copolymerization exhibit amphiphilic properties and form well-organized monolayers at the air–water interface. LB deposition of the floating particle film allows to prepare fairly densely packed monolayers on solid substrates, while electrostatic self-assembly leads to less dense surface coatings. Further LB-type manipulation of the particles allows to build up multilayers although their structure is less regular than that of monolayers. The interfacial arrangement of the particles draws some analogy to the behavior of amphiphilic molecules at interfaces. Particle monolayers may therefore be used as a simple model for the study of molecular aggregation at interfaces. Particle monolayers on solid substrates are also excellent models for a study of particle-electrolyte and -polyelectrolyte interaction. Electrolyte adsorption at the particle surface can be easily studied by quantitative spectroscopic measurements. Moreover, adsorption of inorganic (e.g. Fe^{3+}, $PdCl_4^-$) and organic ions (e.g. copper-phthalocyanine tetrasulfonic acid) at the particle surface is an easy method to introduce functional properties in the particle films such as redox activity, catalytic activity or light emission, for example. Detailed studies are underway and will be published shortly [16].

Acknowledgment The authors gratefully acknowledge Dr. H. Gottschalk, II. Physikalisches Institut, and Prof. Th. Kruck, Institut für Anorganische Chemie, for kindly allowing to use their electron microscopy facilities. E. Gusek is thanked for taking one of the SEM pictures. Financial support by the Deutsche Forschungsgemeinschaft (Project II C 1-Ti219/1-1 and 1-2) is also gratefully acknowledged.

References

1. Schuller H (1966) Kolloid-Z u Z Polym 211:113–121
2. Sheppard E, Tcheurekdjian N (1967) Kolloid-Z u Z Polym 225:162–163
3. Sheppard E, Tcheurekdjian N (1968) J Colloid Interf Sci 28:481–486
4. Garvey MJ, Mitchell D, Smith AL (1979) Colloid Polym Sci 257:70–74
5. Schuller H (1967) Kolloid-Z u Z Polym 216/217:380–383
6. Robinson DJ, Earnshaw JC (1992) Phys Rev A 46:2045–2054
7. Stankiewicz J, Cabrerizo Vilchez MA, Hidalgo Alvarez R (1993) Phys Rev E 47:2663–2668
8. Fulda KU, Tieke B (1994) Adv Mater 6:288–290
9. Ceska GW (1974) J Appl Polym Sci 18:427–437
10. Sakota K, Okaya T (1976) J Appl Polym Sci 20:1745–1752
11. Kamei S, Okubo M, Matsumoto T (1986) J Polym Sci Polym Chem Ed 24:3109–3116
12. Ohtsuka Y, Kawaguchi H, Sugi Y (1981) J Appl Polym Sci 26:1637–1647
13. Lvov Y, Decher G, Möhwald H (1993) Langmuir 9:481–486
14. Kumaki J (1986) Macromolecules 19: 2258–263
15. Tamai H, Hakozaki T, Suzawa T (1980) Colloid Polym Sci 258:870–876
16. Fulda KU, Yarmohammadipour H, Tieke B, to be published

Progr Colloid Polym Sci (1996) 101:184–188
© Steinkopff Verlag 1996

F. Simon
H.-J. Jacobasch
D. Pleul
P. Uhlmann

Complex surface characterization of modified carbon fibers by means of spectroscopic and thermodynamic methods

F. Simon · Prof. H.-J. Jacobasch (✉)
D. Pleul · P. Uhlmann
Institute of Polymer Research Dresden
Hohe Straße 6
01069 Dresden, FRG

Abstract The wish to optimize the interactions between carbon fiber surface and polymer matrix and the basic interest in the fundamentals of adhesion required fiber surface characterization by various physico-chemical measuring methods. This paper discusses the results of different physico-chemical measuring methods, e.g. inverse gas chromatography and electrokinetic measurements, to estimate the acid-base surface properties of carbon fibers. The discussion has been focused on the changes in the Brønsted and Lewis acid-base properties after a surface treatment.

In addition, photoelectron spectroscopic investigations of carbon fibers, which show the elemental surface composition, are a successful tool to demonstrate the molecular reasons of the thermodynamic surface properties.

The combination of different physico-chemical and spectroscopic measuring methods allowed to get comprehensive knowledge of the fiber surface chemistry.

Key words Acid-base properties – carbon fibers – inverse gas chromatography – streaming potential – surface characterization – surface modification – XPS

Introduction

Composites with carbon fibers as reinforced materials show extraordinary properties which make them increasingly more important for technological applications. Especially, low-weight construction components undergoing high mechanical stress are manufactured by carbon fiber reinforced composites [1]. The mechanical properties of carbon fiber composite materials are mainly determined by the adhesion forces between the fiber surface and the polymer matrix surrounding the fiber. From this point of view it is very important to characterize the surface of carbon fibers and draw conclusions with regard to their surface properties and reactivity.

Most of the carbon fiber surface investigations published show either the spectroscopically determined atomic and molecular surface structure [2–6] or thermodynamic methods applied to characterize the carbon fiber surface [7–9]. However, it is well known that the adhesion phenomena base on acid-base interactions between the adhering substances [10]. On the other hand, the acid-base properties have to base on atomic and molecular composition of the surface region. Hence a complex surface characterization requires a combination of spectroscopic and thermodynamic methods and correlation of the data obtained by the measuring methods used.

Materials and methods

An unmodified carbon fiber and three with differently modified surfaces have been investigated. All fibers were commercially manufactured and modified ones. Table 1 contains the trade names and the kinds of surface modification.

Table 1 Carbon fibers used for complex surface characterization

Sample	Trade name	Kind of surface modification
CAR 1	Tenax HTA 5000	untreated, unmodified
CAR 2	Besfight HTA-7	oxidized
CAR 3	Tenax HTA 5131	acidically sized
CAR 4	Torayca	basically sized

X-ray photoelectron spectra of carbon fiber bundles were measured on an ESCAlab 220i spectrometer (Fisons Instruments) using monochromatized Al ($K\alpha$) radiation ($h \cdot \nu = 1486.6$ eV). The kinetic energy of the emitted photoelectrons has been measured by a 180° hemispheric analyser in constant analyzer energy mode ($E_{pass} = 80$ eV [survey spectra], $E_{pass} = 25$ eV [resolved spectra]).

The elemental surface composition was determined from the peak areas found in the survey spectra using Wagner's sensitivity factors [11] and the knowledge of the spectrometer transmission function. Before, the inelastic background was subtracted according to the Shirley method [12].

The electrokinetic measurements have been carried out as streaming potential measurements by means of an Electrokinetic Analyser EKA (Anton Paar, Graz). Aqueous 10^{-3} mol $\cdot$ l^{-1} KCl solutions of different pH values have been pumped through a fiber diaphragm. The potential has been measured in dependence on the applied pressure. The zeta potential was calculated from the measured streaming potential by means of the Smoluchowski equation [13]. To get thermodynamic values like adsorption potentials of H^+ and OH^- ions and pK_b and pK_a values, respectively, a data analysis of the function $\zeta = \zeta$ (pH) was carried out according to a Gouy-Chapman-Stern-Grahame model [14].

Inverse gas chromatography (IGC) measurements have been carried out by means of a Hewlett Packard gas chromatograph (HP 5890). Different test liquids were injected into a carbon fiber packed chromatography colon at 40 °C. The dispersive term of the free surface energy (γ_S^{dis}) has been determined from the retention volume of different homologous alkanes (from C_6H_{14} to $C_{12}H_{26}$). The Lewis acidity ($\tilde{K}_a$) and the Lewis basicity ($\tilde{K}_b$) parameters were determined by different test liquids taken from the Gutmann donor and acceptor scales [15]. More details of the IGC method are given in [16].

Results and discussion

The untreated carbon fiber surface

Figure 1 shows the XPS survey spectra of the differently treated carbon fiber surfaces which have been used to calculate the elemental surface composition listed in Table 2.

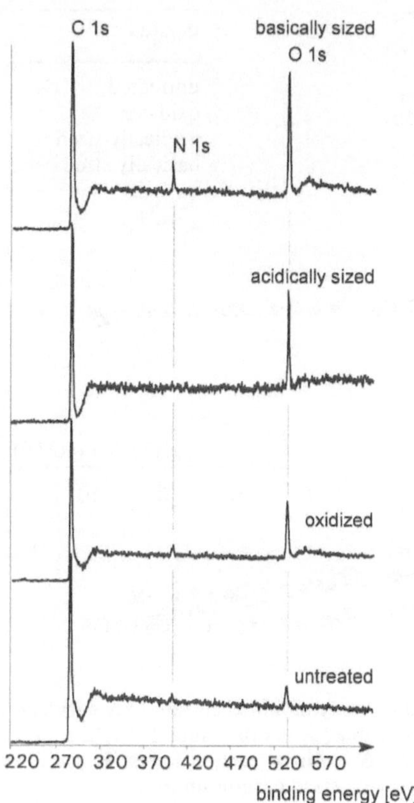

Fig. 1 Survey X-ray photoelectron spectra of differently treated carbon fibre surfaces

The fiber surface of sample CAR 1 contains oxygen and nitrogen in addition to the expected carbon. Traces of nitrogen may have remained from the manufacturing process and oxygen may have been introduced by surface oxidation reactions. The resolved C1s spectrum of CAR 1 shows the presence of alcohol and small traces of carbonyl and carboxyl groups in the surface region. Furthermore, a wide shake-up peak has been observed in the C1s spectrum which indicates conjugated delocalized p_π electrons.

The functional surface groups containing oxygen and nitrogen determine the Brønsted and Lewis acid-base properties and the wetting behavior.

Figure 2 shows the dependence of the zeta potential on pH values reflecting the Brønsted acid-base surface properties.

It can be seen that the surface of CAR 1 acts as a Brønsted base at pH < 4.5. The measured positive zeta potential may be the result of an adsorption of H^+ ions on Brønsted basic surface centers like amino groups.

At pH higher 4.5 Brønsted acid surface groups may have undergone dissociation reactions. In addition, the slope of

Table 2 Elemental surface composition of differently modified carbon fibers determined by X-ray photoelectron spectroscopy

Sample	[O]:[C]	[N]:[C]	[C1s shake-up]:[C]
untreated	0.024	0.018	0.055
oxidized	0.035	0.028	0.024
acidically sized	0.143	0	0.016
basically sized	0.141	0.039	0.014

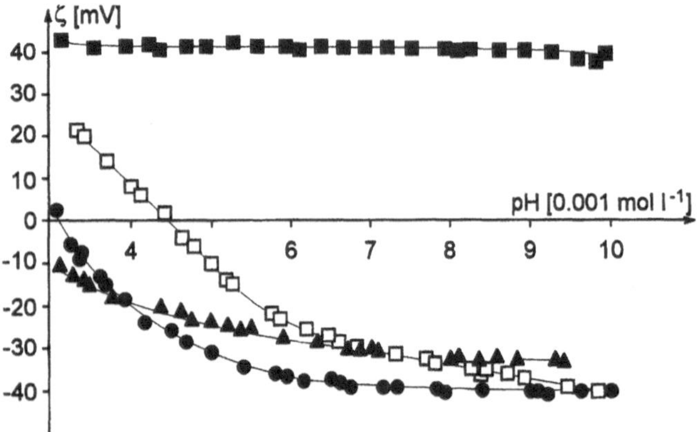

Fig. 2 Zeta potential of differently modified carbon fibre surfaces in dependence on pH values of an aqueous 10^{-3} mol · l^{-1} KCl solution □ untreated carbon fibres, ▲ oxidized carbon fibres, ● acidically sized carbon fibres, ■ basically sized carbon fibres

the function $\zeta = \zeta$ (pH) at pH > 7.5 indicates a preferred OH^- adsorption on the hydrophobic surface centers. The negative surface charges formed by the two described mechanisms account for the observed negative zeta potential values.

The curve analysis of the ζ-pH plot shows the weak Brønsted acid and moderate Brønsted base properties of CAR 1 surface (Table 3).

The Lewis acid base parameters of the CAR 1 surface, as determined by means of inverse gas chromatography (IGC), are shown in Table 4.

The Lewis acid parameter $\tilde{k}_a$ and the Lewis base parameter $\tilde{k}_b$ represent weak Lewis acid base surface properties. These results and the relatively high value of the dispersive term of the free surface energy measured by means of IGC are in good agreement with the XPS results showing a low number of functional groups in the surface region of untreated carbon fibers.

The oxidized carbon fibre surface

After an oxidation treatment (CAR 2) the number of surface groups containing oxygen and nitrogen has increased (Table 2). In contrast, the intensity of the observed shake-up peaks has decreased. Obviously the π electron system of the carbon fibre surface has been destroyed partially during the oxidation reaction.

We assume that the oxidation reaction attacks the carbon rings in the surface region, open a few of the carbon rings by splitting of the C ⋯ C bonds and introduce groups containing oxygen, like OH and COOH, in the surface region.

The increased number of surface groups containing oxygen changes the zeta potential vs. pH plot (Fig. 2). So, the isoelectrical point has been shifted to a lower pH value. The shift of the isoelectrical point shows an increased number of Brønsted acid surface sites. The shape of the zeta potential plot shows a plateau phase in the higher pH range indicating a preferred dissociation reaction of Brønsted acid surface groups. In relation to the unmodified carbon fiber surface, the curve analysis of the oxidized fiber surface gives an increased amount of the OH^- adsorption potential (Table 3). The pK_a calculated from the Φ_{OH^-} value shows a higher Brønsted acid strength than the untreated fiber surface.

In contrast to the strongly increased Brønsted acid strength the increase of the Lewis acid properties, as determined by IGC measurements, has been rather weak.

This result allows to conclude that the oxygen introduced in the surface region is preferably bonded as typical Brønsted acid groups like carboxyl groups. However, it is important to note that the surface oxidation decreases the dispersive term of the free surface energy.

The acidically sized carbon fiber surface

Figure 3 shows the C 1s X-ray photoelectron spectra of the samples CAR 1 and CAR 3. The spectrum of the acidically sized carbon fiber may be divided into two component peaks at the lower binding energy (BE = 284.8 eV) represents C_xH_y species and the other component peak at BE = 286.6 eV shows the presence of $\underline{C}$-OH or epoxy groups in the surface region. In addition, Fig. 1 evidences a loss in the nitrogen peak. This shows that the sizing of carbon fibers leads to quite a new surface composition and that new surface properties can be expected.

The electrokinetic measurements of acidically sized carbon fibers give rather a flat plot of the zeta potential versus pH. The isoelectrical point has been found at pH < 3. That

Table 3 Adsorption potentials (Φ_i) and pK_a and pK_b values of differently modified carbon fiber surfaces determined from the $\zeta = \zeta$ (pH) function by means of a Gouy–Chapman–Stern–Grahame model [14]

Sample	Φ_{OH^-} [kJ·mol^{-1}]	pK_a	$\Phi_{H_3O^+}$ [kJ·mol^{-1}]	pK_b
CAR 1	− 60.7	5.0	− 31.0	10.2
CAR 2	− 73.7	3.3	− 30.8	10.3
CAR 3	− 61.0	5.1	− 27.7	10.8
CAR 4	− 20.9	11.8	− 45.7	7.5

Table 4 Dispersion and Lewis acid ($\tilde{k}_a$) and Lewis base ($\tilde{k}_b$) parameters of differently treated carbon fiber surfaces determined by means of IGC

Sample	γ_s^{dis} [mJ·m^{-2}]	$\tilde{k}_a$	$\tilde{k}_b$
CAR 1	79	0.1	0.7
CAR 2	62	0.5	0.7
CAR 3	32	1.1	0.8

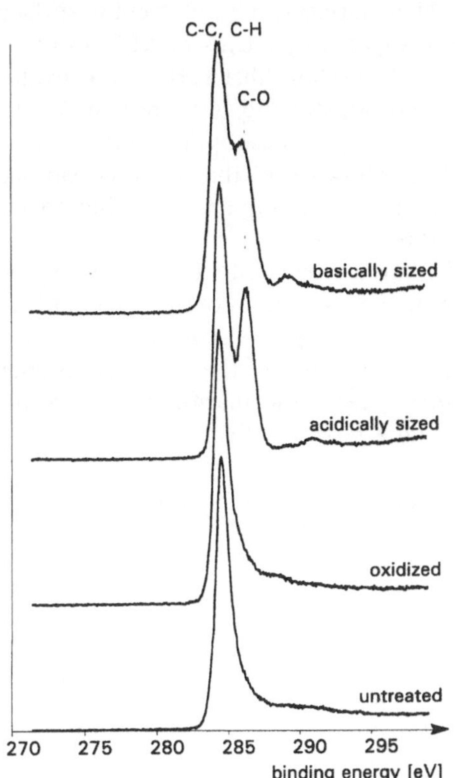

Fig. 3 Resolved C 1s X-ray photoelectron spectra of differently treated carbon fibre surfaces

indicates that the surface of the acidically sized carbon fiber contains a high number of Brønsted acid groups. However, the analysis of the $\zeta = \zeta$ (pH) plot does not show an increased Brønsted acid strength of such groups.

Concerning the increased surface polarity as a consequence of the sizing process, data received from IGC measurements confirm the results of the other methods described above. The dispersive term of the free surface energy has strongly decreased after the sizing treatment. In contrast, the parameters of the Lewis acidity $\tilde{k}_a$ is increased by the influence of the increased amount of surface groups containing oxygen.

The high Lewis acidity and the rather low Brønsted acidity confirm the existence of epoxy or OH surface groups observed by means of XPS.

The basically sized carbon fibre surface

The basic sizing process introduces groups containing nitrogen and oxygen into the surface region (Fig. 1). The elemental ratio [N] : [C] is strongly increased (Table 2). The high amount of oxygen and nitrogen and the structure of the C 1s resolved spectrum (Fig. 3) indicate that the original carbon fiber surface is fully covered by the size.

The Brønsted basic character of functional containing groups nitrogen can be reflected by electrokinetic measurements. The function $\zeta = \zeta$ (pH) is nearly constant over the whole pH scale and all determined zeta potential values are positive (Fig. 2). In contrast to the other samples investigated the pK_b value of the basically sized carbon fiber surface is much lower and represents the surface as a moderate Brønsted base. Obviously, the high oxygen content observed by XPS does not influence the Brønsted acid base properties. A similar result has been found in case of the electrokinetic investigation of sample CAR 3 where the high number of epoxy and OH groups, respectively, did not change the Brønsted acid strength of the fiber surface.

The surface groups containing nitrogen let expect Lewis basic surface properties expressed by the increased Lewis base parameter $\tilde{k}_b$ determined by IGC measurements. The surface groups containing oxygen enhance the Lewis surface acidity and lead to a $\tilde{k}_a$ value similar to the one of sample CAR 3 where nearly the same oxygen content in the surface region has been found.

Prediction of surface reactivity

The discussed acid base properties of the functional surface groups are very important to predict the results of the

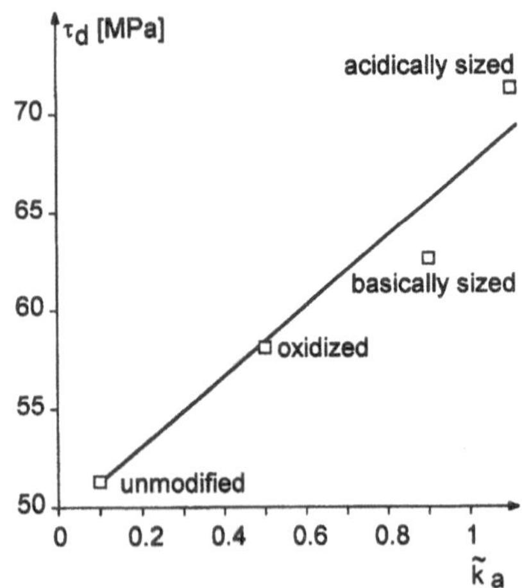

Fig. 4 Dependence of the debonding shear strength τ_d on the Lewis acidity parameter of the differently treated carbon fibre surfaces determined by means of IGC measurements

reaction of carbon fiber surfaces with other components like, for example, resins.

Composites were made from the carbon fiber samples described above and a resin containing epoxide groups (oligo-(bisphenol-A-diglycidether)). The pull-out test served to determine the debonding shear strengths τ_d. Figure 4 shows the results of the pull-out test in dependence on the Lewis acidity parameter determined from IGC measurements.

The debonding shear strength can be seen to increase with increasing Lewis acid properties in the surface region. The oxidized carbon fiber surface with the highest Brønsted acidity shows a significantly lower debonding shear strength than the basically sized carbon fiber surface with very weak Brønsted acid surface properties. Obviously, the reactivity between the fiber surface groups and the epoxy groups of the resin is only determined by the Lewis acid character of the fiber surfaces. Due to the absence of water during the reaction Brønsted acid base interaction plays a minor role.

The results obtained are in good agreement with the well known reaction mechanism of the ring opening reaction of epoxides. In presence of a Lewis acid, the epoxy ring may undergo a nucleophilic ring opening reaction, and in a second step a covalent C-O-C bond may be formed. With increasing number of Lewis acid surface centers, the number of covalent bonds between carbon fiber surface and resin increases. Of course, the increased number of chemical bonds between the two phases enhance the adhesion force.

Conclusions

Various methods have been used to characterize differently treated carbon fiber surfaces. The elemental surface composition was investigated by means of XPS. In addition, the reactivity of the carbon fiber surface was evaluated by means of electrokinetic and IGC measurements. The received results give a good description of the surface properties. It has been shown that the observed surface reactivity must have been determined according to elemental and structural reasons.

Evaluation of the surface reactivity was the basis for predictions concerning the interaction of the carbon fibers with a resin matrix. The examples discussed above have shown the particular importance of the Lewis properties on the adhesion force in case of formation of composites from carbon fibers and a resin matrix.

Acknowledgment The authors are grateful to Dr. Edith Mäder for the determination of the debonding shear strength.

References

1. Heißler H (1981) In: Processing and use of carbon fibre reinforced plastics VDI-Verlag, Düsseldorf, pp 1–4
2. Kozlowski C, Sherwood PMA (1984) J Chem Soc Faraday Tran I 80: 2099–2107
3. Xie Y, Sherwood PMA (1990) Chem Mater 2:293–299
4. Desimoni F, Casella G, Salvi AM, Cataldi TR, Morone A (1992) Carbon 30:527–531
5. King JA, Buttry DA, Adams DF (1993) Polymer Composites 14:292–300
6. Bhardwaj A, Bhardwaj IS (1994) J Appl Polym Sci 51:2015–2020
7. Seo KS, Fornes RE, Gilbert RD, Memory JD (1988) J Polymer Sci, Part B: Polymer Physics 26:245–255
8. Mäder E, Grundke K, Jacobasch HJ, Wachinger G (1994) Composites 25:739–744
9. Mäder E, Grundke K, Janke A (1995) In: Jacobasch HJ (ed) Proc. Workshop Interfaces in Carbon, Glass and Polymer Fibre Reinforced Polymer Composites. Institute of Polymer Research, Dresden, Germany, pp 32–43
10. Marmo MJ, Mostafa MA, Jinnal H, Fowkes FM, Manson JA (1976) Ind Eng Chem Prod Res Dev 15:206–211
11. Moulder JF, Stickle WF, Sobol PE, Bomben KD (1992) Handbook of X-ray Photoelectron Spectroscopy, Perkin-Elmer Corporation, pp 252–253
12. Shirley DA (1972) Phys Rev B5: 4709–4714
13. Jacobasch HJ, Schurz J (1988) Progr Colloid Polymer Sci 77:40–48
14. Börner M, Jacobasch HJ, Simon F, Churaev NV, Segeeva IP, Sobolev VD (1994) Colloids and Surfaces A: Physicochemical and Engineering Aspects A 85:9–17
15. Gutmann V (1978) The donor-acceptor approach to molecular interaction, Plenum Press, New York, London
16. Jacobasch HJ, Grundke K, Uhlmann P, Simon F, Mäder E (1995) J. Adhesion Science and Technology 9:327–350

Progr Colloid Polym Sci (1996) 101:189–193

S. Voronov
V. Tokarev
V. Datsyuk
M. Kozar

Peroxidation of the interface of colloidal systems as new possibilities for design of compounds

Prof. Dr. V. Voronov (✉) · V. Tokarev
V. Datsyuk · M. Kozar
Department of Organical Chemistry
"Lvivska Polytechnika" State University
12 S. Bandera Street
Lviv-13, 290646 Ukraine

Abstract Present work is devoted to the peroxidic modification of dispersed phase surface of various polymeric colloidal systems: emulsions, latexes, polymer–polymer mixtures, polymers reinforced and filled by dispersion, etc. by heterofunctional polymeric peroxides (HEPP).

HEPP are carbonchain polymers which have statistically located peroxidic (R–OO–R′) and highly polar functional groups such as carboxylic, anhydride, pyridine and others along the main chain.

Sharp difference in the polarity of these groups gives to the macromolecules of HFPP the capability of adsorption at the interface in various polymeric colloidal systems. The reactions of the functional groups provide chemical bonding of the macromolecules to the interfacial surface. Thus, in consequence of physical adsorption or chemisorption of HFPP the peroxidation of the interfacial surface, that is the localization of active –OO– groups on the surface, is attained.

Further such –OO– groups can provide the grafting of matrix polymer macromolecules to the dispersed phase surface. The progress of grafting reactions can be carried out in the processes of polymerizational filling, composites' curing or rubber mixture vulcanization.

The activation of latex particle surface by surface-active HFPP is of special interest. Thereby, one can solve both environmental problems and approach the modification of latexes and latex polymers by absolutely new ways.

Key words Polymeric peroxides – immobilization of reactive groups – grafted polymers

Introduction

Polymeric filled materials, polymer–polymer mixtures and aqueous polymeric dispersions are typical colloidal systems with highly developed interface. This causes the peculiarities of the formation and behaviour of such systems and considerable influence of interfacial layer structure on their properties. From our point of view, the adsorption of polymers with reactive groups, in particular with –OO– groups, capable to generate free radicals and initiate radical processes, is a promising method of interfacial surface activation.

Present work is devoted to the peroxidic modification of dispersed phase surface by heterofunctional polymeric peroxides (HFPP).

Principles of the peroxidic modification of the interface of polymeric colloidal systems

Currently, the basic problems of the formation of latexes, filled polymers, and blends of polymers are: purposeful

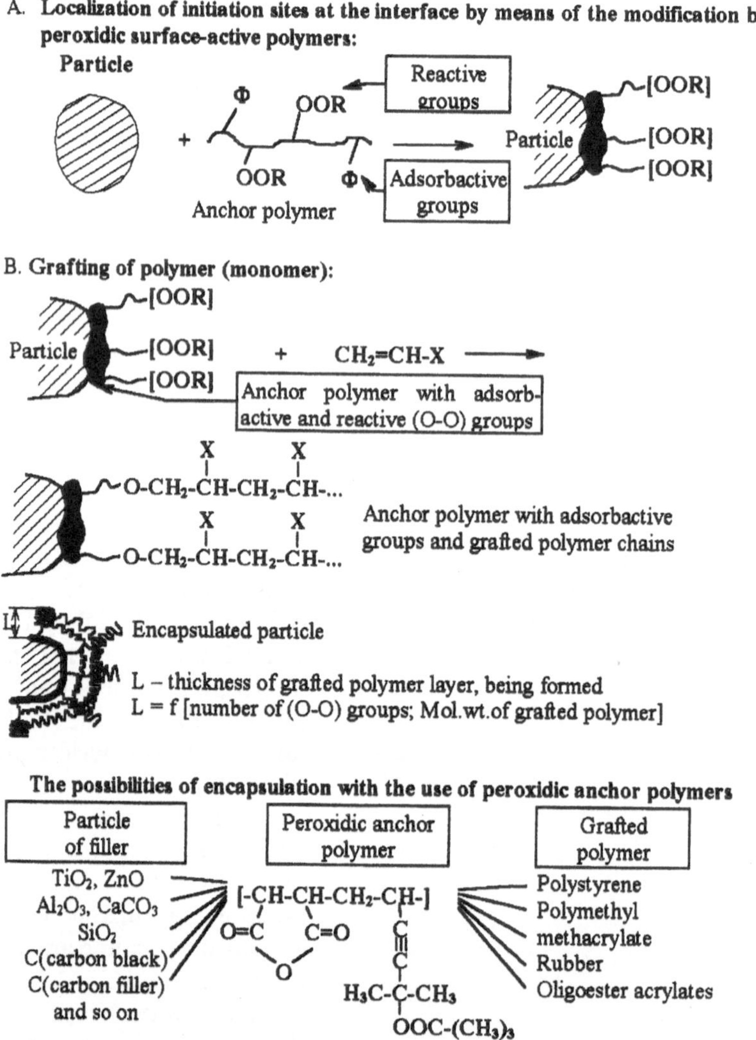

Fig. 1 New approach to solving the problem: Formation of Colloidal Systems

forming of transition (interfacial) layers; stabilization of particles and its control; increase and optimization of adhesion energy value at the interface dispersed phase – dispersive medium (filler – polymeric matrix); deagglomeration of filler particles (lowering of the agglomeration of filler particles); kinetic inhomogeneity of polymerization on the surface of solid, and influence of surface energy of solid on polymer reactivity.

Solution of these problems is possible, in our opinion, by the way of peroxidation of interface of colloidal systems [1–3]. Main conception of this approach is shown in Fig. 1.

This is attained by means of the surface's activation by polymeric surface-active polyperoxides. They act as anchor reactive polymers. It is possible to fulfill stepped or consecutive grafting of polymers and to regulate concentration and nature of peroxidic groups.

Surface-active polyperoxides, "anchor" peroxidic polymers, were synthesized by means of copolymerization of peroxidic and functional monomers.

In addition, the presence of the groups, which strongly differ in polarity, in HFPP macromolecules causes the presence of surface-active properties in them and gives them the ability of sorption at the interface of various polymeric colloidal systems: emulsions, latexes, polymer – polymer mixtures, reinforced polymers filled by dispersion, etc. HFPP adsorption at the interface in different colloidal systems provides the localization of initiation sites and proceeding of reactions in the interfacial layers. The localization can be carried out at the expense of chemical interaction with functional groups of adsorbent, as well as at the expense of physical adsorption of HFPP macromolecules.

Irrespective of the method of surface modification: physical adsorption, chemisorption or microcapsulation – the application of polyperoxides allows to reach the concentration of peroxidic groups of the surface by 2 ... 3 orders of magnitude higher than when it is processed by low-molecular peroxides.

Prospect of practical application of interface peroxidation

Heterofunctional polyperoxides are basically a new type of modifiers, because they have the properties of peroxides and functional polymers. This causes the presence of the complex of specific properties – the ability of sorption at the interface, biphility, and surface-active properties, ability of polymer analogical transformations. These properties are combined at the same time with the ability to initiate homolytical reactions. This combination allows to apply them as chemically-active adhesives and finishing agents; polymeric emulsifier-initiators; active modifiers of mineral filler surface; structurization agents for heterogeneous polymer systems, etc. – i.e., as multifunctional initiators of radical processes.

The adhesives for the fixation of reinforcing fibrous and textile materials (for example, tire cord for elastomers) are the most important area of heterofunctional polyperoxides' application. In comparison with adhesives on the basis of latex, which do not contain –OO– groups, the rise of bond strength of viscosic, amide and anide cords with rubber, treated by peroxidic adhesives, is achieved (Table 1). This phenomenon is explained by increase of concentration of interphase chemical bonds in the area of contacts of fiber-rubber as a result of grafting polymers being bonded to the macromolecules of the adhesive with the help of peroxidic groups. Probably, it is caused by the formation of the optimum proportion of chemical, including strong covalent, and faint physical bonds in the rubber-cord system (Fig. 2). HFPP are of interest for the stabilization of heterophase, thermodynamically incompatible polymers. Due to the surface activity and biphility, at the stage of mixing they act as dispersants, and at the stage of processing, at the expense of forming interfacial bonds, they fix microheterogeneity.

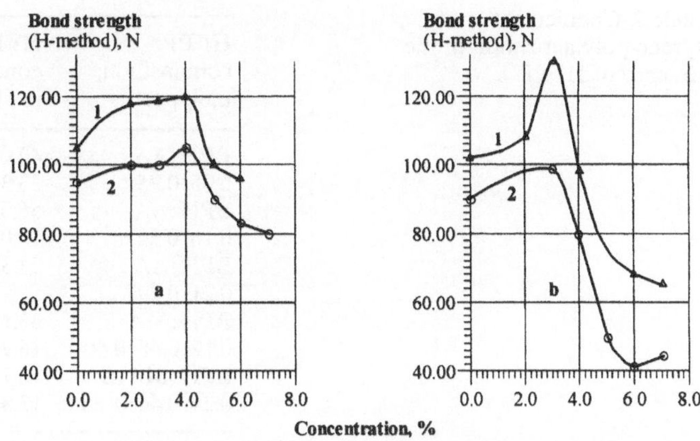

Fig. 2 Plot of bond strength of (a) viscose and (b) polyamide cords with rubber as a function of peroxidic monomer units' content in the latex HFPP adhesive at: 1–293 K, 2–393 K

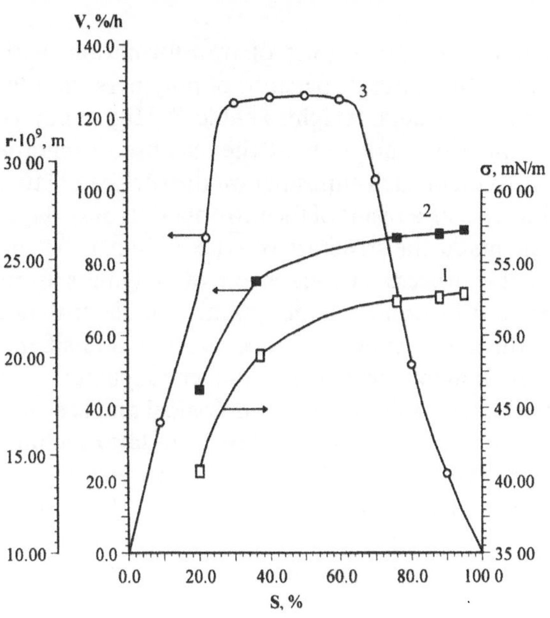

Fig. 3 Dependence of (1) surface tension (σ) of the latex, (2) radius (**r**) of polymer–monomer particles and (3) polymerization rate (**V**) on the conversion for water emulsion polymerization of styrene in the presence of HFPP as an emulsifier and an initiator at 353 K

Table 1 Adhesion properties of peroxidic impregnating composition

Cord	Base* of impregnating composition	Bond strength between cord and rubber SKI-3 by H-method, N		Endurance by multiple compression, thousand cycles
		293 K	393 K	
Viscose 22B	BMVPP-10/4	132	120	1500
	BMVP-10	117	103	1370
Polyamide 23 KHTC	BMVPP-10/4	119	92	324
	BMVP-10	107	78	308
Anide CBM	BMVPP-10/4	132	108	
	Polysar 781 (standard)	128	115	

* BMVP-*poly*(butadiene-*co*-2-methyl-5-vinylpyridine) latex; BMVPP-peroxide-containing *poly*(butadiene-*co*-2-methyl-5-vinylpyridine) latex; Polysar-781-*poly*(butadiene-*co*-styrene-*co*-4-vinylpyridine) latex.

Water-soluble heterofunctional polyperoxides are perspective for application as emulsifier-initiators of emulsion polymerization of water-insoluble monomers (styrene, butadiene, methyl methacrylate, vinyl chloride etc.) (Fig. 3). Both at the thermal and at redox-initiation by peroxidic groups of HFPP, the emulsion polymerization proceeds

Table 2 Characteristics of styrene polymerization in the presence of HFPP

HFPP* composition, mol. parts	HFPP concentration in H_2O, kg/m^3	T, K	Polymerization time, min	S, %	$M\eta \cdot 10^{-6}$ of polystyrene	Radius of latex particles, nm
BPH:AA	42.0	363	90	91	2.62	35
0.05:0.95	33.3	363	60	58	2.25	40
BPH:AA	55.0	353	90	96	3.38	26
0.10:0.90	50.0	353	100	89	2.33	28
BPH:AA	33.3	353	90	89	3.64	24
0.21:0.79						
BPH:MA:S	33.4	363	90	97	2.90	27
0.12:0.48:0.40	16.7	363	90	81	3.43	35
BPH:MA:S	16.7	353	90	83	2.47	32
0.34:0.46:0.20	13.3	353	120	82	2.80	

*BPH – 2-methyl-2-tert-butylperoxy-5-hexene-3-in: AA – acrylic acid; MA – maleic acid; S – styrene.

with high rates to high degrees of transformation, with formation of stable water dispersions of polymers, characterized by high molecular weights (Table 2). High stability of latexes is provided not only at the expense of macromolecule sorption of such emulsifier on the surface of latex particles, but also as a result of the introduction of surface-active fragments in the structure of latex polymer. At the same time, the process of separation of polymers from latexes does not demand a large quantity of electrolytes, and serum do not contain SAS, because the emulsifiers-initiators are bonded chemically and cocoagulate with latex polymers, which has great technological and ecological importance. The application of HFPP in latex technology allows to introduce in the structure and (or) to sorb on the surface of latex particles a predetermined quantity of peroxidic, chromophoric, adhesive, and surface-active groups, thus providing the production and modification of polymers.

Heterofunctional polyperoxides are of special interest for the peroxidation of different fillers, because this allows to increase substantially the degree of composite filling while simultaneously achieving high physical, mechanical and other operating properties. Application of HFPP for surface modification is of great importance, first of all, for the peroxidation of coarse-dispersed fillers: oxides, carbides and nitrides of the metals, etc. These fillers are most widely applied for the production of composites with specific properties: thermo- or electro-conductive, magnetic and so on (Fig. 4). Modification of the fillers' surface by HFPP provides high polymer-phility, due to that they are easily dispersed in organic and polymeric systems (Fig. 5), but the main advantage is that they form grafted polymers and are chemically bound with polymer matrix.

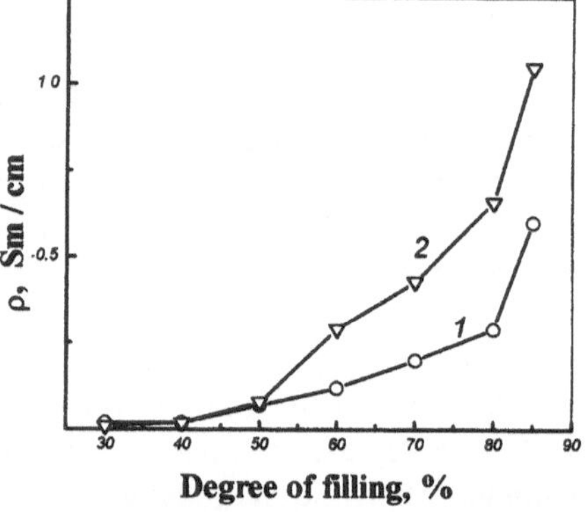

Fig. 4 Plot of rubbers' electro-conductivity (ρ) as a function of the degree of filling by technical carbon: 1 – non-modified filler; 2 – filler modified by HFPP

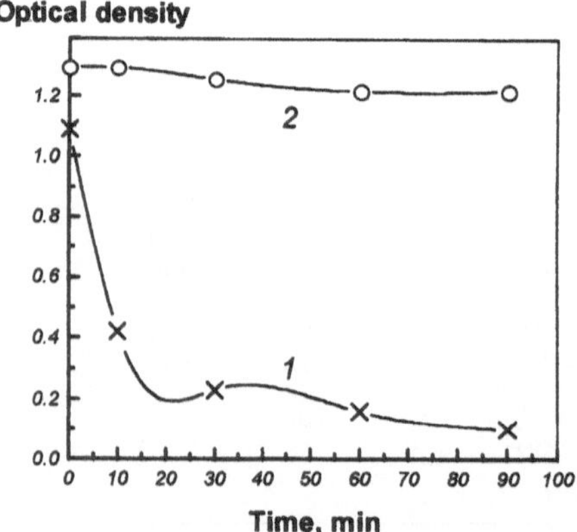

Fig. 5 Kinetic curves of clarification of FeO(OH) dispersion in toluene: 1 – without modifier; 2 – modified by HFPP

Conclusions

1) Adsorptionally active, reactive heterofunctional polyperoxides of different nature were synthesized, offering a new type of surface-active modificators of colloid systems' surface. They afford to pursue new approaches in the problem of creating composite materials.

2) Adsorption or chemisorption of the heterofunctional polyperoxides affords to immobilize polymer initiator of radical polymerization quasi-irreversibly on the surface of metals, salts, polymers, latex particles, etc.

3) For the creation of filled polymers, steric stabilization of dispersive systems, imparting compatibility to polymer mixture, creation of "core-shell" systems, and the grafting of monomers and polymers of appropriate nature is carried out on the peroxidized surface. There are no restrictions as to the nature of backing and grafted polymer.

4) In aqueous solutions heterofunctional oligoperoxides form micellae, have solubilizing properties, and provide the formation of stable emulsions. Water-emulsion polymerization in their presence has peculiarities which are analogical to the known ones. However, their application affords to implement ecologically pure technology and to carry out the modification of latex polymers.

5) The proposed methods for modifying the colloidal systems' surface afford to refine the technologies of composite materials' production.

6) Surface-active polyperoxides are unique patterns for the investigation of surface influence on the polymer reactivity, for the study of the mechanisms of reactions in the interfacial layers, structure of interfacial layer, and formation of adhesive compounds.

References

1. Tokarev V, Kucher R, Voronov S et al. (1987) Doklady AN SSSR 293:166
2. Voronov S, Tokarev V, Petrovska G (1994) Heterofunctional polyperoxides. Theoretical basis of their synthesis and application in compounds. – Lviv: State University Lvivska Polytechnica
3. Minko S, Voronov S, Lusinov I, Tokarev V (1994) Polymer at interphase synthesis, adsorption, conformation and reactivity. – Lviv: State University Lvivska Polytechnica

Progr Colloid Polym Sci (1996) 101:194–198
© Steinkopff Verlag 1996 INTERFACES

J. Gähde
J. Friedrich
T. Fischer
W. Unger
A. Lippitz
J. Falkenhagen

Reactions of polymer analogous model substances with metals and oxides

J. Gähde (✉) · T. Fischer
J. Falkenhagen
Institut für Angewandte Chemie
Berlin-Adlershof
12484 Berlin, FRG

J. Friedrich · W. Unger · A. Lippitz
Bundesanstalt für Materialforschung
und-prüfung (BAM)
12200 Berlin, FRG

Key words Chemical interaction – model compounds – Al_2O_3 and chromium – molecular orientation – formation of complexes – plasma

Introduction

Polymers may physically and chemically interact with functional groups of an inorganic material [1]. In order to study chemical process at polymer-metal or metal oxide interfaces monolayers of low molecular model compounds were used to model a related polymer. As an example, dihexylcarbonate (DHC) and phenylcarbamide acid-*n*-butylester (PCnBE) were used to model polyester-polyurethanes. In that case an interaction with aluminum oxide including kinetic aspects was investigated by means of gas chromatography (GC). In another study XPS and NEXAFS (Near Edge x-ray Absorption Fine Structure) spectroscopy were applied to have more insight into the interaction between chromium atoms and plasma introduced oxygen functionalities at the surface of octadecyltrichlorosilane (OTS) self-assembling layers.

Experimental

Dihexylcarbonate (DHC) or phenylcarbamide acid-*n*-butylester (PCnBE) was dissolved in methylene chloride, aluminum oxide (Aldrich, p_H 9.0, specific surface 150 m^2/g) was added and the solvent was evaporated from the suspension under rotation at 100 Pa and 35 °C. 1 g of the coated pigment was filled into a glass test tube fitted out with a 14.5 normal cut and closed with a glass plug. After storage at 100 °C and different times dioxane was added. The unchanged model compound was desorbed by a 3 min ultrasonic treatment. GC was used (GC Varian 3400) for measuring the content of the model compound in the suspension with 1.5 dibromopentane as an internal standard.

Monolayers were prepared by adsorption of OTS on a hydroxylated Si wafer surface by dipping it into a solution [2]. They were oxidized in an oxygen d.c. glow discharge as described earlier [3]. The deposition of chromium atoms was performed **in situ** with the help of a home-made evaporator. XPS measurements were carried out with a VG Scientific ESCALAB 200X electron spectrometer. Narrow scan spectra were recorded using Mg Kα excitation (300 W) in the constant retarding ratio mode. NEXAFS experiments were carried out in partial electron yield mode on the HE-PGM2 beamline of the synchrotron storage ring BESSY (Berlin, Germany) at a resolution better than 0.8 eV at the C K-edge. The angle of incidence of the linearely polarized x-ray light beam was varied between 90° (E-vector in the surface plane, "normal incidence") and 20° (E-vector nearly parallel to the surface normal, "grazing incidence").

Results

The chemical structure of an aluminum oxide at the surface and the formulae of DHC and PCnBE are shown in Fig. 1. The AlOH groups and $[AlO_4]^-$ Lewis centers as well as adsorbed water at the Al_2O_3 surface cause a hydrolytic degradation of the model compounds. The behavior of the decomposition rate vs. time fulfills the conditions of a first order kinetics. Aluminum oxide has a water content of 1.11 mmol H_2O/g Al_2O_3 measured by Karl Fischer Titration. The water content of the sample is sufficient for the complete hydrolytic degradation of the DHC because 0.25 mmol water are needed only and the degradation proceeds with a high velocity constant of $k = 8.82 \times 10^{-4}\,s^{-1}$ as shown in Fig. 2. *n*-Butanol as the volatile product of the hydrolysis was detected by GC. It is interesting that the addition of water to the coated pigment and storing the sample at 100 °C decreases the decomposition rate constant very strongly as seen in Fig. 2. The added water form a H_2O layer on the pigment which is then overlaid by the DHC [4]. That means that the direct interaction between the functional groups of the Al_2O_3 surface and the DHC molecules is impossible.

A similar behavior was observed with the PCnBE coated Al_2O_3 but the hydrolytic degradation initially starts with a lower velocity in comparison to DHC. The degradation is finished after 4 h, where 80% of the initial amount of the PCnBE is decomposed (Fig. 3). It is supposed that the products of hydrolysis as phenylcarbamide acid and aniline are adsorbed at the Al_2O_3 surface and hinder the chemical interaction with the functional groups of the Al_2O_3. If the storage of the coated Al_2O_3 is carried out in presence of a 12 times molar amount of water then the decomposition rate is very low (Fig. 3).

The adsorption of 0.25 mmol di-2-ethylhexyl phosphoric acid/1 g Al_2O_3 before coating the pigment with DHC reduces the decomposition rate of the DHC very strongly (Fig. 4). This can be explained by blocking of the active centers at the Al_2O_3 surface, for example, by formation of a P–O–Al bond.

In order to study the interaction of condensing chromium atoms with the O-functionalized OTS film, OTS films were oxidized to different degrees by exposure to oxygen plasma. Up to ~28 at % oxygen were detected by

Aluminium oxide surface (Al2O3)

$CH_3-(CH_2)_5-O-CO-O-(CH_2)_5-CH_3$

Dihexylcarbonate (DHC)

$C_6H_5-NH-COO-CH_2-CH_2-CH_2-CH_3$

Phenylcarbamide acid - n - butylester (PCnBE)

Fig. 1 Chemical structure of the Al_2O_3 surface and the polyurethane model compounds DHC and PCnBE

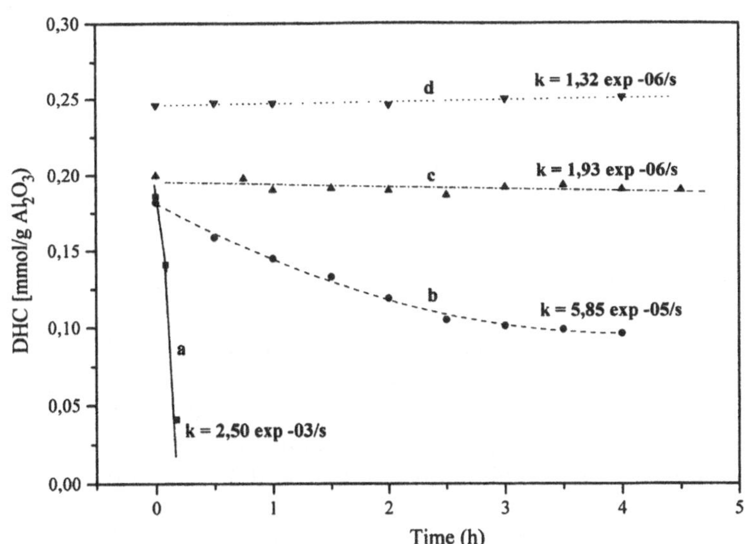

Fig. 2 Degradation kinetics of dihexylcarbonate as coating on Al_2O_3 at 100 °C; a: without addition of water (only adsorbed water); b: addition of a 12 times molar amount of water; c: addition of a 100 times molar amount of water; d: DHC with a 100 times molar amount of water without Al_2O_3

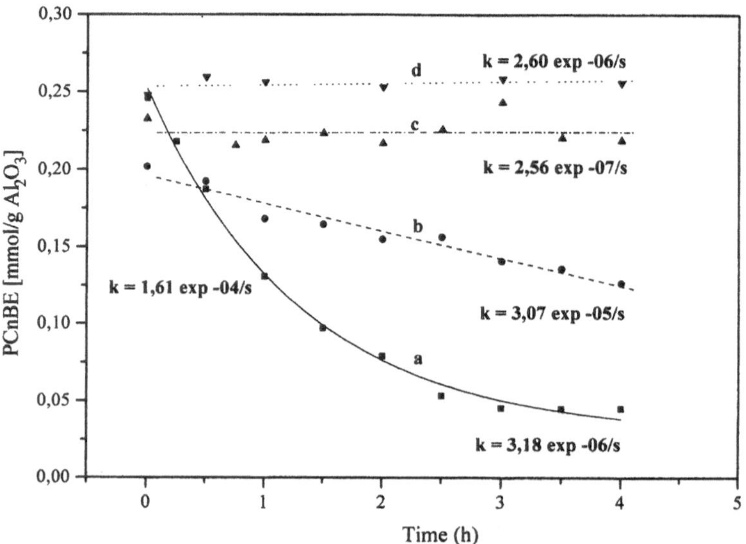

Fig. 3 Degradation kinetics of phenylcarbamide acid-*n*-butylester as coating on Al_2O_3 at 100 °C; a: without addition of water (only adsorbed water); b: addition of a 12 times molar amount of water; c: addition of a 100 times molar amount of water; d: PCnBE with a 100 times molar amount of water without Al_2O_3

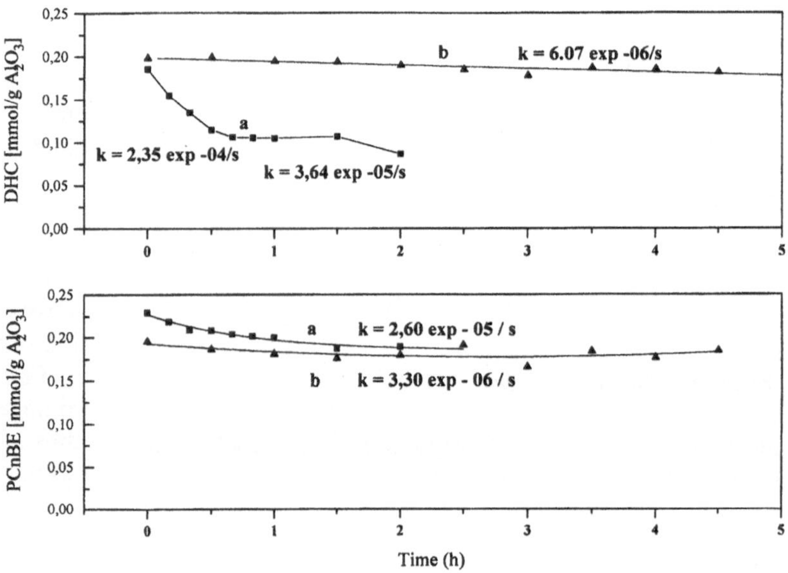

Fig. 4 Degradation kinetics of DHC and PCnBE as coating on Al_2O_3 at 100 °C after pretreatment the Al_2O_3 with di-2-ethylhexylphosphoric acid (DEHP); a: 0.25 mmol DEHP/g Al_2O_3; b: 0.50 mmol DEHP/g Al_2O_3

XPS in the surface layer [5]. NEXAFS at the same time reveals an increasing entanglement of the alkyl chains forming the originally highly ordered OTS layer (cf. Fig. 5, left), to a fully degraded and entangled situation after 4 s plasma treatment. This is verified in Fig. 5 by the total disappearance of any dichroism in the NEXAFS spectra after 4 s plasma treatment.

Deposition of ~1 monolayer of chromium atoms onto an untreated OTS film gives only a small entanglement of the originally well aligned alkyl chains (Fig. 5, right). Deposition of the same amount of Cr onto the 2 s oxygen plasma treated OTS film, which originally shows still a remarkable dichroism in the NEXAFS spectra, results in a full disappearance of any preferential orientation of the degraded and functionalized alkyl chains (Fig. 5, right). The conclusion is, that at very short oxygen plasma exposures oxygen functional groups can be attached to the OTS molecules without substantial influence on their orientation within the film, whereas the addition of chromium atoms to the 2 s oxygen plasma treated OTS film leads to a complete entanglement.

Hints on the chemical interaction between oxygen functional groups on OTS films and evaporated chromium can be obtained from XPS results (cf. Fig. 6). At the beginning of Cr evaporation (2 min) the Cr 2p 3/2 peak shows a peak maximum at ~577 eV, which is characteristic of Cr^{3+}, but after 12 min Cr evaporation a dominating Cr 2p 3/2 peak at 574.5 eV, characteristic of Cr^0, was

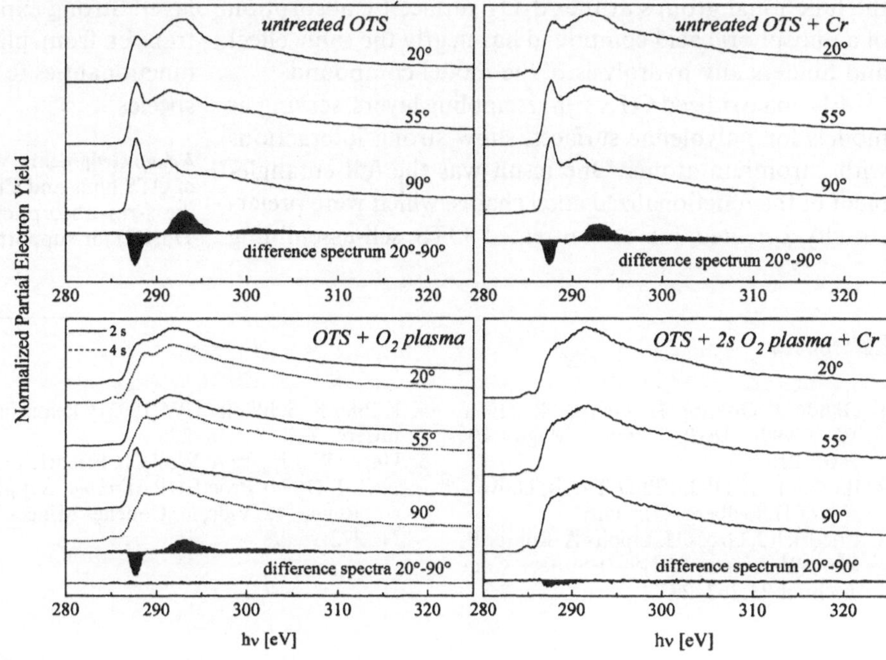

Fig. 5 Original and 20°–90° difference carbon K-edge NEXAFS spectra of OTS self-assembling layers deposited on a Si-wafer before and after different low-pressure oxygen d.c. plasma treatment and/or Cr evaporation

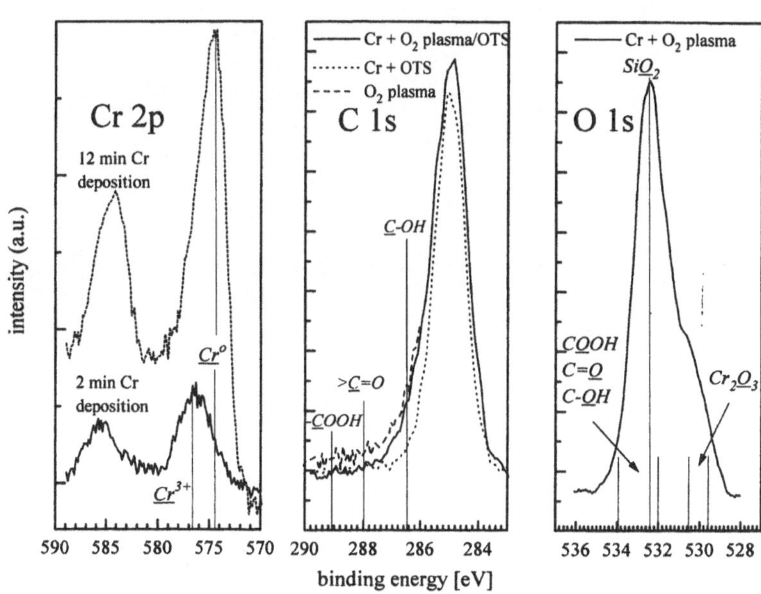

Fig. 6 XPS Cr 2p, C 1s and O 1s spectra of a 2 s oxygen plasma treated OTS subsequently evaporated with Cr

observed. Additionally the C 1 s peak shows a strong intensity reduction at binding energies in a region, where C=O and COO– species are usually observed [6]. Consistently, the O 1 s peak shows a shoulder in a binding energy region, where usually Cr^{3+} oxide oxygen species were observed. The conclusion is that, most probably, an oxygen transfer from O-functional groups (carbonyls, etc.) to condensing chromium atoms was observed. We interpret these experimental results as an interfacial redox process [7].

Summary

DHC and PCnBE as models of polyurethanes show strong chemical interactions with functional groups at the Al_2O_3 surface. Degradation of the model compounds occurs by hydrolysis, but DHC decomposes with a higher rate than PCnBE. Addition of water to the coated Al_2O_3 reduces the decomposition rate by forming a water interlayer, which interrupts the direct interaction at the interface. Blocking the functional groups at the Al_2O_3 surface by adsorption of a phosphoric acid compound has nearly the same effect and hinders any hydrolysis of the model compound.

Plasma oxidized OTS self-assembling layers, serving as models for polyolefine surfaces, show strong interactions with chromium atoms. One result was the full entanglement of the functionalized alkyl chains, which were preferentially oriented for an untreated OTS self-assembling layer. Strong experimental indications exist for an oxygen transfer from plasma-introduced carboxyl and carbonyl functionalities to chromium atoms which form Cr^{3+} oxide species.

Acknowledgments We thank Mrs. K. Bierbaum for the preparation of OTS films and Ch. Wöll and A. Schertel for collaboration during the x-ray absorption experiments at BESSY as well as DFG (Fr. 975/1-2) for supporting this work.

References

1. Gähde J, Goering H, Gehrke R, Hiller W (1993) IEEE Trans Magn 29: 2101–2107
2. Bierbaum K (1995) PhD Thesis, University of Heidelberg, Germany
3. Friedrich J, Gross H, Lippitz A, Rohrer P, Saur W, Unger W (1993) Surface Coat Technol 23:267–273
4. Köhler K, Schläpfer CW (1993) Chemie in unserer Zeit 27:248–255
5. Unger W, Lippitz A, Wöll Ch, Friedrich J, Nick L (1994) Proc EURADH 94, Suppl à la Revue "Le Vide, les Couches Minces" N° 272:95–98
6. Beamson G, Briggs D (1992) High Resolution XPS of Organic Polymers – The Scienta ESCA300 Database, J Wiley & Sons, Chichester
7. Friedrich J, Falk B, Loeschcke I, Rutsch B, Richter Kh, Reiner H-D, Throl U, Raubach H (1985) Acta Polymerica 36:310–320

Progr Colloid Polym Sci (1996) 101:199–202

H. Frenzel
E. Mäder

Influence of different interphases on the mechanical properties of fiber-reinforced polymers

Dr. H. Frenzel (✉) · E. Mäder
Institute of Polymer Research Dresden
Hohe Strasse 6
01069 Dresden, FRG

Abstract In this study the interphase properties of glass fiber polypropylene model composites have been investigated using sized glass fibers with amino silane as coupling agent and varied polymer dispersions as film formers and unmodified (PP) or acidic modified polypropylene (PPM). The interphase was tested by direct melt wetting measurements and single-fiber pull-out tests. The mechanical composite properties influenced by fiber-melt interaction were tested at continuous fiber-reinforced polypropylenes. The direct fiber/melt wetting experiments provide evidence of chemical reactions together with interdiffusion effects in the interphase between sized glass fibers and the modified polypropylene matrices. It was proved that these physical and chemical interactions influence the interfacial shear strength and the macromechanical properties. Differences in the interfacial shear strength were caused by the wetting kinetics due to the different film former formulation.

Key words Composites – interphase – polypropylene – glass fibers – mechanical properties

Introduction

The properties of composite materials are dependent on the properties of the reinforcing fibers, the polymeric matrices and the interphase between fiber and matrix. The interphase is dependent on the surface modification of the reinforcing fiber and the interaction of these modified fibers with the polymeric matrix.

The glass fiber surfaces are commonly modified by sizings which consist of the following main components:

80 … 90% film-former
5 … 10% silane coupling agents
upto 5% auxiliary agents (lubricants, antistatic agents)

The thickness of these sizings is in the range of 0.5 to 1.0 μm for commercially available glass fibers with a diameter between 10 and 17 μm. The silane coupling agents are of the following chemical structure:

$$Y - (CH_2)_n - SiX_3 \quad n = 0\text{–}3 \tag{1}$$

X is a hydrolyzable group responsible for the bond to the glass surface, e.g., methoxy, ethoxy, acetoxy group and Y is an organofunctional group selected to react with the polymeric matrix. Typically, vinyl or methacryloxy groups are used for polyester resins and amino or epoxy groups for epoxy or polyamide matrices.

The function of the silane coupling agents and their interactions with both glass fibers and polymer have been reviewed by Ishida [1]. The generated silanol groups react with the glass fiber surface via hydrogen and covalent bonds (Fig. 1).

Wetting measurements, inverse gas chromatography, electrokinetic measurements (Zeta potential), quantitative adsorption measurements, spectroscopic methods (XPS, TOF-SIMS, FTIR) and microscopic methods (SEM,

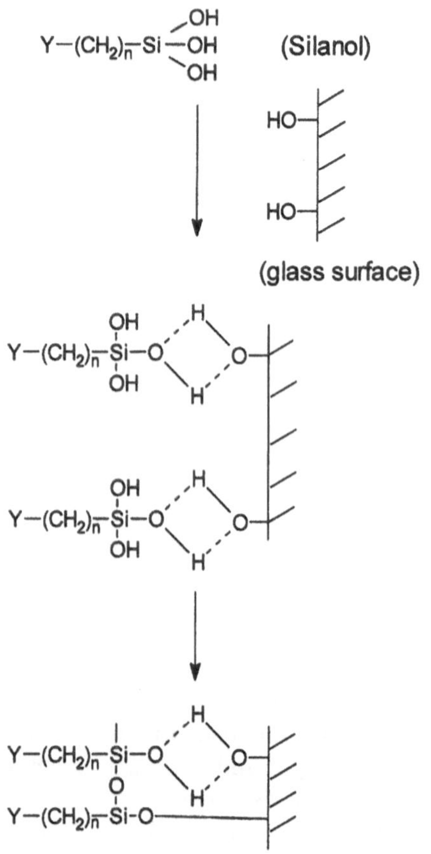

Fig. 1 Bond of the silane coupling agent to the glass fiber surface

Table 1 Characterization of the interphase fiber/matrix

Method	Information
Direct wetting measurements (glass fibers and polymer melt or reacting thermosets)	Kinetics of wetting, mechanism of reaction
Micro-mechanical methods: – single-fiber pull-out test – fiber fragmentation – microindentation test	Damage behavior of the interface (Interfacial shear strength IFSS)
Thermal analysis (DSC)	Reaction sizing/matrix, nucleation
FTIR-microscopy	Reaction sizing/matrix

SFM) are recommended to characterize the state of the modified glass fiber surface [2].

The methods for interphase characterization in composites are listed in Table 1.

The topic of this study is to investigate the interphase properties of glass fiber-PP model composites using differently sized (film former) glass fibers and PP matrices [3]. As experimental techniques wetting measurements with melts, single-fiber pull-out test and mechanical testing of unidirectional composites are used.

Table 2 Summary of sizings studied (variation of film former, coupling agent: aminosilane A 1100)

Sizing (abbr.)	Manufactured by	Dispersion
A	IPF	– 1)
B	IPF	–
C	IPF	1) polyethylene
D	IPF	polyethylene/polyurethane
E	IPF	epoxy
S1	IPF	polypropylene
S2	IPF	polypropylene/polyurethane
S3	GSO	polyethylene/polyurethane
S4	GSO	polypropylene/polyurethane

1) without aminosilane.

Experimental

The materials used in this study were specially sized glass fibers and PP-fibers manufactured at the spinning devices in the Institute of Polymer Research Dresden. Two comparable glass fibers were produced by Glasseiden GmbH Oschatz (GSO). The differently sized glass fibers are listed in Table 2.

Commingled yarns were made from two selected PP and PPM and E-glass filaments (S4) to produce continuous fiber-reinforced UD plates for mechanical testing.

Wetting measurements with test liquids (capillary rise method) were used to characterize the surface energetic behavior in the solid state. Thermodynamic work of adhesion was calculated from the surface free energies of the fibers and the solid polymer matrix determined by contact angle measurements with water and methylene iodide by the geometric mean equation of Owens and Wendt.

A Wilhelmy high temperature wetting apparatus has been developed to study the adhesion between polymer melts and differently sized glass fibers by the determination of the polymer melt surface tension and by direct measurements of molten polymer contact angles on single glass fibers. Details of this experimental technique will be described elsewhere [5]. In the present study, the kinetics of the spreading of molten PPs, unmodified and chemically modified, on unsized and differently sized glass fibers were determined by this technique.

Single-fiber model composites of the glass fibers and matrix polypropylenes were prepared to measure the debonding shear strength by single-fiber pull-out tests. The pull-out tests were carried out by the pull-out apparatus [6] which allows high precision fiber displacement and force measurements of end-embedded fibers. The fibers were embedded at 230 °C under argon atmosphere embedding lengths between 50 and 300 μm and heating/cooling rates of about 50 K/min. The pull-out test was carried out with identical velocities of pulling out the fibers (0.2 μm s^{-1}) at ambient temperature. From each force-displacement-curve the force at debonding F_d and the embedded length l_e are determined and the interfacial shear strength (IFSS) τ_d is calculated. The fiber diameter D is measured microscopically.

$$\tau_d = \frac{F_d}{D\pi le} . \qquad (2)$$

The UD-plates were cut up in test specimens to carry out the following tests:

– tensile test (DIN EN 61)
– three-point bending test and short-beam bending test (DIN 29 971)
– Charpy impact test (by analogy with DIN 53 453) with registration of the forcetime traces
– shear test (compression loading)

Results and discussion

Surface energetic properties

It was found that the surface energetic properties of glass fibers can be changed by sizing with aminosilanes and different film formers (Table 3).

The surface free energy of the solid polypropylenes, determined by contact angle measurements, is lower than the surface free energy of the unsized and sized glass fibers. Thus, it can be concluded that in PP-glass fiber composites complete spreading of the polymer melt on the fibers is thermodynamically favorable. However, the differences in the thermodynamic work of adhesion between PP and PPM and various sizes do not explain the differences in the debonding forces (c.f. Table 4). It should be noted that these calculations consider only physical interactions at the interface. Interdiffusion and chemical interactions are ignored. For that reason we carried out direct wetting measurements with the molten polypropylenes. The results show that PP and PPM indicate very similar surface tension/temperature behavior.

Table 3 Surface energetic properties of glass fibers and polypropylene

Sample	Surface free energy [mJ/m²] γ_s^d	γ_s^p	γ_s	Work of adhesion [mJ/m²] W_a PP	PPM
A	15.7	27.4	43.1	50.8	54.8
B	23.5	8.5	32	55.8	55.6
D	12.2	19.8	32	44.4	47.8
E	13.9	46.2	60.1	50.9	57.7
PP	27.2	0.8	28		
PPM	22.6	2.7	25.3		

The experimental results [7] show that the untreated and differently sized glass fibers were completely wetted by the unmodified PP melt after the fiber is immersed into the melt. The wetting behavior of the PPM melt differs from that of the unmodified PP. Unexpectedly, the PPM melt did not completely wet the glass fibers immediately after contacting the fiber surface. From a thermodynamical point of view this is only possible if the polymer melt surface tension is higher than the surface free energy of the glass fibers. This is obviously caused by the accumulation of the polar component of the chemically modified PP in the interphase region. When the fiber has been immersed in the polymer melt for several minutes the measured cos Θ increases (that means Θ decreases) in dependence on the sizing. Finally, zero contact angle is obtained while the fiber is held stationary in the chemically modified polymer melt. It is assumed that this behavior can be due to the expected strong acid-base interaction between the basic amine groups of the sizing and the acidic groups of the PPM melt leading to a chemical reaction in the interphase. A further proof for this conclusion is the completely different wetting kinetics of the untreated glass fibers with the

Table 4 Interfacial shear strengths (IFSS) of PP homopolymer (PP) and acid-grafted PP (PPM) with different polymeric dispersions and aminosilane

Sizing	Manuf.	Film former	IFSS [MPa] PP	IFSS [MPa] PPM
A	IPF	– 1)	5.9 ± 0.6	8.6 ± 2.6
B	IPF	–	5.2 ± 0.4	
C	IPF	1) polyethylene	5.3 ± 0.5	
D	IPF	polyethylene/polyurethane	6.7 ± 0.9	11.6 ± 1.4
E	IPF	epoxy	6.0 ± 1.0	6.6 ± 0.3
S1	IPF	polypropylene	7.2 ± 0.8	12.4 ± 3.7
S2	IPF	polypropylene/polyurethane	7.2 ± 1.5	11.9 ± 2.3
S3	GSO	polyethylene/polyurethane	8.1 ± 1.5	8.6 ± 1.7
S4	GSO	polypropylene/polyurethane	8.0 ± 1.0	13.2 ± 2.1

1) without aminosilane.

PPM melt. Again, the PPM melt cannot completely wet the untreated glass fiber. But in contrast to the sized fibers, the contact angle does not decrease when the untreated fiber is held in the PPM melt because there are only acidic surface groups which do not interact with the acidic groups of the modified polymer melt.

Single-fiber pull-out test

To study the results of physical and chemical interaction of the polymer melt with the sized fibers single-fiber pull-out tests have been carried out with model composites of differently sized glass fibers as well as modified and unmodified PP. The interfacial shear strengths (IFSS) were calculated as characterization of fiber-matrix adhesion from the force-displacement curves. The IFSS with PPM are higher than that with unmodified ones. Different film formers show lower changes in the IFSS (Table 4). Comparing the differences in IFSS with the different wetting kinetics, it can be concluded that size *E* inhibits the complete spreading of PPM more than size *D* due to the compatibility of the polymeric binder with the polymer matrix resulting in lower IFSS.

Both unmodified and modified PP represent the highest IFSS-values of the compared sizings if the sizing S4 was used.

This effect is not only due to the silane coupling agent but can be concluded by a combination with the film former (PP/PU dispersion) that seems to contribute to a better compatibility with the polymer matrix than, for example, the used epoxy dispersion in size E. A proof for this assumption shows the comparison of the IFSS of glass fiber S4 with PP- and PPM-matrix (Table 5).

Macromechanical properties

The fiber-matrix interaction of the real composites was studied in comparison to the single-fiber pull-out test. The data show that the micro- as well as the macromechanical properties of glass-fiber PP composites are strongly influenced by the matrix modification that leads to a strong chemical bonding in the interphase (Table 5). The experimental results show in the case of polypropylene-glass fiber composites with expected ductile fracture behavior that the higher the adhesion the higher are the IFSS and also shear, impact, 3PB and tensile strengths of the real composites. PP/S4 properties are calculated as 100%.

Table 5 Comparison of micro- and macromechanical properties of PP/S4 and PPM/S4 in continuous-fiber composites

Characteristic	PP/S4	PPM/S4
IFSS [MPa]	8.0	13.2 (165%)
Interlaminar shear strength [MPa] (short beam bending)	no del.	18.7
Shear strength [MPa]	11.0	19.7 (179%)
Tensile strength [MPa]	464.9	620.6 (133%)
3 PB strength [MPa]	245.5	434.4 (177%)
Impact strength [MPa]	479.0	724.0 (151%)

The difference in the absolute values is due to the different loading and on the other hand to those bulk composite effects, e.g., influence of adjacent fibers, fiber-orientation defects.

Bending, tensile and impact strength of these composites show obvious differences due to the influence of the modifier and therefore higher fiber-matrix interaction.

Conclusions

It was proved that the chemical interactions in the interphase caused by the chemical modification of the polypropylene matrix in connection with a compatible sizing increase the interfacial shear strengths of the glass fiber-polypropylene composites. These effects could be confirmed by mechanical testing of continuous fiber-reinforced composites. The use of the direct fiber/melt wetting method can be recommended to characterize the wetting kinetics of PP melts and sized glass fibers.

References

1. Ishida H (1984) Polym Composites 5:101
2. Grundke K, Jacobasch H-J, Simon F, Schneider St (1995) J Adhesion Sci Technol 9:327
3. Mäder E, Grundke K, Jacobasch H-J, Wachinger G (1994) Composites 25:739
4. Mäder E, Bunzel U, Hofmann H, Wulff D Germ Pat applied for June 4, 1993
5. Grundke K, Uhlmann P, Giezelt Th, Jacobasch H-J Colloids & Surfaces, submitted for publication
6. Mäder E, Grundke K, Jacobasch H-J, Panzer U Proc 31. Internat Man-made fibre congress, Dornbirn, Austria, Sept 23–25, 1992
7. Mäder E, Grundke K, Janke A Proc Workshop on interfaces in carbon, glass and polymer fibre reinforced composites, Kaiserslautern, Febr 1–2, 1995, pp 32–43

AUTHOR INDEX

Progr Colloid Polym Sci (1996) 101:204
© Steinkopff Verlag 1996

SUBJECT INDEX